New Media and Visual Communication in Social Networks

Serpil Kır
Hatay Mustafa Kemal University, Turkey

A volume in the Advances in Multimedia and Interactive Technologies (AMIT) Book Series

IGI Global
DISSEMINATOR OF KNOWLEDGE

Published in the United States of America by
IGI Global
Information Science Reference (an imprint of IGI Global)
701 E. Chocolate Avenue
Hershey PA, USA 17033
Tel: 717-533-8845
Fax: 717-533-8661
E-mail: cust@igi-global.com
Web site: http://www.igi-global.com

Library of Congress Cataloging-in-Publication Data

Names: Kir, Serpil, 1985- editor.
Title: New media and visual communication in social networks / edited by Serpil Kir.
Description: Hershey, PA : Information Science Reference, 2019. | Includes bibliographical references and index. | Summary: "This book examines communication strategies in the context of social media and new digital media platforms and explores the effects of visual communication on social networks, visual identity, television, magazines, newspapers, and more"-- Provided by publisher.
Identifiers: LCCN 2019024164 (print) | LCCN 2019024165 (ebook) | ISBN 9781799810414 (hardcover) | ISBN 9781799810476 (paperback) | ISBN 9781799810452 (ebook)
Subjects: LCSH: Social media. | Online social networks. | Visual communication.
Classification: LCC PN4552 .N49 2019 (print) | LCC PN4552 (ebook) | DDC 302.23/1--dc23
LC record available at https://lccn.loc.gov/2019024164
LC ebook record available at https://lccn.loc.gov/2019024165

This book is published in the IGI Global book series Advances in Multimedia and Interactive Technologies (AMIT) (ISSN: 2327-929X; eISSN: 2327-9303)

British Cataloguing in Publication Data
A Cataloguing in Publication record for this book is available from the British Library.

Advances in Multimedia and Interactive Technologies (AMIT) Book Series

Joel J.P.C. Rodrigues
National Institute of Telecommunications (Inatel), Brazil & Instituto de Telecomunicações, University of Beira Interior, Portugal

ISSN:2327-929X
EISSN:2327-9303

Mission

Traditional forms of media communications are continuously being challenged. The emergence of user-friendly web-based applications such as social media and Web 2.0 has expanded into everyday society, providing an interactive structure to media content such as images, audio, video, and text.

The **Advances in Multimedia and Interactive Technologies (AMIT) Book Series** investigates the relationship between multimedia technology and the usability of web applications. This series aims to highlight evolving research on interactive communication systems, tools, applications, and techniques to provide researchers, practitioners, and students of information technology, communication science, media studies, and many more with a comprehensive examination of these multimedia technology trends.

Coverage

- Multimedia Services
- Digital Games
- Digital Watermarking
- Web Technologies
- Digital Technology
- Internet Technologies
- Digital Images
- Audio Signals
- Digital Communications
- Mobile Learning

IGI Global is currently accepting manuscripts for publication within this series. To submit a proposal for a volume in this series, please contact our Acquisition Editors at Acquisitions@igi-global.com or visit: http://www.igi-global.com/publish/.

The Advances in Multimedia and Interactive Technologies (AMIT) Book Series (ISSN 2327-929X) is published by IGI Global, 701 E. Chocolate Avenue, Hershey, PA 17033-1240, USA, www.igi-global.com. This series is composed of titles available for purchase individually; each title is edited to be contextually exclusive from any other title within the series. For pricing and ordering information please visit http://www.igi-global.com/book-series/advances-multimedia-interactive-technologies/73683. Postmaster: Send all address changes to above address.

Titles in this Series

For a list of additional titles in this series, please visit: www.igi-global.com/book-series

Handbook of Research on Media Literacy Research and Applications Across Dsciplines
Melda N. Yildiz (New York Institute of Technology, USA) Minaz Fazal (New York Institute of Technology, USA) Meesuk Ahn (New York Institute of Technology, USA) Robert Feirsen (New York Institute of Technology, USA) and Sebnem Ozdemir (Istinye University, Turkey)
Information Science Reference • ©2019 • 433pp • H/C (ISBN: 9781522592617) • US $265.00

Cases on Immersive Virtual Reality Techniques
Kenneth C.C. Yang (The University of Texas at El Paso, USA)
Engineering Science Reference • ©2019 • 349pp • H/C (ISBN: 9781522559122) • US $215.00

Smart Devices, Applications, and Protocols for the IoT
Joel J. P. C. Rodrigues (National Institute of Telecommunications (Inatel), Brazil & Instituto de Telecomunicações, Portugal & Federal University of Piauí (UFPI), Brazil) Amjad Gawanmeh (Khalifa University, UAE) Kashif Saleem (King Saud University, Saudi Arabia) and Sazia Parvin (Melbourne Polytechnic, Australia)
Engineering Science Reference • ©2019 • 317pp • H/C (ISBN: 9781522578116) • US $225.00

Handbook of Research on Examining Cultural Policies Through Digital Communication
Betül Önay Dogan (Istanbul University, Turkey) and Derya Gül Ünlü (Istanbul University, Turkey)
Information Science Reference • ©2019 • 447pp • H/C (ISBN: 9781522569985) • US $265.00

Advanced Methodologies and Technologies in Media and Communications
Mehdi Khosrow-Pour, D.B.A. (Information Resources Management Association, USA)
Information Science Reference • ©2019 • 752pp • H/C (ISBN: 9781522576013) • US $295.00

Interface Support for Creativity, Productivity, and Expression in Computer Graphics
Anna Ursyn (University of Northern Colorado, USA)
Information Science Reference • ©2019 • 355pp • H/C (ISBN: 9781522573715) • US $195.00

Trends, Experiences, and Perspectives in Immersive Multimedia and Augmented Reality
Emília Simão (Escola Superior Gallaecia University (ESG), Portugal) and Celia Soares (University Institute of Maia (ISMAI), Portugal & Polytechnic Institute of Maia (IPMAIA), Portugal)
Information Science Reference • ©2019 • 277pp • H/C (ISBN: 9781522556961) • US $185.00

IGI Global
DISSEMINATOR OF KNOWLEDGE

701 East Chocolate Avenue, Hershey, PA 17033, USA
Tel: 717-533-8845 x100 • Fax: 717-533-8661
E-Mail: cust@igi-global.com • www.igi-global.com

Editorial Advisory Board

Table of Contents

Detailed Table of Contents

unseen (i.e. what visual messages say and show). Advertising is a public and massive myth-poetic and logo-poetic device and an increasingly multiform, omnipresent, seductive and visually persuasive. It is important to understand the elements of (explicit or implicit) meaning and the corresponding processes and mechanisms through which the meanings produce effects. This chapter assumes itself as a contribution to a desideratum that may be called visual advertising literacy.

One of the most important areas where visual communication is prominent today is marketing. Brands try to adopt to the visual world of today in order to make their communication with their target audience more meaningful and effective. This way, organizations, and therefore brands, take significant steps for differentiation from their competitors by forming their visual identity. Additionally, considering the current advertisements of brands, it is seen that visual narratives are highly abundant. In this context, brands which are starting to use visual communication effectively are gaining a broader place in the memories of their target audience by increasing brand awareness. As a result, it is believed that the significance of visual communication and identity is increasingly higher in terms of influencing existing and potential masses by being integrated into the visual world of today.

This chapter is aimed to put the matter of how is a country image able to be constructed in hand through the medium of the online press, by evaluating The New York Times (USA), The Daily Express (England), Spiegel Online International (Germany), and Le Monde Diplomatique (France) in terms of "Turkish image and identity" throughout four months (January-April) in 2019. The author uses Van Dijk's discourse analysis approach that is based on two main principles, macro and micro discourse analysis, and the content analysis technique. It is possible with this evaluation to determine how Turkish image and identity is established and what kinds of images, expressions, and representations are used by the foreign press, as well as their approach to Turkish identity.

This chapter emphasizes that the changes and developments in information and communication Technologies are reflected in the field of journalism being integrated into new media. The transformation and development in the new media have changed the traditional practices of the perception and communication design of journalism, leaving its place to the elements that include information-communication and communication design. An effective communication design on human beings shows that information-target, mass-message-communication pieces are the interaction of today's consumer's taste and consumption desire.

An inescapable part of our everyday lives, visual communication is a key driver of engagement on social media. These are redesigned their news feed to allow greater emphasis on visual content, resulting in greater interaction. This chapter discusses the current scenario of cyber and social media crime in India and how the government has incorporated the necessities to fight against it. It will also include the types of social media crime enumerating the provisions of Information Technology Act, Indian Penal Code. Through this chapter, the author discusses the various types of cybercrimes, which are cyber defamation, cyber pornography, cyber stalking, fraudulent transaction and misrepresentation, hacking. The author laid emphasis on what legislations are in action to deal with such crimes and how strictly the offenders are punished. The author also discusses the competency of the present legislation and how the loopholes, if any, can be filled to make the virtual world a better place for everyone.

In this chapter, results obtained from a longitudinal study on Social Media (SM) use are reported. Previous studies have mostly carried out contextualized research and not a lot of it has been done in Kenya and especially with the emerging mobile application SM platforms. The key objective of the study was to understand the general aspects of emerging SM platforms with a view of mapping out study areas going forward. The study used mixed method approach for an extended period. To effectively carry out the study, seven themes were identified through a preliminary study and literature review. A summary of the results show that mobile app SM platforms are gaining popularity among users. SM uses are majorly socialization, but other uses such as political campaigns, fundraising, and religious uses are taking root. SM groups are dominating; even though SM is reach in functionality, users expect more. There exist various challenges associated with social media use and SM study methodological challenges. Finally, the study established seven key themes which can frame SM studies.

This chapter examines social media relations, which build virtual Erdogans as two opposite realities, with netnography method because of community composition and cultural sharing contents. It will be analyzed visual 'Erdogan' productions in Anti-Tayyip (Opponent) and Erdoğan Sevdalıları-Lovers of Erdogan (Fan/Supporter) communities and it will be drawn post-truth biography of a leader in visual culture of social media. Two different/opposite virtual realities of Erdogan, which are reproduced in social media sociality every day, lead to expansion of polarized political climate in the context of organic society and absorb the political identity of Erdogan.

One of the concepts that have a strong and dominant effect in transforming the culture, individual, and society of social media has been privacy. Everything that belongs to our domestic space in modern times, which should not be known/seen by others, is made public by ourselves in the postmodern age with new media tools. In social networks focusing on vision and surveillance, privacy is restricted, eliminated, or stretched by individuals themselves for the creation of ideal profiles. The privacy settings that a person thinks are under his control seriously affect the way he uses social media. This chapter will try to determine which subject/situation/images are perceived as intimate among university students, and how the boundaries of social media and privacy are drawn and transformed. The study is based on the assumption that the level of privacy awareness and the level of knowledge control influence the quality and frequency of social media sharing of users.

Çiftlik Bank (Farm Bank) is an investment system based on fraud that may be described to be a Ponzi scheme as a commercial term. It reached thousands of investors in Turkey via the service it provided over the internet and attracted attention by leaving a high profit margin for its investors for some time and using some abstract concepts that are held sacred by the majority. Çiftlik Bank created an earning-oriented exclusivity for its investors, but also created suitable scapegoats for the community, along with fraud. This chapter focuses on the rhetorical conflict between the scapegoat virtual group organized with the name "Çiftlik Bank Victims" on Facebook, and the society, as well as the activities of regaining the contingent/select identity.

Identity emerges as a flexible, multidimensional, variable, and slippery concept that cannot be defined through the processes of discussion and understanding. The new construction area of this concept, which is regarded as a process constructed on the social plane, is the social networking platforms. This is because these platforms are the most common communication environments where people and their lifestyles are presented to the outside world, in addition to the cheap and rapid satisfaction of their needs for information and entertainment. Face-to-face communication and language practices are not sufficient enough in the identity presentation anymore. Individuals choose to design and update their identities through social networks and to perform an image-based identity manifestation. This chapter examines how identity was established and manifested through social networks, and analyzes the identities the popular people in these networks designed and exhibited.

Nowadays, in order for brands to respond to consumer expectations, digital media efforts need to be involved in the brand communication process. Brands have a unique way to remind their names in a consumer's mind with real-time marketing. In addition, real-time marketing offers a way to make it easier for marketers to reach their target audiences at a low cost when increasing the speed and functionality of information. In this chapter, real-time marketing posts that have high user interaction on Twitter are handled in the context of their process, content features, and message appeal. Examined were 185 tweets. According to the data obtained, brands are trying to create positive brand image for consumers. Besides, it has been determined that both informational and emotional appeals are used intensively in order to create brand awareness.

The development of technology at an incredible speed today and the fact that the internet has become an important area of social life has led to differentiation in the structure of mass communication and content production, too. This differentiation has stimulated advertisers and companies to reach the target audience through social networks with many users and different characteristics. Companies employ different strategies to be effective in these platforms. One of these strategies is collaboration with social media phenomenon. The relationship between the social networks considered as the new medium of advertising, social media phenomenon identified as influencer in these networks, and advertising is examined within the scope of this chapter. In this context, data obtained from interviews with 50 Instagram phenomenon by using semi-structured interview technique, which is a qualitative research method, were analyzed and advertising collaborations with influencers in social networks were evaluated.

The owners of Facebook and WeChat repeatedly promote their media as the preferred platform for people to connect. Improving social relationships was marketed as the reason for their innovation. But users' urge to unite on these OSN services alone cannot explain the success of these media in the US and China. There is a different or rather new business approach underpinning these OSN services that contribute to their success. The author argues that there is an implication of owners' profit-based interest in ensuring the popularity of their online platforms. Audience commodity analysis as discussed by Dallas W. Smythe and Christian Fuchs is employed in the contexts of the US and China to comprehend the complex factors related to online social media owners' interest and their negotiation with the government in online media's prosperity. Through archival research including examination of newspapers, policy documents from OSN-based companies, and survey results from 2015 to mid-2018, this chapter demonstrates the political economy of Facebook and WeChat.

Foreword

Starting with the publication of the first newspaper in the 17th century, the history of modern media is moving along an axis that has more and more control over people's worlds and their lives. Coming with the internet, one of the most common tools of the information age is social network, which dominates every field from human psychology to social environment and connects everyone without boundaries.

In the early ages, people were communicating and understanding each other through the simplest, but most effective, form of communication—face-to-face communication. They developed various tools and ways to communicate more easily as the ages progressed. The communication that one needs to convey his impressions and views, share life, and solve problems has led people to new inventions and developments as a fundamental need.

Having started with transmitters, followed by telephone, radio, and television, these developments reached the age of informatics with the spread of the internet, which was actually developed for military purposes. After nearly 300 years of traditional media, the new media, which was improved with computer-based techniques, created the modern media; especially with the internet, an age of message flow that is not easy to control has begun. The internet is the beginning of a technological revolution in world history.

With its features like wieldy, high, and easy level of accessibility and participation, the social network that provided people with opportunities such as simultaneous interaction, message transfer, and sharing and getting information, has become the most used media, leaving behind all traditional media.

Developing with new technologies on a daily basis, social networks are platforms with high interaction power and versatile technology opportunities, where users communicate through a profile they create. In addition to these characteristics, social networks not only provide more satisfaction because of the amount of people who use them and the amount of people the messages reach, but also attract attention because of the thoughts and characters they include.

However, although social networks include true and accurate messages, they sometimes might be the targets of criticism with their virtual content being far from reality. In addition to social networks, people who plan to go beyond the limitations of real life aim to create a new identity and get away from reality with the help of technological reality systems such as virtual reality applications, digital broadcasting, and digital games where the user is involved and can gain experience. This type of communication, which aims to create a kind of interaction with visual elements, is called visual communication. In this type of communication, messages sent to people are transmitted through visual materials, so emotions or thoughts are transmitted through symbols instead of words in communication.

In this context, today's technological developments and visual communication elements gain importance. This book establishes the relationship between technology, new media, and visual communication and explores the effects of visual communication on social networks, visual identity, television, magazines, newspapers, cinema, and digital pollution.

It focuses on the development of technology and media over the ages, the networks—based on technology—that twine around the world and surround humanity and the people who are captivated to it within modern life. The book also includes invaluable research in the field of visual communication and social networks, including opportunities, challenges, and current trends facing researchers and practitioners. It was prepared with a large composition consisting of 16 chapters and the collaboration of many academicians. It includes a detailed and comprehensive review of the topics covered. Since most of them are empirical studies, their contribution to the field is very valuable. First of all, I would like to congratulate Assistant Professor Serpil Kır, who edited and contributed to the book with her selfless work, and all the academicians who contributed to the book with their original works. I thank each of them for their valuable contributions to the field of communication.

Veysel Eren
Hatay Mustafa Kemal University, Turkey

Preface

Considering the transformation in communication infrastructure that penetrates the exchanges of social meaning in parallel with the change of ages, it is not possible to separate visuality and visualization from digitalization. By reason of the transition from analogue to digital age, radical differences emerged in the ways in which picture practice form the reality. The flexible and coded structure of digital world has led to the reorganization of the shooting, printing, reproduction and distribution processes of the image. Visuality that is increasingly surrounding the media, communication infrastructure and culture renewed in the digital age has been joined in a hybrid production area which is more open to intervention. The new digital system includes a wide range of text and graphic designs, picture, illustration, adding-removing-creating in computer and new printing forms (Manovich, 2006a, p. 5). Visuality integrated into computer mediated communication systems has been renewed in its structural features in the shadow of a set where the live image is gradually losing its importance and being manipulated by software programs. Because reality effect on digitally recorded images can be edited in computer by means of various cut-add-subtract or color correction settings without any obligation to stay with the original image. This re-editing process is, to transfer the capabilities of the camera and picture to the computer, along with the person. Such that, by means of copy and variability features of new media, a new composition can be produced by building a large number of pieces of picture and organic images can be obtained by adding the same image to different collages. The questionable dimensions of reality become more evident in such a case.

The fact that changes in the essence of visual communication do not cause too much tension can be explained by the effect of absorbed reality, the relationship between post-modernism and social forms becoming abstract and dislocation in the axis of visual esthetic elements. Because, unlike many features of digital culture, the fact that the change in visual culture is not criticized too much can be described as a Velvet Revolution (Manovich, 2006b, p. 1). After this dislocation that does not disturb the communities and does not cause any concern, a cultural mechanism produced and served by software programs of effect images emerged. Typical traces of effect culture can be found in photo-oriented social media platforms such as Instagram etc. The special effects that George Meliés brings to the visual culture are now among the routines of a new generation of visual communication practice (Manovich, 2001, pp. 176-178). Therefore, a large audience, including online users, naturally accepts color and format changes on images. Moreover, some art movements ground on direct manipulation. Manipulation, which is a different case from abstraction through salt signifiers of reality, is the work of creating and using the non-absolute signifiers of reality. Computerization, which is at the center of visual production processes at the present day, increases the use of CGI (Computer Generated Imagery) technologies. Many online users use Illustrator for vector-based images, Photoshop for pixel-based images, Wavefront for 3D modelling and animation and After Effect for visualization and 2D modelling (Manovich, 1998, p. 3).

Visual surrounding of individual and collective representation actions in digital platforms means that a new language based on mutual agreement and meaning sharing is in effect. According to Hall (1997, p. 5), who establishes the theory of representation parallel to structuralist theories of language and accepts meaning as a symbolic representation produced in the structural mobility of language, there is a direct connection between the represented ones and language. Representation is established in association of language and meaning. The representation power of visual is an unspecified and endless experience that users build with mutual visual conversations. This eternal language extends from the moment of production of the image to the infinite interpretation and expansion. The spoken language built by the different fractions of the image is universal and includes a non-verbal symbolic order. In the digital age, in which visual culture rises and eye re-establishes its dominance on a different level, the number of people who speak the visual language has increased and the usage patterns of social areas where visual compositions such as Instagram or YouTube are produced have grown. Anymore, a supranational form of society that only communicate through visual fictions and sophisticated presentations without any written statements has arisen. Although the stories told by the visual codes digress from their artistic and realistic aspects in this period, the affinity of embodying practice to popular culture is a stubborn fact. Because the character of representation is concerned with the production of a meaning pointing to other than itself. Regardless of the meaning produced in the content, it always carries the language of the image in its potential.

Digital visuality has an appearance shaped in the abstract and concrete complexity in the traditional context in which it cannot prove it by direct links. The real-like sensation in the perception of organic real-like appearances requires a distinction between the reality and phenomena. Because reality is a fluid motif that can be reconstructed in many styles and interpretations, that can be stretched and whose ties to the original can be ignored. Further ahead, reality codes that show the behaviors of establishing with images in the post-modern era have opened the way to produce infinite reality in the specificity of a single image. Keyes (2004) argues that in the new world, where he defines as 'Post-Truth Era', the functions of visuality provide evidence for reality and the system of truth is connected to the signifiers manipulated through emotions and beliefs. The impact on the reality dimension of this era, in which visuality is purely contrary to the aim of creating images and documentation, often includes the meaning of what is not 'as it appears'. In this new phase in which the appearances are not as they appear, visuality is a force that controls the effects of those who are not as they appear. Because the apparent reality has reached the dimensions to be replaced for the actual one.

The relationship between images and reality, which Baudrillard discusses within the framework of Simulation theory, is shaped in a critical collision where visual communication is also the subject of intense moments of contact. The simulacra, which Baudrillard (1998) describes as unreal even though it has the signs of reality, mediates the hyperrealities, that is, the simulations that feed human phantasms by carrying meta-meanings. It is possible to interpret the reality representations of visuality as a world of simulacra. Because, although the signs composed in the visual carry the surrogates and symptoms of reality, they evoke other realities.

The narrative position of the visual cannot be reduced to a superficial recording and time-related context. Limiting the visual as a mechanical copying, a documentation process of writing histories and recording moments, should be regarded as a large-scale injustice against the designs that break and twist history and save history from the relative conditions of time and present it to universal eternity. Because, the visual communication language on digital platforms cannot be thought of with the plain meanings of the objects in its content. While visuality in digital media seems to be inherent to time and space in this sense, it contains the provision of timelessness and placelessness in its background. While describing the instrumental action, Deleuze (2013) argues that mechanical categories including cinema and photography are more effective than traditional methods of understanding culture. According to him (pp. 78-80), many cultural contents in visual roaming are vagrant and this idea contributed to Manuel Castells' theoretical consciousness.

This book consists of academic studies organized within the framework of the theoretical perspective outlined above. The studies in the book contribute to the multidirectional reading of the subject by evaluating communication actions in visual culture with different themes. The book covers a wide spectrum of studies consisting of individual use motivations, community models, professional practices and technological transformations.

OBJECTIVES, IMPACT, AND VALUE

The mission of the this book is to provide broad and comprehensive international coverage of subjects, issues and current trends relating to all areas of visual communications. Emphasis is highly placed on publishing research articles, case studies and book reviews that seek to connect theory with application, identifying best practices in visual communication and social networks. In this respect, book links both theoretical and practical approaches of visual communication and social networks to make a proactive contribution to the field.

This book is an applied research, refereed, international book that provides complete coverage on the opportunities, challenges, and current trends encountered by researchers and practitioners in the field of visual communication and social networks. This book offers an important and critical platform for researchers, practitioners, entrepreneurs, policymakers, and educators to present and discuss their experiences and perspectives on important issues and current trends related to marketing activities and research in an online context.

ORGANIZATION OF THE BOOK

In the study named 'Presentation of Visual Culture Elements in Digital Environments With Special Effect Technologies', which is the first chapter of the book, transformation of narrative structure in new generation visual fiction processes and effectiveness of special effects technologies in design are examined. In the study, which draws attention to the fact that special effect has become a constituent element in CGI technologies following a graphic that is distant from image alone, it is explained that organic elements are not needed as much as before in the fiction of digital visual reality. The study tries to prove that this situation determines attitudes and perceptions about reality and that the narrative position of the special effect is effective on amateur practices as well as professional practices. The second chapter, named as 'Visual Literacy and Visual Rhetoric Images of Ideology Between the Seen and the Unseen in Advertising', examines the surreal and poetic narrative world of advertising within the framework of visual rhetoric. The theoretical approach in the study is based on the fact that visualization techniques that create the magical world of advertising and narrative elements shape the structure of visual literacy and that visuals loaded with ideological images contain explicit and hidden semantic codes. In the third study named 'Visuality in Corporate Communication', the visual usage intensity of brands in corporate communication and marketing practices and the contribution of trends in visual culture in terms of branding is examined. The study shows that functional interaction between visual communication and corporate identity is producing more impacts to influence current and potential masses by being integrated into today's visual world. In the fourth chapter of the book named 'How Is a Country Image and Identity Construction Reflected via Discourses in Press?' discursive forms of construction of Turkey image through online international press is examined. The study, where the platforms of New York Times, The Daily Express, Spiegel Online International and Le Monde Diplomatique investigated by Van Dijk's critical discourse analysis, focuses on the cyclical relationship in different countries and Turkey's identity building depending on the ideological point of view. In the fifth chapter of the book, named 'Journalism and Communication Design in New Media', the reflections of changes and developments in information and communication technologies in the field of journalism integrated into new media are examined. In the study, the assumption that continuous transformations and developments in the new media change the perception of journalism and traditional applications of communication design is emphasized.

In the sixth chapter of the book, named 'A Critical Appraisal of Crime Over Social Networking Sites in the Context of India Social Networking Sites', the legal regulations on cybercrime and the types of digital activities defined as cybercrime are examined in terms of the practices in India. The study discusses various cybercrimes such as Cyber Insult, Cyber Pornography, Cyber Stalking, False Transactions and Misinformation and Hacking in the context of social media crime types regulating the provisions of the Indian Criminal Code and the Information Technologies Law. In the seventh chapter named 'Mobile-Based Social Media, What Is Cutting? Mobile-Based Social Media: Extensive Study Findings', the usage areas and increasing effects of mobile social media applications are examined. The study, which focuses on mobile applications in Kenya, defines these applications in seven themes and explains the usage patterns in mixed method. The study also explores the impulses behind the popularity of mobile social media applications and the social media motivation of users. The eighth chapter of the book, named 'Erdogan vs. Erdogan: A Polarized Post-Truth Case in Social Media Reality', examines the relation between the political polarization and social mediatic representation in Turkey around the concept of Post-Truth and biographical representation of Erdoğan, an effective political subject, in virtual communities. According to the study, which presents examples of distortions of facts through the representation practices in two large Facebook groups consisting of supporters and dissidents of Erdogan and maps post-truth relationships, Erdoğan's original personality moves away from the original also in social media as in the system of images in traditional media representation and he is absorbed in the fantastic representation practices of the two opposite poles. In the ninth chapter, named 'Is Somebody Spying on Us? Social Media Users' Privacy Awareness', the extent of the confidentiality problem is examined through the field study, depending on the boundaries of the private and public spheres that transform with social media. The study, which aims to understand the privacy criteria of university students depending on the concepts of trust and sincerity, focuses on the determinants of the relationship between the sharing person and the people/followers or non-follower but witnesses in cases of privacy violations. In the study, with the social media platform itself becoming a problem about privacy in recent years, attitudes towards information privacy are increasing. In the tenth chapter, named 'Virtual Resistance of "*Çiftlik* (Farm) Bank Scapegoats" and Discursive Atonement of "Being Scammed"', the processes of transforming social media into a purification mechanism as a compensation of the victims of *Çiftlik* (Farm) Bank, which defrauds the masses with the investment system, known as the Ponzi Scheme, is examined. In the study, which aims to understand the *Çiftlik* (Farm) Bank investors', declared as scapegoats by the society, virtual community organizations on Facebook through the critical discourse analysis approach, a theoretical framework is drawn to the efforts of the victims to justify themselves and to avoid the gloom of scapegoating by using various rhetorical instruments. The deceived masses of the system that collects investors with religious and political rhetoric by deviating from the aims of its establishment adopt an incriminating attack attitude with the opposite modification of the same rhetoric to get rid of scapegoat and to purify.

In the eleventh chapter of the book, named 'Identity Design and Identities Exhibited in Social Networks: A Review-Based on Instagram Influencers', the relations of sharing in social networks and the relations of self-representations with identity building are examined. The study, which questions the mass prevalence of trend types as a form of relationship and construction under the leadership of popular social media users/Influencers, assumes social media as an appropriate, flexible, variable and slippery social base in terms of identities. The twelfth chapter, named 'Real-Time Marketing as a New Marketing Approach in the Digital Age a Study on the Brands' Social Media Sharing in Turkey', examines the reflections of marketing activities to the example of the real-time application brands in Turkey. The study, measuring the interaction metrics in real-time marketing actions in brand accounts on Twitter, says that brands create positive images in this way. According to the study, real-time marketing activities on digital platforms are used extensively to build brand awareness in an emotional and rational context. In the thirteenth chapter of the book, named 'Social Networks: The New Medium of Advertising – Instagram Case', the social media considered as the new medium of advertising and the social media phenomena that exist in these networks and defined as influencer and the relationship between advertising are examined. In the study, considering the joint studies with social media phenomena that have become a very important social media strategy in today's advertising world, semi-structured interview technique was used. Data obtained from interviews with 50 Instagram phenomena were analyzed and advertising collaborations with influencers in social networks were evaluated. The fourteenth chapter named 'Rise of Facebook in the USA and WeChat in China Commodification of Users' provides a critical economic-political review of the online applications, Facebook and WeChat. The study, which examines the profit and ownership relations of social networks within the framework of the 'Audience Commodity Analysis' model of critical theorists, shows the political economy of Facebook and WeChat through an archive survey of newspaper reviews, OSN-based companies policy documents and survey results. In the fifteenth chapter of the book, named 'Voyeurism in Social Networks and Changing the Perception of Privacy on the Example of Instagram', the monitoring relations in social networks are examined in terms of visual self-presentations and peeping relations on Instagram. The study, which provides a description in terms of Turkey, focuses on the digital representation of transformations in the culture of privacy. The study, which interprets the relationship between traditional culture and digital culture from the perspective of Voyeurism, considers this conflict as a problem, providing the infrastructure for cultural crises and identity complexities. In the sixteenth chapter named 'Carnivalesque Theory and Social Networks: A Qualitative Research on Twitter Accounts in Turkey', the public activities and sharing events of Turkish users on the Twitter social network were examined in the context of Carnivalesque Theory and a structural description is made. In the study using Content Analysis technique, the prominent concepts and markers in theory are searched for the equivalents in popular Twitter accounts. In the study, it is stated that the structure of Twitter, which is one of Turkey's most popular social network, is very similar to the structure of Carnival and Carnivalesque Theory.

CONCLUSION

This book is valuable as it contains works that explain the nature of visual communication with practical cultural examples and theoretical approaches. The value of the book, as well as the scientific work in the content, is also related to an understanding of a publication that plans reader interests. Because in a wide and multi-part field such as visual communication, theoretical descriptions are expected to give more weight to some problems and to be in close relationship with the forms influenced by popular culture. In accordance with the assumption that constitutes the theme and core of this book, the studies were prepared by considering concrete cultural equivalents in daily digital practices. This book, which examines theory in daily life, or rather, in the non-daily and continuous life of digital media, collects the explanatory energies of a collective that has scientifically witnessed the transformation of visual communication.

REFERENCES

Baudrillard, J. (1998). *Simulakrlar ve Simülasyon* (O. Adanır, Trans.). İzmir: Dokuz Eylül University Press.

Hall, S. (1997). The Work of Representation. In S. Hall (Ed.), Representations. Cultural Representations and Signifying (pp. 13–74). London: Sage Publications.

Keyes, R. (2004). *Post-Truth Era: Dishonesty and Deception in Contemporary Life*. New York: St. Martin's Press.

Manovich, L. (1998). *Database as A Symbolic Form*. Retrieved from http: //manovich.net/index.php/projects/database-as-a-symbolic-form

Manovich, L. (2001). *The Language of New Media*. MIT Press.

Manovich, L. (2006a). *After Effect, or Velvet Revolution (Part 2)*. Retrieved from http: //manovich.net/index.php/projects/after-effects-part-2

Manovich, L. (2006b). *What Is New Media? In* R. Hassan & J. Thomas, (Eds.), *The New Media Theory Reader*. Berkshire: Open University Press.

Acknowledgment

As an editor, I owe a great debt of gratitude to my dear academics who wrote chapters for this book. This book will shed light on researchers, students, and academics interested in the field, thanks to the contributions of valuable authors. I would also like to express my gratitude to the academics of the editorial advisory board who provided full support in the constitution of the book.

I would like to thank IGI Global Editors, who answered my questions in every detail about publishing the book and did not refrain to support.

I am grateful to my parents who have supported me throughout my education life and career. I am very grateful to my beloved fiancée, who has given countenance to the development of the book. Finally, I would like to thank the IGI Global publishing house for helping me to publish my first edited book.

Chapter 1
Presentation of Visual Culture Elements in Digital Environments With Special Effect Technologies

Türker Elitaş
Kyrgyz-Turkish Manas University, Kyrgyzstan & Hatay Mustafa Kemal University, Turkey

ABSTRACT

This chapter is important in terms of the visualization of the meaningful relationships of the digital media which emerge with the integration of technology into the communication processes and the special effects applications which are the other technology that consecrate it. In addition, visual culture is a phenomenon that is shaped by vision-based elements in the attitudes and behaviors that arise from a simple point of view through evaluation and reception process. In this process, there is a dialectical relationship between perceiver and perceived. The most important argument of this dialectical relationship is the images. In this context, this chapter is important in terms of revealing the fact that special effects applications, which are the constituent element of the image in the digital media, which presents the relationship between perceiver and perceived, are presented with simulaks and the individual determines the cultural attitudes and behaviors in these digital environments with these unrealistic images.

DOI: 10.4018/978-1-7998-1041-4.ch001

INTRODUCTİON

The understanding of communication, which has reached a different dimension in the focus of technological developments, has become a discipline in which the messages are mass-consuming, especially with the mass media spreading on the social plane. The subjective structure of communication has evolved into objectivity produced under the influence of social collectivity with mass communication tools. The most widely used mass communication tools are undoubtedly new communication technologies. Addressing the majority of the world's population, these technologies and the basic indicator of these technologies, the internet, and computer, during a period of mass-oriented transmission processes, has undertaken the task of producing information. The introduction of individuals' participation in social processes through new Communication Technologies has led the internet to become a source of communication that produces meanings for social reality. New communication technologies, which gather many different sources of information on a holistic basis, address different audiences with heterogeneous content. As a matter of fact, the reality produced in New Communication Technologies, which dominate both visual and auditory sensations, has a high level of credibility.

The new functions that are added to the communication tools together with the developments in communication technologies create changes in the functionality of the vehicle to which it is added. Manipulative operations on the information used to produce meanings of reality are very important in the context of creating a new perception of reality. Visual content used in digital media is altered by manipulative processes and is intended to be ambiguous.

The manipulation process, called special effects, refers to the manipulation of visual and audio content. First, with the special effects that began to be used in cinema, television's rise and with the transformation into a home-type cinema, it has also been used in television content, and in recent times, new communication technologies have become the founding element of the virtual space. Special effects applied as analog in the early days, today with the development of digital-based technologies, mostly applied in the computer environment. Special effects, which are the constituent elements of visual culture, have become one of the essential elements of the narrative structure of the image, reinforcing the reality produced by connecting the visual and auditory content of the image. The start of building the organic bond of the image with special effects has resulted in the presentation of encoded meanings in the content with special effects.

In this study, a theoretical and conceptual discussion on visual culture and individual relation is made on the relation of visual culture elements with special effect technologies in digital environments. In this context, study data analysis is based on the theory of meaning, which is a hermeneutic (interpretation) approach that emerged in the 19th century. The hermeneutic approach makes it possible to read and study texts composed of written texts or images in human sciences, especially in philosophy, art history, religious studies, linguistics, and literature.

ON VISUAL CULTURE

Visual culture is an interdisciplinary concept. Therefore, it is possible to encounter different definitions by different disciplines. However, the concept at the center and the common point of these different definitions is the image element, which is the constituent elements of the image. According to Dunjum, which explains visual culture through this constituent element, *"visual culture is the objects and images we encounter in our daily lives, such as TV, movies, books, magazines, advertisements, home, and clothing design, shopping mall and amusement park, show arts and other visual products and communication forms"* (2002, p. 19). Again, Tavin, who takes the image to the center and explains the visual culture from a postmodern point of view, evaluates the visual culture as a result of the effects of the images produced by communication technologies on human attitudes and behavior. In addition, visual culture is a situation that focuses on various visual practices; this is indeed every visual that is made up of various objects and images in motion (2009, p. 22).

Visual culture, which is a description of the acquisition and attitudes gained with the sense of sight, is a re-interpretation of the codes attributed to culture through the experience of seeing. This interpretation of the visual act is particularly closely related to the depth of the relationship between the image and the individual and to the positioning of the image. In this ordinary movement, the image is an important element in the process of interpreting and internalizing what an individual sees and showing variation from individual to individual. So understanding the visual culture in full is a good way to read the concept of image.

In this culture where images are dominated, individual forms attitudes and behaviors under the roof of the image. The image, a projection that the senses transmit to the brain, can be explained as the object's response to the brain. Thoughts on the image go back to ancient times. The image especially tried to be explained in a philosophical plane by Plato and Aristotle, has been transformed from past to present as an expression of the relationship between perceiver and perceived, and the interactivity between perceiver and perceived has been tried to be explained as an image in the present age. The image of the object perceived through the sensory organs is the imaginary image of the object in the mind (Bolay, 2009, p. 44). Akarsu, on the other hand, described as *"a tangible or intellectual copy of something that is perceived by the objects and the bodies of senses, which are seen in a way that can be directly re-introduced to an object"* (1988, p. 104). The image is not just a trace of visual activity in the brain. It is also an image designed by memory without warning. In this context, images can be created by the projection of the human brain in the brain without being perceived. In this context, it is possible to search for visual cultural arguments, which are a process that is woven with images, everywhere the image exists.

Visual culture, which is an image process, requires the effective use of visual action. In this process, especially the individual's vision of the images and analysis of these images is one of the most important characteristics of visual culture. In this context, developing communication technologies, especially the means of transmitting images from writing to the internet, triggered the rise of visual culture, as well as the fact that the individual perceives the images according to global values. In the period when new communication technologies, which McLuhan described as the electronic age today, were introduced to social and cultural structures, Bernard's claim that the visual cannot be independent of culture (2002, p. 27) has lost importance with the formation of global culture created by global images.

It is possible to divide the historical development process of visual culture into two periods. The first of these periods is the period starting with the written tradition in which the instinct of seeing and being affected was dominant and the cave paintings were used as a narrator, and the second is the process that begins with the written tradition of culture and the written tradition in which seeing is more important than other senses, and that communication technologies are channeled between perceiver and perceived.

The stage of modern visual culture, which begins with the written tradition, finally extends to digital elongation. From a modern point of view, Mirzoeff points out that visual culture was created in the hegemony of new communication technologies after the 1960s and that images were a tool of manipulation over perceptions (1998, p. 45). In this context, the phenomenon of visual culture has entered into a transformation in parallel with the development of technology, and the image has only moved away from the projection in the mind of the perpetrator and became a tool for the construction of perception.

SPECIAL EFFECTS TECHNOLOGIES AND DIGITAL CONSTRUCTION OF THE IMAGE

IIn parallel with the development of technology, special effect applications have been a major design technique in many areas, from theatre to TV, computer games to entertainment and from cinema to radio programs, showing great improvement since the day they were first used. This design technique is the technological builder of a construction process for visual and auditory senses. While this construction process is aimed at reproduction of reality, it is circulated through the reproduction of real communication technologies. Special effect technologies direct the individual's perceptual processes by interfering with or manipulating the constituent elements, especially the image, of the fact that enters the circulation.

"Special effects" refers to techniques based on visual and auditory effects" (Yurdigul & Zinderen, 2013, p. 11). The special effect that emerged in France at the end of the 19th century, which has become an invention of the computer adapted to modern life, was first applied in cinema in 1895 (Yurdigul & Zinderen, 2013, p. 13). "The Execution of Mary Stuart" is a special effects application, which enters the world of the show and is frequently mentioned in many projects in the following process, which means that the image will be changed and the image and sound effects of the imagination will be processed.

Special effects technologies, which are widely used in cinema, are used in the content of mass media which are of aesthetic concern and aim to reach the desired level of interest. In this context, digital media, which is organized by the media paradigm of the dominant ideology and which is an effective producer for values inherent in this paradigm, has become a communication space with the content of system-oriented representations and the structure that nourishes the social legitimacy and derived sub-systems, and the continuity of ideological processes and building the conscious perceptions.

The determinant position of the simulations in digital media representations has led to hyper-reality, which makes reality more attractive in societies that make the show phenomenon a way of life. As a necessity created by this orientation, the concept of special effects has silenced qualitative differences between media content. The use of special effects in digital environments, as in cinema, *"makes it very easy to create scenes that are impossible to create in natural ways or that are too risky or costly to create"* (Yurdigul & Zinderen, 2013, p. 12).

Especially after the 1990s, both video and audio play played on the viewer, and the lure of digital media combined with special effects applications has become the reason of choice at the individual level. In particular, individuals who develop practices for visual perception in digital environments live the possibilities offered by the internet without limits and consumes them by seeing and internalizing the content from a modern perspective.

While the individual consumes the content under the roof of the attractiveness offered by digital, it also reshapes the visual texts prepared within the visual feast and the visual culture memory from the past. The individual who is updating his visual culture at the modern level, which began to emerge with the printing press, is facing the manipulation of the image.

According to Manovich, special effect applications or visual manipulation techniques that have become very popular in cinema and feed the film industries have almost completely acquired a computer derivative, with communication technologies developed from the 19th century to offer advanced opportunities in the 20th century (Keskin, 2018, p. 337). This feature gives special effect technologies an integral part of a massive composite, and it has become an essential part of participation and interest in every environment in which it is used.

The visual culture, which has reached a different dimension in the focus of technological developments, has become a modern phenomenon, especially in the mass consumption of the messages along with the mass distribution of mass media on the social plane. In this transformation process, the subjective structure of the object has become objectivity produced under the influence of social collectivity with mass media. The most widely used mass media is digital media. Addressing the majority of the world's population, this communication technology has undertaken the task of producing mass-focused information. The initiation of the participation of individuals in social processes via the internet has led to the rise of digital media to be a source of communication that produces meanings for social reality. New communication technologies, which gather many different sources of information on a holistic basis, address different audiences with heterogeneous content. As a matter of fact, the new communication technologies that dominate both the visual and auditory sensations processes have a high level of credibility.

The efforts of individuals to understand the events and phenomena that are developing around them have always existed in the historical process as a result of their urge to recognize, learn and share. The phenomenon of an image, which produces indirect meanings related to social reality and spreads its meanings to the social plane, provides clarity and demonstrates continuity in terms of its need for meaning. The image phenomenon at the center of visual culture affects social processes as a source of information that produces meanings for social realities and produces a wide range of journals. In this regard, the image is seen as a priority for individuals to obtain information about reality.

With advances in communication technologies, new functions that are added to the communication tools create changes in the functionality of the tool it is added to. Manipulative operations on the information used to produce meanings of reality are very important in the context of creating a new perception of reality. Visual content used in digital environments is intended to be vague by manipulative processes.

The manipulation process, called special effects, refers to the manipulation of visual and audio content. First, with the special effects that began to be used in cinema, television's rise and with the transformation into a home-type cinema, it has also been used in television content, and in recent times, new communication technologies have become the founding element of the virtual space. Special effects applied as analog in the early days, today with the development of digital-based technologies, mostly applied in the computer environment. The special effects that reinforce the reality created by connecting the visual and auditory content of the image have become one of the essential elements of the creation or consolidation of perceptual processes. The start of building the organic bond of the image with special effects has resulted in the presentation of encoded meanings in the content with special effects.

The continuous development of technology triggers a parallel change in visual culture. Especially the developments in communication technologies have brought the individual's relationship with communication tools to high levels. The individual who puts his / her relationship with communication technologies to high levels is constantly exposed to messages created with the image and sound element which is an output of these technologies. The individual who constructs his ideas about the real life, including the cultural phenomenon, through the image and sound, has to update every manipulation of the image and sound or the image process in his mind in the output of every new communication tool.

Special effects are already one of the most important applications in the presentation of the image, which is one of the main constituent elements of the image and supporting the image, by manipulating the sound. Special effects, which are subject to many classifications, vary according to the intended use and the vehicle used. Yurdigul and Zinderen (2013, p. 12), who hold all variables in a pot and make a general reading, classify special effects technologies as follows.

Optical effects
- Slow motion
- Dunning Pameroy processing
- Schufftan
- Optical drive
- Scene transition
- Rear projection

Practical effects
- Atmospheric effects
- Pyrotechnic effects

Miniature effects(models/constructions)

Matte painting

Make-up effects
- Makeup and prosthetic practices
- Animatronics

Sound effects

Digital effects (CGI)
- Animation (3D)
- Chromakey
- Bullet time
- Motion capture
- Embarkation
- Image Manipulation

These effects, especially on the image and sound as a technological development is included in the literature. Because there is no scene that can no longer be explained by these effects. All the images that will be introduced into the mass circulation have been transformed into a visual and audio feast with special effect technologies in our age. However, these technologies not only eliminated the technical disadvantages, but also interfered with the individual's semantic plane. This intervention, according to the content prepared and the purpose of the communication tool on the legitimacy and negativity of the debate is still ongoing, the existence of an intervention based on seeing is not open to discussion.

The most commonly used special effects in digital environments are CGI-based effects, of course. These effects that are prepared in the computer environment are also being circulated through the computer.

Animation (3D): creating models for 3D animation in a computer environment, the impression of motion is realized with computer-based animation. While space or object models can be made using various geometric shapes, the process becomes a bit more difficult when it comes to human or creature modeling. It may be necessary to make a physical sculpture before it is designed in a digital environment. The physical model or any object to be scanned is transferred to the computer by the laser scanning process called 'cyber-scan' (Yurdigul & Zinderen, 2013, p. 51). "*Cyber-scan method is a feature that allows the entire human body to be scanned and transferred to the digital environment.*"(Yurdigul & Zinderen, 2013, p. 52).

Chromakey is the process of creating impossible, expensive and danger scenes through the computer. This process is a process of compositing the virtual environment or real environments as a background, which means combining multiple images (Yurdigul & Yurdigul, 2014, p. 131). In other words, it is the process of creating a background.

Bullet time: this technique, successfully applied during the shooting phase of the Matrix film, was carried out with 124 cameras placed in the studio environment and a green box background. The film was placed and shot in such a way that the cameras were placed to create a ring, and the camera turned around the player in the scenes where the time was frozen and the player was hanging in the air. With this technique, the camera rotated around the player, creating a more streamlined narrative structure with the effects of acceleration and deceleration of movements, with the effect of the green box, another scene was placed in the background.

Motion capture: digital motion capture technique (Rickitt, 2000, p. 311) is defined as the recording of the movements made by a live actor with the devices connected to the computer. This motion data transferred to the computer is applied to character models designed in 3D to allow the model to simulate one-to-one movements of the live actor. When these devices are applied to the 3D model, the face and the physical movements of the model are realistic, such as a human or an animal movement.

Trap: the most common technique used in digital environments in special effect variants. In this method, the main paradigm is the incorporation of many images into the image. In particular, even in devices that are described as smartphones, a simple application can be prepared with the trap method, visual cultural elements discussed as colors, symbols, and signs in the sense and perceptual strengthen intervention.

Image manipulation: digital technology allows to play different kinds of images. One of them is 'color-contrast' arrangements (Rickitt, 2000:82). This means that after transferring the captured image to the digital environment, adjustments should be made at color and light levels.

VISUAL CULTURE AND SPECIAL EFFECTS TECHNOLOGIES FROM A POSTMODERN PERSPECTIVE

Visual culture is a process in which images and pictures reveal meanings through the combination of images. In this context, this culture is an important element in two different areas of activity in which the individual physically participates and takes part with his digital identity through communication technologies. In these two different activities, using the act of seeing an individual effectively creates judgments, positions its membership and assimilates the truth.

The basic elements of visual culture are colors, symbols, and signs. While these visual arguments serve as a representation of perception and meaning in the process of meaning the truth of the individual, they also assume the role of language and codes of the new culture, which at the same time is considered as postmodern. In particular, individuals who participate in digital media with their digital IDs make a real difference over these languages and codes. However, these elements, which are often prepared with special effect technologies, are not only presenting the truth but also presenting the image that is close to the truth. Baudrillard considers this as a transition to a symbolic society. Baudrillard, who makes symbolic social inference through the relationship between image and reality, speaks of the existence of the imaginary image of itself that has no relation to reality (Baudrillard, 2011, p. 19).

The image is the only representative of reality. In this context, the signs of visual culture in digital space are actually a realistic simulation. In this simulation axis, the individual who builds his / her own reality and environment, as Baudrillard states, declares his / her membership to a symbolic society and interprets visual cultural elements in the magic of the visual feast prepared with special effects.

Colors, symbols, signs and other metaphorical meanings attributed to visual culture draw the boundaries of the image. Especially when these limits are evaluated by postmodern reading, they refer to the limitlessness and the copy of the truth.

From a postmodern point of view, visual culture is an artificial culture. This is particularly noticeable when a vehicle enters between perceiver and perceived. In digital environments, this tool is the new communication technologies and the internet that promise timeliness and uncertainty. Especially in this environment where all arguments about visual culture are circulated effectively, the designer of artificiality is also a special effect technology. In fact, this artificiality is almost the same as the simulations and simulation approaches of Baudrillard. The visual cultural elements prepared during the dawn of Digitalism are the postmodern definition of this illusion, presented under the title of the claim of reality, but which shows variation according to perception or is constructed as real perception.

Special effect applications, which are the most commonly referenced technology in all image environments, are frequently used in digital environments. Especially in these environments where CGI-based effects are used too much, special effect technologies make the reproduction of reality in the illusion surrounded by simulations and also simulate the environment in which the individual physically participates in the simulation in digital space.

It is impossible to make a conciliatory definition for the concept of postmodernism, which has frequently been debated since the twentieth century. In addition to the theoreticians who claim that this concept refers to the final phase of modernism as a process, after the First World War, modernism is now replaced by postmodernism and in this process the concept of theoreticians who claim that the mass media provoked a change in the social, cultural and economic level.

Whether the post is after modernization or the final phase of modernism. (Mongu, 2013, p. 29), but there is one thing that is common. This is the fact that a change is taking place. Harvey, who claims that this change is basically moving towards the ontological field, talks about the same change *"as opposed to the perspective of modernism, which allows the modernist to better understand the meaning of a complex but yet singular reality, the question of how radically different realities can coexist, touch each other and intertwine"* (1997, p. 57). Harvey argues that postmodernism and modernism have deviated meaning and purpose. Hassan explains these changes and new formations by comparing modernism and postmodernism with each other and argues that the postmodernization phase is temporary, discontinuous and differentiating and that it promises individual pleasures by feeding on a totally disastrous environment (Hassan, 1985, pp. 123-4).

In the postmodernist paradigm, media, communication tools, virtual reality, top reality are important tools of democracy and individuality and myth. It recommends that everyone is equal, and that even the individual has equal rights in the face of institutions and the state, and that the individual should be strengthened in the face of sovereign ideologies such as the state. According to postmodernism, modernism has many problematic areas. Modernism cannot produce solutions to these problematic areas. These problems are humanism, peace, tolerance, human rights, democracy, and majority repression. Postmodernism is a process that seeks solutions to these problems and does not reject modernism but feeds on it.

Looking at all these developments regarding postmodernism, postmodernism is always ready for new formations. Cultural breakthroughs and new togetherness in this paradigm, especially in support of locality and individuality, are proof that the concepts of modernism, which postmodernism promises to solve, have changed as desired.

Mirzoeff, who states that visual culture is a culture of postmodernism, considers this culture as one of the consequences of the capitalist system, stating that it is the field of understanding, learning, and perception acquired by those interested in visual technology (1998, p. 3). In this context, as visual technology Mirzoeff emphasizes the importance of cinema, television and the internet in visual culture.

CULTURE OF VİSUAL MANİPULATİON AND THE 'NATURALNESS' OF SPECİAL EFFECT

The participatory culture of the new media has broken the monopoly authority in production processes and the flow of information that media cartels dominate. The production of content in the monopoly of the info-capitalists has become a synergistic collective where participants are invited in the new media. Manovich links the source of increasing visualization in the media and the fundamental innovation that builds new media, the rise of software and the use of personal versions that are available to everyone (2012). The reality underlying the content in the new media culture is software technologies, and these technologies enable the user to control the content with its specificity. The coded structure of software technologies requires the digital representation of the expected instruments, the ability to flex and gain 'variability' features that allow them to be played on (Manovich, 2001, p. 22). When visual content is transferred to the system with numerical representation, it acquires the qualities that can be reworked infinitely at the same base. This creates a visualization practice dominated by effective interventions by destroying the possibilities for the image to remain faithful to its origin.

The internet system that builds the infrastructure of the new media and this system is required for the production of cultural code interface designs, social media platforms, web sites, mobile applications, games, and interactive shopping environments are a software (Manovich, 2008, p. 94). Software, which is a prerequisite for defining organic reality in the system, creates virtual social centers that are adapted to everyday life with structural designs that simulate sharing relationships and creates space for multiple relationship representations. Because new media leads to new ways of accessing manipulative information, techniques, search engines, hypermedia, databases, visualization, and simulation (Manovich, 1999, p. 15).

Image technology shows a radical change towards the end of the 1990s. Media formats such as cinematography, graphics, 3D animation, typography are represented in each other and combined infinitely. At the beginning of the millennium, linear media was obsolete and hybrid media became normative culture (2007, p. 1). Although the variety of coding content is closely related to the segmentation of the media, the coding cycle and modularity features have an effect on visual variation. Because the variable code structure of an image contributes to the addition of modeled designs in different software. It is not a coincidence that the new media has changed the visual, which is defined in the system by means of the basis of trans narratives and the multi-media elements, and that it is open to changes in color, light, and parts, even with a technique which is called "pop-art" in the art. A visual can switch to a completely different representation as a result of the concrete additions of the designer, apart from the linear meaning it carries. Indeed, in the new visual humor culture called Caps/Memes, it is constructed in a similar way to the functions carried by caricatures and chromosomes. The photo, which is recorded from anywhere in everyday life and which is viral in the common sharing culture, gets the special effect structure with the addition of a written recipe. Special effects, which are seen in news, movies, shows, television programs and all types of publications where software is used, help to establish a culture of entertainment in the new media. In this regard, Caps/Memes applications are evaluated as an amateur humor product, as well as the technical dimension and special effects should also be evaluated as a variant. Because the basic foundation of Caps/Memes culture is software. At the beginning of this software comes the package called Photoshop. Manovich (2011), in his article on Photoshop, points out that this software is the most important production center of the effect culture. Because Photoshop, which is created for direct visual manipulation, prepares the context for adding content as well as for manipulating the image. Caps/Meme applications often occur through the addition of a text strip that describes the content, not directly through the replacement of existing images. This process is very simple in terms of Photoshop techniques, the priority of speed and aesthetic concern in design is not carried in the sharing system, making it easy to enter in a short period of time. It includes a number of production techniques such as design, typography, cell animation, 3D computer technologies, animations, painting, and cinematography, as well as new expression styles that diversify the expression beyond Photoshop (Manovich, 2006b, p. 3).

While talking about the revolutionary qualities of the mass communication infrastructure, it can be said that the new media controls a low level of tension at many points. The cultural form that shows the most change with the new media is included by the visual aesthetic understanding. Throughout history, aesthetic culture, which guided the desires of societies and their perceptions of beauty, has also determined the forms of presentation of the truth, which visual representation is based on. As a form of art and documentation, the volume of photography gained in social life has questioned the manipulative and interventionist world of painting. However, the revolution with the new media has resulted in a loss of social justifications for the photography being faithful to the 'live image'. According to Manovich, the interesting thing is that unlike many aspects of digital culture, the transformation of the visual aesthetics and the shifting of the visual culture in the axis does not take much criticism (2006c, p. 1). After this shift, which does not disturb the communities and does not concern them, a cultural mechanism has been created and served by the software programs for special effects. Instagram and similar photographic-oriented social media can have significant traces of the effect of culture. George Meliés' special effects on visual culture (Manovich, 2001, pp. 176-178) are now among the routines of photography practice. Interest directed to visual effects without the need for a mere image of photography, while making reality more questionable, the reason why it is not questioned can be explained by The Velvet Revolution. After this digital revolution that surrounded the millennium era, pure image is a matter of choice and exception, and effects and hybrid images are normalized (Manovich, 2007, p. 1). In addition, there is a special effect orientation in the relationships that everyone creates with their images, from professional media content producers and desktop publishers to amateur social media practitioners. In fact, it is difficult to come across an image that has no effect on everyday social media. Design-oriented productions, such as Caps/Meme, require that the photo be transformed through software and reworked with manipulative interventions. Instagram, which is a photo-based social network, is the reason behind the ever-increasing user profile, the prevalence of visualization culture, the development of mass communication infrastructure and the use of tools and visual construction to manipulate users ' visual content (Manovich, 2016A, p. 4). The creativity features that are important to the participation culture and prosumer type users, allow the image to be transformed into a visual feast with human intervention, thereby saving reality from monotonous typesetting. Because photographic vision is technical and mechanical creativity that controls the relationship of the eye with the object. The fact that the content presents a different visual feast than the one seen with the machine is considered among the basic merits expected from the users (Sontag, 1993, p. 103). The fact that users are a multi-media content producer together with the new media also included the possibility of interacting with the visual feast. For more visual aesthetics and feast, users who compete to produce a different composition have constantly expanded the boundaries of intervention in the image.

According to Manovich, it is necessary to make a separate space in the history of the new media during the stages in which the historical flow is interrupted. Because the new media culture is an intermediate form, that is, the transition culture, where the visual is dominant. Because in the practice of reading, the distribution of written symbols has been reduced, and the tradition of establishing the narrative in relation to visual elements has started to settle (2016a). Manovich (2016a), who works on a new kind of cultural analytical called instagrammism, argues that the visual representation system that creates this culture is a software-oriented manipulation. The visual content block, which is evaluated in a wide spectrum from normal photographs to special-produced content (caps/memes, graphic, etc.), is perceived as a natural and usual production technique rather than a shame of playing on the image.

Manovich (2010), who discusses what visualization means in the digital network system, says that the image transfer recorded with cameras is over and that "objective image" is not a necessary requirement in computer-generated images. Visualization in the digital system is not the job of adapting the photo, but rather the job of creating the image. From the moment a visual is taken with digital cameras, passing through computer processes it becomes a completely different form. For this reason, traditional photography culture should be evaluated separately from digital photography culture (Manovich, 2016b). While photography had a lot of informatics and documentary functions, such as "presenting evidence to reality" in previous periods, today's digital culture does not have much importance on the connection of a photograph with reality. The photo has lost its traditional features in terms of many aesthetic and design elements in reflecting the content, apart from including real objects and individuals.

Computer Generated Imagery technology, which is at the heart of today's visual production processes, increases the use of computers. Many social media users use Illustrator for vector-based images, such as industry professionals or amateur photographers, Photoshop for pixel-based, Wavefront for 3D modeling and animation, After Effect for visual effects and 2D modeling (Manovich, 1998, p. 3). The design called Caps/Meme has an info-graphics feature, so it needs the design process. With mobile phones adapting to the growing culture of special effects, it is normal for the photo menus to offer automatic effect options. The visual effects that are automatically applied as soon as the picture was taken have also brought convenience to sharing. Manovich's new media 'automation' feature, which is among the principles of the new media, describes a system that is calculated by computers and which processes without human intervention. Visual sharing sets, which are described as image culture, have a special effect network called 'automation aesthetics' (Manovich, 2017) which is cliched and a relationship practice in which all content in this culture is organized by computers. Manovich (1997) discusses photographs taken by computers in the context of an automation system when the new media began to be debated for the first time. The software provides the individual with the possibility of technical intervention, but each process has a part that is completed and calculated by computers. When Caps/memes design is carried out, the design moves of the individual are calculated by writing the program and the commands defined in the code system are activated automatically. Red strips to be added to the photo are written as code in the software base. The rendering calculation is automatically performed by the software when converting the ribbon, JPEG, PNG, or other formats applied with the command. A similar logic works on mobile phones in effect applications before and after photography. The effect applied to the photo by instant commands has quickly reflected the screen to complete a pre-defined algorithm and visual matrix.

CONCLUSION

The phenomenon of digitization is a kind of response to technology in the present century. This concept makes itself felt in almost every area in the social life dynamics of the individual, as well as in the case of communication as the editor of social life practices, becomes a processual and instrumentally effective subject. In particular, the individual who is positioned at the heart of their lives in this era where the communication action is now based on technology, has introduced new media, social media, media content providers, file sharing programs, etc., and new socialization, sharing, learning, and transferring media without physical participation and has been involved in a digital-oriented interaction and communication process by developing relationship practices. The internet, which is an alternative to the places where it is physically involved, invites the individual with the facilities and facilities it offers and directs the individual to question all the learned and gained performances by providing physical participation, primarily belonging to the individual.

As a projection of reality, the individual who begins to position himself in digital environments is interacting with digital identities like themselves through their digital identity and even building a digital perception based on reality, which is a reflection of reality. In this process of perception, the individual is actively using his or her vision as in routine social life practices.

With the pleasure of being in many places at the same time with digital media, the individual traveling on a simultaneous axis by developing different types of communication and codes is also interacting with the technology itself. This interaction with technology is the transmitter of the image, which is the constituent element of the individual and visual culture that develops attitudes and behavior through the action of vision.

By offering the pleasure of spatial limitedness to the digitality of which technology is blessed, the place ends its hegemony over belonging; it redefines many dynamics, especially culture, which are restrictive and organizer of space, under the umbrella of limitless and physical belonging. The individual who positions himself over space and defines his language, religion, attitude, and behavior through spatial commitment uses the power of the visual action during this definition process. However, the individual who interprets his actions based on this view under the shadow of the norms and values of the place to which he belongs is also included in the social process by accepting what he sees.

Throughout the history of mankind, the individual who contributes to continuous visual production processes and who is affected by them realizes cultural construction on this axis. During this construction process, the space that is only physically involved is no longer an important actor, but instead, digital spaces designed by digital reality are important. It is also part of a digital-centric entity that has established its presence in this media with its digital identities. The most important of these formations is undoubtedly cultural environments based on the logic of digital-based reconstruction of visual culture.

Visual culture is a way of understanding objects and images that we encounter in everyday life using visual practice. However, with this understanding and interpretation based on the visual act, it points to a problematic area in the digital environment. Because the individual interprets the image with the reality conveyed by the instrument in the digital environment. In a digital environment, this reality is simulated, manipulated, modified simulation.

In this context, it is discussed that special effects are used in digital environments as an essential element in the scope of the study and special effects are effective in the context of creating a perception of reality, it is discussed that the use of CGI-based special effects in the construction process of cultural codes is becoming widespread and saves time and cost, and the use of CGI-based special effects is an effective narrative tool for cultural content that takes the form of a demonstration, and that special effects are an element of inter-text transitions.

REFERENCES

Akarsu, B. (1988). *Felsefe Terimleri Sözlüğü*. İstanbul, Turkey: İnkılap Kitabevi.

Barnard, M. (2002). *Sanat, Tasarım ve Görsel Kültür. (Güliz Korkmaz, Çev.)*. Ankara, Turkey: Ütopya Yayınları.

Bolay, S. H. (2009). *Felsefe Doktrinleri Ve Terimleri Sözlüğü*. Ankara, Turkey: Nobel Yayın Dağıtım.

David, H. (1997). *Postmodernliğin Durumu. (Sungur Savran, Çev)*. İstanbul, Turkey: Metis Yayınları.

Duncum, P. (2002). Visual Culture Art Education: Why, What and How? *Journal Art and Design Education*, *21*(1), 14–23. doi:10.1111/1468-5949.00292

Keskin, S. (2018). Reklam Gerçeğinin Dijital Failleri Olarak "Efekt Kimlikler": Turkcell'in Emocanları Üzerinden Bir Kimlik Okuması. *Global Media Journal TR Edition.*, *8*(16), 328–353.

Manovich, L. (1997). Automation of Sight from Photography to Computer. Retrieved from http://manovich.net/index.php/projects/automation-of-sight-from-photography-to-computer-vision

Manovich, L. (1998) Database as A Symbolic Form. Retrieved from http://manovich.net/index.php/projects/database-as-a-symbolic-form

Manovich, L. (1999). Avant-Garde as Software. In M. Revolutions (Ed.), *S. Kovats*. Frankfurt, Germany: Campus Verlag.

Manovich, L. (2001). *The Language of New Media*. Cambridge, MA: MIT Press.

Manovich, L. (2006a). After Effect or Velvet Revolution (Part-1). Retrieved from http://manovich.net/index.php/projects/after-effects-part-1

Manovich, L. (2006b). After Effect or Velvet Revolution (Part-2). Retrieved from http://manovich.net/index.php/projects/after-effects-part-2

Manovich, L. (2006c). Import/Export: Design Workflow and Contemporary Aesthetics. Retrieved from http://manovich.net/content/04-projects/051-import-export/48_article_2006.pdf

Manovich, L. (2007). Understanding Hybrid Media. Retrieved from http://manovich.net/index.php/projects/understanding-hybrid-media

Manovich, L. (2010). What is Visualization. Retrieved from http://manovich.net/index.php/projects/what-is-visualization

ManovichL. (2011). Inside Photoshop. Retrieved From http://manovich.net/index.php/projects/inside-photoshop

Manovich, L. (2012). Media After Software. Retrieved from http://manovich.net/index.php/projects/article-2012

Manovich, L. (2016a). Instagrammism and Contemporary Cultural Identity. Retrieved from http://manovich.net/index.php/projects/notes-on-instagrammism-and-mechanisms-of-contemporary-cultural-identity

Manovich, L. (2016b). What Makes Photo Cultures Different? Retrieved from http://manovich.net/index.php/projects/what-makes-photo-cultures-different

Manovich, L. (2017). Automating Aesthetics: Artificial Intelligence and Image Culture. Retrieved from http://manovich.net/index.php/projects/automating-aesthetics-artificial-intelligence-and-image-culture

Mirzoeff, N. (1998). What is Visual Culture? (Nicholas Mirzoeff, Ed.), The Visual Culture Reader. (s: 3-13). USA: Routledge.

Möngü, B. (2013). Postmodernizm ve Postmodern Kimlik Anlayışı, Atatürk Üniversitesi Sosyal Bilimler Enstitüsü Dergisi, 2/17.

Rickitt, R. (2000). *Special Effects: The History and Technique*. New York, NY: Billboard Books.

Sontag, S. (1993). *Fotoğraf Üzerine. (Trans.) Reha Akçakaya*. Istanbul, Turkey: Altıkırkbeş Publications.

Tavin, K. (2009). Engaging Visuality: Developing a University Course on Visual Culture. *The International Journal of the Arts in Society*, *4*(3), 115–123. doi:10.18848/1833-1866/CGP/v04i03/35641

Yurdigül, Y. & Yurdigül, A. (2014).Tv Haberlerinin Anlatı yapısının Oluşturulması Sürecinde Özel Efekt Teknolojileri: NTV ve CNN Türk Ana Haber Bültenleri üzerinden Bir İnceleme. İstanbul Üniversitesi İletişim Fakültesi Dergisi. 46, 123-148.

Yurdigül, Y., & Zinderen, İ. E. (2013). *Sinema ve Televizyonda Özel Efekt*. İstanbul, Turkey: Doğu Kitapevi.

ADDITIONAL READİNG

Dereli, A. (2006). Görsel Efekt Teknikleri. Videograph. Sayı: 1. İstanbul.

Elitas, T. (2018). Televizyon Reklamlarında Özel Efekt Teknolojileri ile Sunulan Marka İmajı. *Social Sciences Studies.*, *4*(17), 1426–1437. doi:10.26449ssj.503

Fieling, R. (1985). *The Technique of Special Effects Cinematography*. London: Focal Press.

Miller, R. (2006). Special Effects: An İntroduction to Movie Magic.Twenty-first Century Books: Minesota.

Warren, C. (1959). *Modern News Repording*. New York: Harper and Row.

KEY TERMS AND DEFINITIONS

Culture: The culture that creates a sense of belongingness by directing the attitudes and behaviors of the individual is also an important element in the transfer of the social memory to the new generations as a means of transference. This content is the reflection of the general who determines the attitudes and behaviors of the individual as part of the social structure.

Digital Media: It refers to the change of traditional media with the development of communication technologies.

Image: Image is theunderstanding and impression that the person or institution has left in the minds of other people or institutions. Your image, your clothing, your behavior, your ability to speak, your manners and courtesy rules as a whole and the way you are perceived by the society.

Internet: It is a worldwide and widely growing network of communication, where many computer systems are interconnected.

New Communication Technologies: New communication technologies resulting from the infrastructural integration of traditional communication technologies are the technologies that use computer skills and interact between users and information.

Simulacr: Indicators representing reality.

Special Effect Technologies: It is a way of creating scenes that are commonly used in the film, television and entertainment sectors, which cannot be created by normal means or are too risky.

Visual Culture: The individual entering into a new reading process against the environment and the social memory based on the visual culture makes an interrogation on the attitudes and behaviors that he / she considers outside of the acceptability.

Chapter 2
Visual Literacy and Visual Rhetoric:
Images of Ideology Between the Seen and the Unseen in Advertising

Paulo M. Barroso
Escola Superior de Educação de Viseu, Portugal

ABSTRACT

Advertising imposes ways of seeing, thinking, feeling and acting; it leads consumers to act without them noticing; it creates an ideal social imaginary of a "perfect world" or "happy ending" for the daily needs and problems of consumers. Advertising does this by formulating a proposal for a collective and ideal good. Following a theoretical strategy and a critical analysis, it is an approach intended to relate rhetoric, ideology, and literacy of advertising image, exploring the implied ways of the seen and the unseen (i.e. what visual messages say and show). Advertising is a public and massive myth-poetic and logo-poetic device and an increasingly multiform, omnipresent, seductive and visually persuasive. It is important to understand the elements of (explicit or implicit) meaning and the corresponding processes and mechanisms through which the meanings produce effects. This chapter assumes itself as a contribution to a desideratum that may be called visual advertising literacy.

DOI: 10.4018/978-1-7998-1041-4.ch002

INTRODUCTION

Modern and Western cultures are increasingly visual and rhetorical. In these cultures, ubiquitous images convey messages appealing everything all the time. These messages are not innocent. On the contrary, they follow planned strategies of seduction and persuasion, imparting ideas, ideals, values and imposed ways of seeing, thinking, feeling and acting (ideologies) mixed with seemingly simple and understandable information.

The most common, inevitable and influential messages are those of advertising. Advertising is everywhere, showing images as command words, such as "buy it now", "try", "drink", or "enjoy". "Advertising has become an accepted part of everyday life" and "the symbolic attributes of goods, as well as the characters, situations, imagery, and jokes of advertising discourse, are now fully integrated into our cultural repertoire", argues W. Leiss et al. (2005, p. 3). Using an imperative form and a seductive image, the effectiveness of advertising messages is mostly due to the visual impact. The images are simply to understand and to follow their commands.

For example, advertising is strategically designed to highlight the characteristics of the product, using images of certain elements (sun, heat, summer) and colors (red, yellow, blue), so that these characteristics, thus evidenced, raise the audience's desire of consumption. Therefore, beer advertisements use visuals to suggest the taste and texture of the product, i.e. to call to mind the heat (the consumer's need or problem) and the refreshing sensation (the consumer's satisfaction or solution) drinking the beer.

In this chapter, the visual literacy applied in advertising is focused in the potentialities of the image, on what the image shows without showing (the implicit). The advertising's visual literacy is an ability to read / understand the rhetorical strategies of advertising messages. It is necessary to identify the practices of meaning production that makes up the advertisements and dismantle its engine to see how advertising works, i.e. to break down the advertising strategy itself and to understand the forms of advertising in the societies in which they are inserted, as well as the effects of meaning.

However, the application of this strategy of understanding and observing advertisements requires (such as advertising literacy) the ability to critically analyze and interpret content or statements, interrelating psychological, social, symbolic and ideological processes.

The main objective of this chapter is to understand and critically analyze the strategies and mechanisms of the meaning of visual representation as a socio-cultural construct. To understand visual literacy as a cultural and practical competence is also to recognize the importance of these skills of cultural understanding and visual hermeneutics against the visual rhetoric used by advertising strategies.

This chapter aims to analyze the advertising rhetoric, i.e. the strategic exploration and creation of myths or mythical meanings expressed publicly in advertising images. It is argued the increasingly secular, tautological and paradoxical semiocracy and iconocracy in the public space, because of a screen-society based on public and massive (advertising) speeches, which are myth-poetic and logo-poetic devices, i.e. paradoxical public discourses (namely screen images) with argumentative fallacies and a sophisticated visual rhetoric. This public space becomes both a social and mythological imaginary and a mode to express ideologies.

Meanings are constantly produced and influence us everywhere. Every day we receive and make use of a large variety of advertising signs. Thus, literacy is relevant to let us know or find a way of understanding how these signs and meanings are as expressive as influential. The power of signs (to create representations) increases the persuasive force of advertising and the pertinence of literacy approach lies in its awareness about the wider field of meaning-making. For this reason, the primary focus of this chapter is on how literacy can be used in the study of the advertising, considering that meaning in the advertising messages are conveyed by signs and literacy is concerned with the ways of how signs work and may be read and understood. If "language is the most fundamental and pervasive medium for human communication", as Jonathan Bignell (2002, p. 6) says, rhetorical advertising language is even more pervasive and inflowing. Thus, literacy is a useful skill to our perception and understanding of such pervasiveness and influence, since our perception and understanding of reality is constructed by signs (words and images) which we use every day.

However, our relationship with reality and signs that name real or unreal things is not so simple. Our relationship with reality and signs depends on our perception and understanding of reality, which is constructed by the signs. A sign is a medium and a medium is conventionally something which acts as a channel, passing something from one place to another. Any sign has the function *sine qua non* of representing something, i.e. "to be instead of" or "to be in the place of". This is the replacement function or the semantic transitivity *aliquid stat pro aliquo* (Eco, 1984, p. 213), i.e. "something is in the place of something else" (Barroso, 2017, p. 343). There must be enough literacy to perceive this semantic transitivity; otherwise, communicative interaction does not result in mutual understanding. The use, perception and understanding of signs requires literacy, because a sign always disclosure anything latent by its representation. In the advertising language, signs belong to sign systems and require the consumer's literacy for the perception and understanding of the respective semiosis process, the recognition or grasp of something that functions as a sign in the message advertised. This semiosis process is like a mental construction of the reality through the signs of our language. As Jonathan Bignell (2002, p. 7) says, "signs and media are the only means of access to thought or reality which we have".

We live surrounded by signs (particularly advertising signs) and sign systems, like advertising. These signs shape us, they are sometimes ideological and shape our ways of seeing, thinking, feeling and acting. We use and consume them constantly and everywhere. Our consciousness and experience "are built out of language and the other sign systems circulating in society that have existed before we take them up and use them." (Bignell, 2002, p. 7).

Advertising is one of the most powerful sign systems we have. Accordingly, for Saussure (1959, p. 14), language "is the social side of speech, outside the individual who can never create nor modify it by himself; it exists only by virtue of a sort of contract signed by the members of a community". As per Saussure (1959, p. 130), "in reality the idea evokes not a form but a whole latent system that makes possible the oppositions necessary for the formation of the sign", because "by itself the sign would have no signification". Therefore, Saussure's perspective admits that our thought, our sense of social identity, our collective and individual experiences, our perception and understanding depend on the systems of signs, like the sign system of advertising, which codify our social life and give shape and meaning to our consciousness and interpretation of reality.

Considering that "there is no perfect analytical method for studying the media", according to Jonathan Bignell (2002, p. 3), "different theoretical approaches define their tasks, the objects they study, or the questions they ask in different ways". Therefore, following a theoretical and reflexive strategy and a critical analysis of the role of rhetoric and literacy in advertising, an approach to relate rhetoric, ideology and literacy of advertising image is intended, exploring the implied ways of seeing and not seeing (i.e. what advertising visual messages are saying / not-saying and showing / not showing).

If rhetoric plays a key role in the production of effective messages, literacy shows its usefulness, allowing to receive these messages in a more comprehensive way from the public perspective. This methodology serves to conceptualize and problematize rhetoric and literacy in advertising and, more specifically, in the realm of the image. The theoretical approach is supported by the bibliographical resource of authors and reference works. After understanding the specific and delimited fields of visual rhetoric and literacy, the strategy focuses on the recognition, identification and characterization of the meaningful elements usually present in advertising messages, since they influence both the production and the reception of the meanings. Then, the approach is more practical, using examples to demonstrate the opposite (but also complementary) fields of rhetoric and literacy, i.e., the production (coding) and the reception (decoding) of the meanings. There is an inevitable dialectic between the visible and the invisible in the images, a language-game regarding the seen and the unseen, based on what is shown and what is not shown. Visual literacy and visual rhetoric are both recognized as belonging to a visual culture, where images convey ideology between the seen and the unseen in advertising.

The concept of literacy has changed with the new visual, global and digital cultures. Therefore, it is important to reflect on the appropriateness of media literacy, starting with the curricula (Burn & Durran, 2007, p. 95). Given the influence of the media in today's information societies, it is surprising how little research and attention has been paid to advertising literacy. On this point, we must agree with Malmelin (2010, p. 130). As advertising is omnipresent and multiform in today's societies and it is also an important component of media literacy (Silverblatt et al., 2014; Potter, 1998), advertising literacy is a training or a tool for a cognitive ability; it is useful for everyone, because we are all consumers in act or, at least, in potency.

BACKGROUND

Advertising's principal task is to increase sales, but it also communicates, explicitly or implicitly, social values, ideas and ideals. Advertising is a rhetorical device to shape and disseminate cultural standards. Advertising uses persuasion (*Pheitó*, as the name of the Greek goddess that personifies seduction and persuasion) to promote an idea or motive to the audience. Advertising attempts to persuade the audience changing behaviors and attitudes.

According to Silverblatt et al. (2014, p. 32), persuasion is a function in which the communicator's objective is to promote an idea or motivate the audience to change specific behaviors and attitudes. "The ultimate purpose of persuasion is control. Advertising attempts to persuade you to think positively about a product and, ultimately, to purchase the advertiser's brand." (Silverblatt et al., 2014, p. 32).

The audience is "suggested" to think positively about a product and to purchase the advertiser's brand. Using visual (but also textual) rhetorical figures, advertising influences as far as the literacy or illiteracy of the audience allows.

Advertising imposes ways of seeing, thinking, feeling and acting; it leads their targets to act without them noticing; it creates an ideal social imaginary of a "perfect world" or "happy ending" for the needs and everyday problems of consumers. It does this by formulating a proposal for a collective and ideal good. It is this proposal that unites the rhetorical image of advertising, ideology and (il)literacy. Considering that advertising is multiform, omnipresent, seductive and persuasive (i.e. socially influential), it is important to understand the visual elements of meaning (explicit or implicit meanings) and the corresponding processes and mechanisms of producing meaning effects.

In advertisements, the visual elements must be capable of representing concepts, abstractions, actions, metaphors, etc. (Scott, 1994, p. 253). Public discourses must follow Aristotle's recommendation to elaborate an effective discourse in four phases: *inventio* (*πίστις*, *pisteis*), to grasp what needs to be proved; *dispositio* (*τάξις*, *taxis*), to arrange the arguments; *elocutio* (*λέξις*, *lexis*), deciding how to express; and *actio* or *pronuntiatio* (*ὑπόκρισις*, *hypocrisis*), to deliver the speech with appropriate gestures and expressions (Connolly, 2007, p. 148). All these four phases are needed to arrange the visual elements that must carry meanings in different manners and styles. "The rhetorical intention behind a visual message would be communicated by the implicit selection of one view over another, a certain style of illustration versus another style, this layout but not that layout." (Scott, 1994, p. 253).

Remembering Guy Debord's *The Society of the Spectacle*, "the spectacle's function in society is the concrete manufacture of alienation" (1995, p. 23). The screens are all over the public space; their images impose socio-cultural values and representations. There is a sort of cultural imperialism on public images. Notably, in the capitalist societies, the dominant classes (the political, religious, economical and media agents) create useful cultural systems to transmit core values and perpetuate the domination. According to Armant Mattelart (quoted by Espinar et. al., 2006, p. 106; Mattelart & Siegelaub, 1979, p. 57), cultural imperialism is a set of processes by which a society is introduced within the modern world system and forms its management by the induction of fascination, pressure, force or corruption, shaping social institutions to match the values and structures of the dominant system. The mediated imperialism promoting the consumerism using public images is alike, it is the media action as an extension process of cultural imperialism.

According to a reflective and critical approach, starting from Deleuze's "civilization of the image" as a civilization of the cliché (Deleuze, 1997, p. 21), it is possible to understand the iconocracy of the public space (the proliferation of advertising screens) and test the hypothesis of a tautological-society transformed by the epidictic and apodictic visual and public speeches of advertising, i.e. speeches re-meaning and secularizing the public space, the social imaginaries and the strategic ways to express the collective thought.

Visuality is dominant in modern cultures. This is one of the main theses of Giovanni Sartori, in his book *Homo Videns: Teledirected Society*. According to Sartori (1998, p. 45), there is a hegemony of the seen, a primacy of the image, a prevalence of the visible over the intelligible, which leads to a seeing without understanding. We live in a culture of the image (Sartori, 1998, p. 115) and the image is not seen in Chinese, Arabic or English, because it is simply seen. Sartori (1998, p. 45) refers to the impoverishment of the capacity to understand reality on the part of the human being when it is exposed to the images and the effects of the media. Modern visual cultures reverse the progress of the intelligible and change it for the sensible, for the return to seeing pure and simple. This perspective of Sartori is relevant to understand how the seeing overlaps the reading, how the image overlaps the word.

However, literacy is not exclusive of the verbal reading. The seeing also has to do with learning and knowledge than just with the simple transfer of images to the brain. According to Gillian Dyer (1982, p. 75), when we see things or images, we know what is there partly because of knowledge gained from previous experience; we read the image "rather than just absorb it, and it is therefore accurate to talk of visual literacy". The modern, visual, global and digital cultures are reproduced in the increasingly visual and rhetorical advertising, a global phenomenon as a "visual turn", but for W. J. T. Mitchell, this might be called "the fallacy of the pictorial turn". Such development of the pictorial would be "viewed with horror by iconophobes and opponents of mass culture, who see it as the cause of a decline in literacy, and with delight by iconophiles, who see new and higher forms of consciousness emerging from the plethora of visual images and media" (Mitchell, 2005, p. 346). Because "many of us are being influenced and manipulated, far more than we realize, in the patterns of our everyday lives" by media messages (and by advertising in particular), as Vance Packard (2007, p. 31) underlines in his well-known *The Hidden Persuaders*, the need and importance of advertising literacy is notoriously justified. Packard's negative perspective about the manipulative and hidden impacts of advertising is also supported by Wilson Bryan Key (cf. 1973 and 1976). Both authors disassemble and criticize the misleading and unnoticed techniques of subliminal perception in advertising.

Anyway, in the present approach on visual literacy and visual rhetoric, advertising literacy focuses particularly on the image and its potentialities, on what advertising visual messages say without saying or shows without showing. That is the implicit. Advertising literacy is taken as an ability to read / understand the rhetorical strategies of advertisements and what do images really want. Visual literacy and visual rhetoric of advertising invite us to ask:

- What does the advertising image say/show?
- What does the image intend and how does it manifest its pretension?
- How to read/understand the image?
- What is the idea that the image conveys and how it conveys?
- How the qualities of the product are enhanced?
- What feelings, sensations, and manifest values are aroused in the advertising visual message?
- What are the latent values that seem most valued in the advertisement?
- What is the connotation of the product with these values?
- What influences and affects the emotions of the audience?
- Who fits the profile of the consumer according to the advertisement?
- What is the promise that the advertisement makes about the product?
- What can the product effectively guarantee?

The identification of the meaning structure and its significant elements that make up the advertisements allows to dismantle the advertising engine to see how it works, i.e. to break down the advertising strategy itself and understand the forms by which advertising manifests intentions, suggestions, affections, effects of meaning.

However, the application of this strategy of understanding and observing advertisements requires the ability to analyze and interpret information, interrelating psychological, social, symbolic and ideological processes. "Understanding advertising's role requires attention to the context of the production of its messages, to the technology utilized, and to the changing habits of mind and techniques employed by its practitioners", states W. Leiss et al. (2005, p. 19). Advertising offers cultural patterns and proposals for a good life and shows how to achieve personal pleasure and social success, since it is based on fables, fairy tales, and troupes. Therefore, it must be understood as representing a cultural discourse (Leiss et al., 2005, p. 19).

The meaning of advertisements is carried out jointly between who "writes" and who "reads and perceives" the signs in the advertising message. These signs are polysemous and belong to the cultural sign system. They produce meanings of everyday experience, inviting audiences to participate in their ideological ways of seeing, thinking, feeling and acting. That's why many advertisements do not directly invite consumers to buy the products, insisting more on an ideological rather than a commercial approach.

ADVERTISING AND LITERACY

Advertising is as effective as profuse. The effectiveness and the profusion are two characteristics of advertising and both are evident in the daily life of modern societies (Malmelin, 2010, p. 130). Advertising manifests itself in many ways and everywhere; it is omnipresent, in an irrefragable way, in all possible and imaginable spaces. For example: "The United States has arrived at the stage of *ubiquitous advertising,* in which all conceivable public space is dedicated to advertising, including checkout lines, gas pumps, ATM machines, and urinals." (Silverblatt et al., 2014, p. 273). Consequently, "*place-based video screens* show advertisements in public spaces, such as gas stations and doctor's offices. Advertisers also reach consumers in nontraditional ways, including podcasts, blogs, video games, e-mail messages, cell phones, and video on demand." (Silverblatt et al., 2014, p. 273). Advertising is intrusive, seductive and persuasive, mainly in the so-called visual cultures, consumer societies or mass markets, where the levels of production and consumption of material goods are high.

Advertising messages change according to the time and space (culture), in terms of content and form. The subject (what they say) and the approaches (how they say) are now more irreverent, implicit and visual. To interpret these messages, consumers (who also change) need literacy, basic skills of understanding advertising messages (e.g. to be able to recognize, evaluate and understand the manifest intentions in the advertisements and what these say or the actions they propose or persuade to consumers (Malmelin, 2010, p. 130).

Advertising literacy is a kind of cognitive filter, a rational competence against the influences triggered by the stimuli (signs) that advertising uses to elicit responses (favorable and unconscious attitudes and behaviors) in the consumers. Advertising literacy is the ability to analyze and evaluate the seductive and persuasive messages that advertising creates using emotional, rational, or ethical arguments.

Advertising literacy presupposes skills to recognize and identify messages, perceiving their commercial and persuasive goals, or their argumentative strategy. It is a kind of personal mechanism of defense and control of the emotional responses to the advertising messages. According to R. Heath (2012, p. 123), "if emotion in communication increases level of attention, then advertisements that incorporate a lot of emotive content are likely to be paid a lot more attention". Consequently, emotion can work subconsciously. That's why McLuhan argues that advertisements are not meant for conscious consumption. For McLuhan (1994, p. 228), "they are intended as subliminal pills for the subconscious in order to exercise a hypnotic spell, especially on sociologists" and "that is one of the most edifying aspects of the huge educational enterprise that we call advertising."

Advertisers may not let us think about the message nor directly persuade us to buy their products, but they may also transform us and compel us to buy a product without even we knowing why we're buying it, "as a visceral response to a stimulus, not as a conscious decision", states C. A. Hill (2008, p. 39), and "this is best done through images".

Advertising messages are essentially aimed to influence their target for consumption and to increase sales of products or services. However, they also have the function of transmitting social values, in a direct or indirect, conscious or unconscious way. Thus, advertising resorts to rhetorical communication strategies and shapes the patterns of culture (Wicke quoted by Galician & Merskin, 2007, p. 37).

More than texts, images are stimuli that reach us on a more affective or emotional level in advertising techniques. The impact of the image is effective in this circumstance as well as for acting on a more unconscious level. The power of visual impressions to awaken our emotions has been observed and harnessed since classical antiquity (Gombrich, 1974, p. 244). The knowledge and use of images for affective strategies (ways in which the image can affect, whether we want it or not, through emotional appeals / arguments), is a powerful force to create emotions and to arouse attention, interest and desire.

Advertising messages influence public attitudes and behavior by appealing to emotions. For example, subliminal appeals to primary emotions such as guilt or the need for social acceptance through color, light, shape, size of visual signs with meanings and connotations that are intended to produce. Advertising messages capitalize irrational feelings (e.g. fear or guilt) to promote products and brands. Whoever conceives advertising messages is concerned with covering up or triggering meanings using rhetorical figures, that modify the perceptions of the consumers about the product, service or brand.

The persuasive power of advertising messages is not unlimited; it's relative. Despite the use of sophisticated techniques of rhetoric and the placement of visual impulses, advertising cannot persuade people to buy something they don't want or dislike (Silverblatt et al., 2014, p. 274): if someone does not like beer, no advertisement will be able to change the consumer and make him drink beer. However, more than convincing that we want the product, advertisers try to convince that we really need the product, creating false needs. That's why they use the imperative (e.g. "Drink X" or "Try Y now!").

In this perspective, rhetoric, ideology and literacy in advertising are adjacent and complementary. Rhetoric is the art of enunciating and visual rhetoric is the same art, but of showing; ideology is a hidden content (ideas, ideals, values) that affects consumers collective conscience and thought; literacy is the ability to understand or not (illiteracy) the advertisements as commodities, ideological goods that go beyond the promotion of the brand or product, based on certain ways of seeing and thinking social values.

Advertising images obviously don't come to us with an instruction manual to be read / interpreted. Besides that, advertising images have different coding levels and the higher coding levels make it difficult to recognize the various meanings and connotations (e.g. metaphorical meanings, connecting an immediate consumer product, such a soda, with an intangible feeling like happiness), insinuations (seductive or sexual) and intertextualities (e.g. with television programs).

Thus, it is important to understand the rhetorical processes of advertising statements and the explicit or implicit elements of signification, as well as the corresponding processes of unveiling the mechanisms that produce meanings and their effects regarding advertising ideals and values. Advertising literacy is the ability to recognize these hidden meanings, the sign system and the meaning production and their effects. Rhetoric, ideology and literacy are related in the advertising image, exploring the implied modes of showing pleasant signs (colors and forms) that are also psychological, social, cultural and ideological ways of seeing and thought.

ADVERTISING'S RHETORIC, IDEOLOGY AND LITERACY

If the concept of ideology, according to Giddens (2006, p. 605), was only first used in the eighteenth century (by Destutt de Tracy), this term gained greater notoriety through Marx. However, if the first use had the sense of "science of ideas", Marx's use had an essentially critical and pejorative meaning, because it means "false consciousness". Marx refers to influential groups in society. These groups are capable of instilling and controlling the dominant ideas that circulate in society in order to justify their position.

In this conception of Marx, a relation between ideology and power is observed, insofar as ideological systems serve to legitimize the power held by certain groups. According to Giddens (2006, p. 605), ideology is about the exercise of symbolic power, it is about how ideas are used to hide, justify or legitimate the interests of dominant groups in the social order. Ideology is, therefore, a set of ideas or beliefs shared to justify the interests of certain dominant groups. An ideology is a shared way of perceiving reality, which assumes that some false (or inaccurate) and imposed ideas or ways of perceiving and understanding the reality are true. As Barthes (1991, p. 137) points out, such ideas serve the ideological interests of a particular group in society and these ideas may be transmitted by advertisements.

Eco argues that "ideology is a message which starts with a factual description, and then tries to justify it theoretically, gradually being accepted by society through a process of overcoding" (Eco, 1976, p. 290). In Eco's semiotic perspective of codes, "there is no need to establish how the message comes into existence nor for what political or economic reasons"; instead, to establish in what sense this new coding can be called ideological (Eco, 1976, p. 290).

Therefore, there are ideologies in all societies and cultures with inequalities between individuals (Giddens, 2006). The meanings of advertisements are designed to move out of the page (in poster advertisements in magazines) or the screen (in TV commercials) and to shape our perception and understanding. Advertising has the traditional function to sell things to us and ask us to participate in ideological ways of seeing (Williamson, 1978, p. 11). But advertising also creates structures of meaning in which literacy is required for their perception and understanding. The need for an advertising literacy is to unchain the way how advertisements are proposed to be read. Consumers need to notice the indirect, subtle or implied ways of seeing.

According to Bignell (2002, p. 31), "ads often seem more concerned with amusing us, setting a puzzle for us to work out". "The aim of ads is to engage us in their structure of meaning, to encourage us to participate by decoding their linguistic and visual signs to enjoy this decoding activity." (Bignell, 2002, p. 31). Since advertisements belong to a meaning structure and a sign system, they make use of signs, codes, cultural and historical meanings, social myths and ideologies and worldviews. Consequently, consumers must be literally prepared to recognize and decode them.

The relationship between the media and ideology, the one that is interesting to analyze in this chapter, arises with the question about the possible ideological weight of the content transmitted in an imperceptible and influential way. If public discourses, such as advertising, favor or promote one ideal over another, the media diffuse ideology and broaden its scope in society as the messages reach large audiences. In this case, the media diffuse values and beliefs that contribute to securing the domination of more powerful groups over the less powerful (Giddens, 2006).

For Slavoj Zizek, ideology can designate anything, an attitude or belief; it impels to action and it arises inadvertently, even when we avoid it. Ideology is not simply a "false consciousness" or an illusory representation of reality; the fundamental dimension of' ideology is rather the reality itself which is already to be conceived as ideological. Ideology is rather the same reality that is already prepared to be conceived as ideological (Zizek, 2008, p. 16). According to Zizek (2008, p. 24), the most elementary definition of ideology is probably the well-known phrase from Marx's *Capital*: "they do not know it, but they are doing it". The concept of ideology implies a kind of basic, constitutive naivete. Ideology seems to arise exactly when we try to avoid it, failing to appear where it is clearly expected to exist: "It seems to pop up precisely when we attempt to avoid it, while it fails to appear where one would clearly expect it to dwell." (Zizek, 1994, p. 4). This is precisely an essential, but paradoxical, characteristic of ideology: its capacity not to be perceived as such.

The concept of ideology implies a constitutive unconsciousness or a collective illiteracy about ideology itself. Ideology and its presuppositions are not recognized. There is a distance or divergence between reality and distorted representation, i.e, a false consciousness of reality. This is because the concept of ideology implies either a naivety, which is collective, for the non-recognition of assumptions and mechanisms of influence, or distance.

The media, particularly advertising, are diffusers of ideology. As a strategic communication technique, advertising fits into an ideological scenario of collective conceptual or mental construction of reality. It is a system of representations, images, myths, values, ideas or concepts with the practical implications of a given culture. Advertising messages impose ways of persuading their audiences to act and behaving in certain ways.

For example, the image of the model Gisele Bündchen projected at the Empire State Building in 2017 in a promotional action of the 150th anniversary celebration of the *Harper's Bazaar* American magazine. The Empire State Building symbolizes the technological prowess and economic strength of the United States and, in general, of the contemporary capitalist world. In this promotional strategy, the building is a means to project the values and content of *Harper's Bazaar* magazine: fashion, beauty, celebrity, lifestyle. This strategy had already been followed in 1999 with the projection of the image of a woman in lingerie on the facade of the London Parliament, a strategy of the men's magazine FHM.

The strategies go even when the president of the United States is used on a billboard in Times Square. In this case, the outdoor of the brand Weatherproof use the image of Barak Obama while he wore one of his coats, adding the slogan "A Leader in Style". The image was a photograph taken during the visit of the United States president to the Great Wall of China and aimed the promotion of the called "The Obama Jacket", according to *The New York Times* (Jan. 7, 2010).

In advertising, a myth should be explored as a form of a sophisticated discourse aimed at meeting the psychological needs of the consumers. The myth is explored in the advertisements that invite consumers to participate in ideological ways of seeing the world and of seeing themselves. These advertisements come with ideological commitments, because they present and propose a given way of seeing the world.

Myths are used in advertising messages to easily convey the main idea: to sale and consume the product, service and brand. Myths create an optimal atmosphere for the product, service and brand and influence consumers that they should be like the situation presented in the advertisement. A myth in an advertisement may simply promise that the product will bring happiness or success, eternal youth and infinite beauty. Connotative images help to construct the myth in the advertisement. Certain images of beautiful, young and attractive women are most prevalent in modern advertising in order to make the consumers (both men and women) fascinated by pictures of "perfect people" in the present global world. Therefore, the study, understanding, identification and recognition of myths in advertising are useful to critique advertisements and to be aware about what influence do social values have on the success of advertising (Galician & Merskin, 2007, p. 45).

Advertising is a different media from other mass media and, therefore, the effects of advertising messages are also different in its use of stereotypes. Advertising use stereotypes to sell the product (Galician & Merskin, 2007, p. 45). The myth plays an important role in the advertising message. The myth is easily confused with a fabulous narrative popularly built. It is, therefore, an imaginative elaboration of the collective spirit; an allegory or metaphorical representation of a situation taken as exemplary and accepted by all who sustain and share it. Through the myth, something is exposed in a representative and unrealistic way. For Barthes, the myth is "speech stolen and restored" (Barthes, 1991, p. 124). In

Mythologies, Barthes says that the myth has the task of giving an historical intention a natural justification, and making contingency appear eternal. "Now this process is exactly that of bourgeois ideology. If our society is objectively the privileged field of mythical significations, it is because formally myth is the most appropriate instrument for the ideological inversion which defines this society" (Barthes, 1991, p. 142). For Barthes, at all the levels of human communication, "myth operates the inversion of *anti-physis* into *pseudo-physis*".

The concept of literacy has changed and advanced in recent years. The emergence of new media forces this evolution of the concept, which now becomes more adapted to the evolution of societies and the diversity of new communication devices and technologies. From the traditional and limited concept of literacy as the ability to read and understand the meanings of written words the evolution has now reached to a more embracing literacy. Thus, it is more appropriate to speak in different types of literacy.

Advertising literacy is a pedagogical and an educational skill for emotional responses against the appeals and arguments used in advertising messages and the influences triggered by the stimuli (conceptual, rhetorical and aesthetic signs, whether textual or visual) that advertising uses to provoke responses (reinforcement, change or creation of favorable, often unconscious attitudes and behaviors) of consumers. For example, in a subliminal, hidden or covert advertising messages, whose designation refers to the degree of camouflage of meanings, but not of signs (cf. Key, 1973 and 1976).

As an ability, advertising literacy analyze and evaluate the seductive and persuasive messages that advertising creates through the recourse of emotional arguments (e.g. "Because you feel X and desire the product Y or the brand Z"), rational arguments (e.g. "Because X is reasonable" or "Y is the best – cheapest - choice") or ethical arguments ("Because X helps you to save and it is the convenient choice"). Advertising literacy essentially requires recognition and identification skills of advertising messages, as well as the perception of the message and its commercial and persuasive goals or the followed an argumentative strategy.

Advertising is persuasive, it is a technique of effective use of language, in which the most important is not so much what is said; it is the way it is said. This rhetorical mode regarding the way of saying is not generally perceivable by the consumer. The cultural and linguistic structure of current and complex societies is based dichotomies, such as ancient versus modern (or tradition versus contemporary). As Barthes (1994, p. 12-14) recognizes in his book *The Semiotic Challenge*, the world is incredibly full of ancient rhetoric. If the contemporary world is full of ancient rhetoric, according to Barthes, there is a common point to which all connotative systems refer: ideology. All meanings and connotations lead to ideology. Ideology is the form of meanings of connotation and rhetoric are the way to express ideology.

When a commercial says "Lose 30 pounds in one week, eating what you like, when you like", it is a sophism. There is a conscious intention about what is said and what is said is not true. When a commercial says "We have been producing cars for over 100 years", it is a paralogism, because there is no intention to deceive, but a fallacy is used (that of antiquity or *ad antiquitatem*) to give credibility to what is said.

Among the languages used by the media, advertising discourse is characterized primarily by its connotative, suggestive and persuasive domain. Whether focused on the product or the brand (to show that the product or brand are superior to other products and brands through a "unique selling proposition"), or focused on the public / consumer (in a segmentation strategy to accentuate the benefits of consumption, calling the consumer public to a certain attractive consumer community), or focused in the context (showing the product involved in a pleasant and attractive environment), advertising explores connotations and creates specific meaning structures. Therefore, "advertising often relies on connotative words to sell their products" (Silverblatt et al., 2014, p. 187). In other occasions, advertisers often employ euphemisms to change the public perceptions of products (Silverblatt et al., 2014, p. 188).

Advertising languages (especially visual rhetoric, as art or technique of using effectively the visible to persuade, influence, please, or provoke desire) are strategically exploited in advertisements using resources and rhetorical arguments conceived to be not aware and rationally understood by consumers.

CODING LEVELS AND LITERACY

If Confucius's maxim "a picture is worth a thousand words" recognizes the polysemous richness of the image, the deliberate formation of coding levels (Eco, 2001, p. 162) further enriches what the images say and show implicitly without saying and showing it, while at the same time hindering their readings / understandings of their meanings, that is, they make it difficult to recognize the various connotations. The more elaborate, complex and high the coding level is, the smaller and more difficult the reading / understanding level is.

For the analysis of an advertisement composed of text (slogan and argument) and image, for example, Eco distinguishes five coding levels. According to Eco, there are coding levels of the image (the first three, below) and coding levels of the text (the last two on the following scale), all five in order of increasing complexity:

1. Iconic level, the place of visual recognition codes. This level groups the concrete, objective, explicit and denotative information about the image (the minimum units assigned to the visual representation). For example, a smiling woman holding a baby or a silhouette representing a woman, which has the function of stimulating people's desire for the product. In this case, the iconic level would just be the image of the woman meaning "woman" (the general idea of "woman"), which everyone clearly and immediately sees in the advertising image.
2. Iconographic level, the place of visual statements contextualized in a given culture; the iconic cultural knowledge (a kind of visual education of a certain culture where the images belong). This level groups the connotations manifested in the image according to two coding types: a) historical coding, a coding based on conventional meanings (e.g. the aureole to signify holiness); b) modern coding, created by the advertising itself (e.g. the dressing style or being fashionable). In the advertisement example of the previous level, the coding would be "mother's love" based on the same image of the woman.
3. Tropological level, the place where the visual rhetoric appears in a trope form, for example, metaphors, considering that images acquire a representative value (the specimen represents the genre, type or species), based on a culturally acquired competence. This level is confined to the figures of rhetoric, the tropes. In the example of a woman holding a baby, it would be "Be like this mother".

4. Topic level, the place of stereotyped cultural connotations. This level comprises all the socially accepted general ideas and stereotypes used in advertising communication, in order to allow rapid and automatic decoding of the message. In this case, it would be "All mothers are like this one".
5. Enthymeme level, where the arguments are unfolded and are presented in an abbreviated way. This level uses the conventional arguments. It pays attention to the polysemy of words. In the example, it would be: "If all mothers are like that and so are you, you are a good mother".

Considering any current advertisement, we may verify if these coding levels correspond to the complexity of the levels of reading and understanding the messages conveyed. Taking the given example of an advertisement with a smiling woman holding a baby, like the advertisement of Johnson & Johnson with a slogan saying "Skin so gentle, as gentle as a mother's love", which has apparently a coding level of low complexity, the coding levels depend on the literacy of the target.

In the advertisement of Johnson & Johnson, the coding levels are identified and recognizable as:

1. **Iconic Level:** (A close-up image of) a woman holding a baby and smiling at him;
2. **Iconographic Level:** Maternity and mother's love (the meaning of "woman" leads to the meaning of "mother" and "baby", that is "mother's love");
3. **Tropological Level:** Comparing the softness of the skin provided by Johnson & Johnson with a mother's love), the slogan is "Skin so gentle, as gentle as a mother's love"; thus, be like this mother (which is an example for all mothers);
4. **Topic Level:** Mother's love (all mothers are so gentle like this);
5. **Enthymeme Level:** If you are also a mother, you are a good (gentle and lovely) mother to Johnson & Johnson.

When the product is unique, objectively distinct and superior to competitors, or when the product offers a clear benefit to the consumers, the task of advertisers is easy, and the structure of advertising messages is simple. But when the product has no distinct benefit nor a different feature from the other products of competitors brands on the market, advertising messages must be based on rhetorical strategies that make the product seem more seductive and distinctive (Silverblatt et al., 2014, p. 317).

In this case, the coding of the message is more careful and strategic. The understanding level is correlative to the coding level and it is hardly literal (for example, the simple recognition of the placement of the product in a television program) and even more rarely comes to appreciative level, according to the following typology of reading / understanding (literacy) levels.

1. **Literal Literacy Level:** Recognition and memory of the facts established in the text / image (main ideas, details and sequence of events).
2. **Interpretative Literacy Level:** Reconstruction of the meaning of the text / image (obtaining inferential meanings from the act of reading), derivation of generalizations, distinction of the essential, abstraction of the message as a sign system, differentiation of justified conclusions from unjustified conclusions, interconnection of contradictory data, etc.

3. **Evaluative or Critical Literacy Level:** Formation of judgments, expression of own opinions, analysis of the intentions of the text / image, elaborated cognitive processing.
4. **Appreciative Literacy Level:** Affection by the content of the text / image and style of expression, influencing attitudes, behaviors, thoughts, knowledge, ways of seeing and interpreting the reported reality.

The reading / understanding levels are literacy levels, i.e. degree of regarding the capacity of decoding the messages (text and image) and, therefore, these levels are connected to the previous coding levels.

MEANING AND LITERACY OF COLORS IN ADVERTISING

In advertising, the meanings of colors, for example, are important for the construction of meaning in the message: red to connote passion and feelings of desire in messages about perfumes; yellow to establish references to the warm and sunny environment conducive to the suggestion of a refreshing drink; blue (of the sea and the sky) associated with freedom, infinite or lack of limits in the proposals that are based on these approaches. The study of the general effects of color as a persuasive communication tool reveal the color as a significant element (Garber & Hyatt, 2008, p. 314), especially in advertising, considering that the world of advertisements is peopled by fantastic images (Scott, 1994, p. 252). According to L. L. Garber & E. M. Hyatt (2008, p. 314), "as much as color is a powerful and salient persuasive communications tool, it is as well a complex, multidimensional phenomenon, poorly understood yet difficult to examine, making individual response to color exposure notoriously hard to explain or predict".

Colors are signs (signifiers with meanings), elements that produce meaning. Colors are "the product of the brain's interpretation of the visual sensory information that it receives" (Garber & Hyatt, 2008, p. 315; Scott, 1994). The study of advertising images is related to the development of a theory of visual rhetoric, argues L. Scott, for whom "pictures are not merely analogues to visual perception, but symbolic artifacts constructed from the conventions of a particular culture" (Scott, 1994, p. 252).

Colors have natural and conventional, denotative and connotative meanings that go unnoticed to the public and consumer literacy. Colors are used intentionally in advertising to express sensations, feelings, ideas, values, etc., because they seduce, they call our attention and arouse interest, exert psychological influence, stimulate desire for the product, service or brand. Colors are stimuli, because they are signs; they are one of the most decisive factors in the consumer choice process. Consumers are sensitive to the visual appearance and the impact of the products during the decision and purchase moments. Colors increase brand recognition and they are also responsible for the acceptance or rejection of products. The judgment of colors is, as a rule, subconscious. Advertisements with more colors are more appealing than those with less colors, because colors facilitate the perception and the reading of the advertising message and predisposes favorably the consumers.

The meaning of advertisements is not only developed by those who conceive the sign in the advertising message. The meaning is also carried out by the consumers and based on their advertising literacy. If the signs that advertising uses are polysemous and belong to the cultural system, these signs create / form myths. For example, the myth of whiteness, that motivates consumers to demand for clothing whiteness. It is like a Manichean combat against two opposite forces, the good against the evil. In this case, it is the whiteness against dirtiness in advertisements for laundry detergents. The meanings are suggested or induced by the signs present in the advertisements and, then, they are understood by the consumers, who complete them mentally. Consumers have a sort of minimum of literacy; if not advertising literacy, it would be general cultural literacy, which are public vehicles of cultural and ideological meaning.

The meanings of the advertisements are designed in such a way as to overflow the commercial statement or consumerist appeal, as well as the medium of communication where they arise. The meanings also produce and reproduce everyday experiences, inviting consumers to participate in their ideological and collective ways of seeing, thinking, feeling, and acting. That's why many advertisements do not directly invite consumers to buy the products, insisting on a humorous approach rather than a commercial approach. Words and images, signs and symbols, shapes and colors, gestures and odors, everything that surrounds us or everything we say or do may become a sign, a language, whether we want it or not. When this happens, all these signs carry connotative meaning, i.e. manifest an ordinary reality, objects or an abstract or concrete, absent or invisible idea.

The meanings of colors are important for the message coding process. Colors are significant elements; they contribute to the meaning structure built by the advertisements. Colors are used strategically in advertising. They have always been used in other forms of life in the history of humankind, in all cultures and social and communicative interactions, such as body paintings in primitive tribes representing virtues or symbolic powers to express sensations, feelings, states of mind, ideas.

Colors always represent something, but sometimes the influence and meaning of colors go unnoticed. In advertising, colors are used to stimulate desire for the product, service or brand, for example, leading to the choice, purchase and consumption of products.

Colors are sensitive and transmit implied information beyond what is expressly stated, as in the case of a slogan with a black formal color that gives rise to tranquility and reason (the *logos* of the message), a green color of the logo (the *ethos* of the brand and its values of credibility) or a red color that exacerbates and excites the *pathos* of consumers.

CONCLUSION

According to Bill Bernbach (quoted by Yilmaz, 2017, p. 47 and Tungate, 2007, p. 51), advertising is not a science, it is a persuasion and persuasion is an art. Persuasion is a rhetorical art or technique of language use and advertising makes use of this rhetorical art, in which the most important is not so much what is said; it is also important the way it is said. In advertising, what is said is not stated explicitly; it is implicit, connoted. What is said is not expressed in a direct, objective way, but even so the message fulfills its persuasive effect and influences consumer behavior.

Given this strategical and rhetorical nature of advertising, messages require a certain level of codification and consequent understanding. Thus, disciplining the way of seeing images is learning to see structures of meaning. Mainly the most hidden, the syntactic forms and semantic models that participate in advertising, according to an innovative or effective and consistent way.

The connotations that the message establishes are triggered from certain stimuli present in advertisements, but they exist in the form of meaning in the public mind, which completes or forms them based on what is advertised. It is part of advertising literacy, considering that the images are strategically chosen and sometimes are mischievous and misleading, not corresponding to the words of the advertisement or to the properties and benefits of the products.

When the advertisement says "this car flies" or "with this perfume you conquer all women", everyone knows it is false, but understand the idea, because the exaggerations of self-promotion, the poetic falsity and the lack of logic and reasonableness do not usually have great evil, especially in the advertising messages (Neves, 2014, p. 384).

Advertising messages dramatize how products satisfy the needs and solve personal problems (Silverblatt et al., 2014, p. 321). The advertisements show smiling and satisfied people who benefit from the purchase and consumption of the product. Even if they do not say it explicitly and only show pictures of smiling and happy people (preferably attractive, beautiful and young women) enjoying the product, as if it were a cause-and-effect relationship, the advertisements suggest that the product will bring happiness and success to who does the same. It's the happy ending advertising.

Women representing the ideal of feminine beauty, celebrities (e.g. Zsa Zsa Gabor or Orson Welles), animals or children are appealing elements for advertising messages to persuade without public awareness of the strategy. However, a responsible ethics in advertising would make it an ideal communicative action based on four fundamental principles that would guide the messages: comprehensibility, sincerity, legitimacy and truthfulness.

As Jean Baudrillard points out in *Simulacra and Simulation*, the most interesting aspect of advertising is its disappearance, its dilution as a specific form, as a medium, since "advertising is no longer (was it ever?) a means of communication or of information" (Baudrillard, 1997, p. 92). "If at a given moment, the commodity was its own publicity (there was no other)", continues Baudrillard (1997, p. 92), "today publicity has become its own commodity".

In this perspective, considering a) that the world changes permanently and cultures become more visual, global and digital, with more instantaneous and ephemeral collective experiences, and b) that rhetorical advertising strategies are increasingly multiform and effective in the capacity for persuasion and distinction between what is true and lie, media literacy (in general) and advertising literacy (in particular) will have a positive impact by training and empowering consumers with valences that protect them from purely consumerist attacks.

In the curricula, advertising literacy prepares students (while future consumers) for a visual, global and digital culture in their habits and customs, as well as sharpening their approaches to consumption, imposing ideal modes of seeing, thinking, feeling and acting, leading consumers to act unconsciously; creating ideal social imagery. It is therefore necessary to extend literacy studies and practices to all areas of life, especially the new media, which are constantly changing and expanding their power of influence over people.

REFERENCES

Barroso, P. M. (2017). The semiosis of sacred space. Versus – Quaderni di Studi Semiotici, (125), 343-359.

Barthes, R. (1991). *Mythologies* (A. Lavers, Trans.). New York, NY: The Noonday Press.

Barthes, R. (1994). The Semiotic Challenge. (Trans. R. Howard). Berkeley: The University of California Press.

Baudrillard, J. (1997). *Simulacra and Simulation* (S. F. Glaser, Trans.). Michigan: The University of Michigan Press.

Bignell, J. (2002). *Media Semiotics*. Manchester, UK: Manchester University Press.

Burn, A., & Durran, J. (2007). *Media Literacy in Schools: Practice, Production and Progression*. London, UK: Paul Chapman Publishing.

Connolly, J. (2007). The New World Order: Greek Rhetoric in Rome. In I. Worthington (Ed.), *A Companion to Greek Rhetoric* (pp. 139–165). Oxford, UK: Blackwell Publishing. doi:10.1002/9780470997161.ch11

Debord, G. (1995). *The Society of the Spectacle* (D. Nicholson-Smith, Trans.). New York, NY: Zone Books.

Deleuze, G. (1997). *Cinema 2 – The Time-Image* (H. Tomlinson & R. Galeta, Trans.). Minneapolis, MN: University of Minnesota Press.

Dyer, G. (1982). *Advertising as Communication*. London, UK: Routledge. doi:10.4324/9780203328132

Eco, U. (1976). *A Theory of Semiotics*. Bloomington, IN: Indiana University Press. doi:10.1007/978-1-349-15849-2

Eco, U. (1984). *Semiotics and the Philosophy of Language*. London, UK: MacMillan. doi:10.1007/978-1-349-17338-9

Eco, U. (2001). *A Estrutura Ausente. (Trans, Pérola de Carvalho from the original La Struttura Assente)*. São Paulo, Brazil: Editora Perspectiva.

Espinar, E., Frau, C., González, M., & Martínez, R. (2006). *Introducción a la Sociología de la Comunicación*. Alicante, Spain: Publicaciones Universidad de Alicante.

Galician, M.-L., & Merskin, D. L. (2007). *Critical Thinking About Sex, Love, and Romance in the Mass Media – Media Literacy Applications*. New Jersey: Lawrence Erlbaum Associates, Publishers. doi:10.4324/9781410614667

Garber, L. L., & Hyatt, E. M. (2008). Color as a Tool for Visual Persuasion. In L. M. Scott & R. Batra (Eds.), *Persuasive Imagery - A Consumer Response Perspective* (pp. 313–336). New Jersey: Lawrence Erlbaum Associates.

Giddens, A. (2006). *Sociology*. Cambridge, UK: Polity Press.

Gombrich, E. H. (1972). The Visual Image. *Scientific American, 227*(3), 82–97. doi:10.1038cientificamerican0972-82 PMID:4114778

Heath, R. (2012). *Seducing the Unconscious: The Psychology of Emotional Influence in Advertising*. Chichester, UK: Wiley Blackwell. doi:10.1002/9781119967637

Hill, C. A. (2008). The psychology of rhetorical images. In C. A. Hill & M. Helmers (Eds.), *Defining Visual Rhetorics* (pp. 25–40). New Jersey: Lawrence Erlbaum Associates Publishers.

Key, W. B. (1973). *Subliminal Seduction*. Upper Saddle River, NJ: Prentice-Hall.

Key, W. B. (1976). *Media Sexploitation*. Upper Saddle River, NJ: Prentice Hall.

Leiss, W., Kline, S., Jhally, S., & Botterill, J. (2005). *Social Communication in Advertising – Consumption in the Mediated Marketplace*. New York, NY: Routledge.

Malmelin, N. (2010). What is Advertising Literacy? Exploring the Dimensions of Advertising Literacy. *Journal of Visual Literacy, 29*(2), 129–142.

Mattelart, A., & Siegelaub, S. (1979). *Communication and Class Struggle – 1. Capitalism, Imperialism*. New York, NY: International General/International Mass Media Research Center.

McLuhan, M. (1994). *Understanding Media – The Extensions of Man*. Cambridge, MA: The MIT Press.

Mitchell, W. J. T. (2005). *What do Pictures Want? – The Lives and Loves of Images*. Chicago, IL: The University of Chicago Press. doi:10.7208/chicago/9780226245904.001.0001

Neves, J. C. (2014). *Introdução à Ética Empresarial* [*Introduction to Business Ethics*]. Lisbon, Portugal: Principia.

Packard, V. (2007). *The Hidden Persuaders*. New York, NY: Ig Publishing.

Potter, J. W. (1998). *Media Literacy*. London, UK: Sage.

Sartori, G. (1998). *Homo Videns: La Sociedad Teledirigida* [*Homo Videns: Teledirected Society*]. Buenos Aires, Argentina: Taurus.

Saussure, F. (1959). *Course in General Linguistics* (W. Baskin, Trans.). New York, NY: The Philosophical Library.

Scott, L. (1994). Images in advertising: The need for a theory of visual rhetoric. *The Journal of Consumer Research, 21*(2), 252–273. doi:10.1086/209396

Silverblatt, A., Smith, A., Miller, D., Smith, J., & Brown, N. (2014). *Media Literacy - Keys to Interpreting Media Messages*. Santa Barbara, California: Praeger.

Tungate, M. (2007). *Adland – A Global History of Advertising*. London, UK: Kogan Page.

Wicke, J. (1988). *Advertising Fictions: Literature, Advertisement and Social Reading*. New York, NY: Columbia University Press.

Williamson, J. (1978). *Decoding Advertisements: Ideology and Meaning in Advertising*. London, UK: Marion Boyars.

Yilmaz, R. (2017). *Narrative Advertising Models and Conceptualization in the Digital Age*. Hershey, PA: IGI Global. doi:10.4018/978-1-5225-2373-4

ADDITIONAL READING

Baran, S. J. (2013). *Introduction to Mass Communication – Media Literacy and Culture*. New York: McGrawHill.

Gaines, E. (2010). *Media Literacy and Semiotics*. London: Palgrave Macmillan. doi:10.1057/9780230115514

Gil, I. C. (2011). Literacia Visual. Estudos sobre a Inquietude das Imagens [Visual Literacy. Studies on Image Uneasiness]. Lisboa: Edições 70.

Hoechsmann, M., & Poyntz, S. (2012). *Media Literacies – A Critical Introduction*. Sussex: Wiley-Blackwell. doi:10.1002/9781444344158

Kress, G. (2003). *Literacy in the New Media Age*. London: Routledge.

Potter, J., & McDougall, J. (2017). *Digital Media, Culture and Education – Theorising Third Space Literacies*. London: Palgrave Macmillan. doi:10.1057/978-1-137-55315-7

KEY TERMS AND DEFINITIONS

Advertising: From the Latin *advertere*, "to direct the attention of someone to", the action of making public, promoting a product, service or brand.

Communication: From the Latin *comunicationis*, which means "to make common" (the information), communication is a global and social phenomenon based on the transmission of information through verbal or non-verbal messages from an emitter to one or more receivers.

Culture: The set of material or immaterial aspects that define a way of life (values, customs, history, traditions, rituals, beliefs, symbols and languages, instruments and consumer goods, laws, codes and norms, social activities and practices, and institutions).

Ideology: A set of ideas, ideals, beliefs and social values disseminated and shared to hide, justify or legitimate the interests of certain dominant group in the social order.

Literacy: The ability to perceive, read and understand common information, statement or content about everyday life.

Persuasion: The communicative practice, activity, technique of exerting influence over other people.

Visual Culture: A form of life based on the visible; a cultural pattern emphasis toward images and screens, with predominance of visual forms of communication and information.

Visual Literacy: The cultural and practical skill to read / understand what images show according to their rhetorical strategy. Visual literacy is focused on the potentialities of the image, on the suggestive and evocative power of images, on what images shows and suggests (the implicit, the unseen).

Visual Rhetoric: The art and technique of using and exploring effectively the communicative and suggestive power of images to provoke effects and shape people's consciousness and thought, such as persuasion, please, desire.

Chapter 3
Visuality in Corporate Communication

Özen Okat
Ege University, Turkey

Bahadır Burak Solak
Trabzon University, Turkey

ABSTRACT

One of the most important areas where visual communication is prominent today is marketing. Brands try to adopt to the visual world of today in order to make their communication with their target audience more meaningful and effective. This way, organizations, and therefore brands, take significant steps for differentiation from their competitors by forming their visual identity. Additionally, considering the current advertisements of brands, it is seen that visual narratives are highly abundant. In this context, brands which are starting to use visual communication effectively are gaining a broader place in the memories of their target audience by increasing brand awareness. As a result, it is believed that the significance of visual communication and identity is increasingly higher in terms of influencing existing and potential masses by being integrated into the visual world of today.

DOI: 10.4018/978-1-7998-1041-4.ch003

INTRODUCTION

The time allocated for reading by people is increasingly rarer in the intense struggle in life with an accelerating pace. Reading and the process of making sense that starts with combining words are processes that are more complicated and time-consuming in comparison to perceiving and making sense of visual content. For this reason, for a simpler and more comprehensible communication process, people always prefer visual content more. An example of this issue is that the number of Instagram users worldwide is higher than the number of Twitter users, and content that is shared on Instagram receives more interaction (Statista, 2019). Another striking example of this is the rapid increase in the rates of using emojis. Now, instead of expressing thoughts and feelings through text, it is possible to communicate with others faster and in a simpler way by emojis which are visual. In some cases, these visual emojis are used to strengthen the thoughts and feelings that are expressed through texts.

Communication that is achieved with "visible" content is known as visual communication. Even though people do not understand the language of each other, they are able to communicate via visual images and symbols. This is because visual communication is more universal in comparison to written and verbal communication, and therefore, it is easier to perceive and make sense of its content. For instance, when someone who visits London without speaking English sees a traffic sign on the side of a road with the symbol of no parking can comprehend what it means thanks to the symbolic narrative even though they do not understand the language that is spoken in that country. Communicating through gestures may be possible in a country whose language we do not speak. Moreover, in today's world where time is very limited and life is very fast-paced, visual images are preferred by individuals much more in comparison to written texts as they are perceived faster. Therefore, with numerous different advantages it provides and for reasons such as the qualities of human nature, visual communication has a much larger place in the world of communication. Thus, it is clear that visual communication should not be left to coincidences in commercial spaces where very large investments take place. The necessity of all organizations taking part in the world of commerce to pay attention to information on visual identity planning and the perceptual characteristics of their target audience is clear.

VISUAL DIMENSION OF COMMUNICATION

A person is a being that prioritizes what is visual. This is because, although we utilize several of our senses for communication, for all individuals who have the capacity to see, the basis of learning, attitudes and behaviors is in direct interaction with visual content. A person assesses the reflections coming from around them within visual communication systems, and more of the time in their daily life, they direct their actions based on visual communication. This visual attitude is almost in the genes of people (Ketenci and Bilgili, 2006, p. 268). In this context, it may be argued that the sense of sight is one of the most important senses in people.

People identify and try to make sense of the objects, events and situations around them firstly by seeing them. Humanity has spent tens of thousands of years to develop this quality (Uçar, 2004, p. 17). Since the dawn of humanity, the most effective form of expression has included the pictures, shapes and symbols that are drawn. With these symbols, humans have aimed to transmit the thing they want to express in the shortest and clearest way possible. The first examples of the shortest and clearest form of expression, namely visual communication, were encountered with pictures and shapes in cave drawings. These symbols not only allowed communication among people living in those times but also left traces that have reached today in the form of knowledge about those people (Yaban, 2012, p. 973). In this sense, it may be seen that visual communication firstly emerged for the purposes of helping decision-making and transmitting information. Likewise, cave paintings and pictograms which are the oldest forms of visual communication were used to transmit information on marking territory, identifying dangers, finding and showing routes for migration and identifying locations of shelter and sources of food/drinks. Usage of such marks may be traced back to more than 200,000 years ago (Jaenichen, 2017, p. 10).

As in the past, visual contents are prevalently utilized and play an important role in the process of communication in today's modern world. In their daily lives, people frequently encounter stationary and motile images, texts and visual elements that are organized in certain forms. These visual messages are designed to serve various objectives such as informing, explaining things, providing directions, persuasion, education, sales and entertainment. The process of visual communication starts on the level where interaction with various visual constructs start for such purposes (Holsanova, 2014, p. 331). The process of visual communication involves sharing all visible things within a perceptual context (Kavuran & Özpolat, 2016, p. 267). Visual communication takes place at the end of the process that progresses in the form of the eye's function of seeing and interpretation of sensationally obtained information in the brain. Information that is collected from the environment and the visual contents that carry this information are stored in a person's memory. In the process that follows, the mind tries to interpret and recall information by using visual messages in different forms (Onursoy, 2017, p. 50). Visual images enter long-term memory storage faster and easier in comparison to written and verbal texts. This is because visual communication is the oldest and most natural form of human communication. As mentioned before, the history of visual communication may be traced back to the cave paintings of the ancient time. In this context, visual communication is considered to be a significant form of communication for all eras where visuality and visual expression forms have been in the center (Uyan Dur, 2015, p. 444).

Individuals and communities in developed or developing countries obtain their information to a large extent from visual sources such as newspapers, television, magazines, posters and brochures. Other than this type of information, several helping tools that make daily life easier are presented visually. For example, people try to recognize places they are visiting for the first time with maps, they utilize visual information consisting of writings, photographs and other symbols before buying a product from a store, or they resort to the guidance of visual elements for distinguishing men's restroom from women's when they visit a place (Becer, 2006, p. 29).

As a general definition, visual communication is transmission of information, ideas and feelings by using various images, symbols and visuals. In other words, all types of exchanging messages through visual contents are known as visual communication. It is one of the three main forms of communication in addition to verbal and non-verbal communication. It is known that visual communication is the form of communication that is perceived by people in the easiest way. Visual communication includes numerous examples such as various signs, typographies, photographs, graphical designs, animations and movies (Ruhela & Parween, 2018, p. 748).

At times of primitive life, the most effectively used communicational contents were pictures, symbols and marks. Instead of speech, several pictures and shapes used as visual communication tools that symbolize their lives and communicate may be observed on clay tablets or rocks where they lived (Ketenci & Bilgili, 2006, p. 266). Today, with the prevalence of using visuals, it has started to be seen that visuals may be used just like written language to transmit a thought or meaning to people. Just as in the case of written texts, narratives may be constructed on different semantic levels by using visual contents (Öncü Yıldız, 2012, p. 74). Different visual contents may be used for meaning creation and transmission on conventional and digital platforms. One of the most popular ones among such visual contents today are emojis that are frequently used on digital platforms. It is seen that usage of emojis on various digital messaging platforms and social media is increasing fast. The visual language that was used at times of primitive life has been reanimated today in the form of "emoji communication". It has now become almost impossible to encounter content without emojis while viewing posts on social media. Similarly, many people use emojis on digital messaging platforms, and it is even observed that people sometimes communicate by using only emojis. This is evidence that the visual-based simple form of communication used by the ancestors of today's people who lives centuries ago is still going on in the context of contemporary developments and innovations.

We firstly identify and try to understand what is around by seeing it and react to it later. This is why the sense of sight is one of the most important senses of humanity (Ketenci & Bilgili, 2006, p. 265). A study conducted by educational psychologist Jerome Bruner at the University of New York also stated that people remember 10% of what they hear, 30% of what they read and about 80% of what they see (Lester, 2006). In this sense, it was argued that products of visual communication are at the top of factors that motivate learning, grab attention, make comprehension easier, transfer information simply and in a comprehensible way and promote learning while having fun (Mercin, 2017, p. 211). In short, communication that is carried out with visual content is both more effective and more memorable. Thus, brands, one of whose most important objectives is to have a place in the minds of their target audience by effective communication, have used products of visual communication from the past to the present intensively.

IDENTITY AND ITS VISUAL FORM

In today's world of marketing where the level of competition increases unpredictably, firms and thus brands that have to make large investments towards the process of communication also utilize opportunities of visual communication. In this period where the importance of visual communication has been proven in several studies (Bosch, LM, De Jong, & Elving, 2005) (Melewar T. C., Measuring visual identity: a multi-construct study, 2001) (Peng & Hung, 2016) (Rutter, Stephenson, & Dewey, 1981), the most useful method for firms and brands to differentiated from and rise among their competitors is showing their difference by presenting their target audience with a visual integrity. Visuality that is different and striking is a factor that always provides advantages for prominence. The most fundamental way for a firm/brand to be able to present visual integrity is a well-managed corporate visual identity design. However, corporate visual identity should be shaped in the context of the identity of the organization. A practice of corporate visual identity that is independent of corporate identity will fail in presenting an integrated point of view and will collapse like a building without a foundation. Therefore, before talking about corporate visual identity, it would be useful to define the concept of corporate identity where it comes from and its components.

Corporate Identity

Corporate identity is usually considered to be interchangeable with the concept of corporate visual identity. However, just as the character of people is not limited to their appearance, corporate identity is not limited to its visual components. It is also seen in recent studies that the attention in the context of the definition of corporate identity has shifted from design to the own nature of the organization, and the importance paid to visual elements in definition has decreased (Bosch, Elving, & de Jong, 2006). Corporate identity is the self-presentation of an organization, rooted in the behavior of individual organizational members, expressing the organization's 'sameness over time' or continuity, 'distinctiveness' and 'centrality' (van Riel, 1997). Identity refers to individuality that is a means by which others can differentiate one person from another (Melewar T. C., Determinants of the corporate identity construct: a review of the literature, 2003). Corporate identity can thus be defined as the picture of the organization in terms of how this is presented to various audiences (Cornelissen, 2004). With this explanation, Cornelissen (2004) focused on how corporate identity is perceived by the audience of the organization and defined identity as "a picture of the organization".

Corporate identity (Figure 1) consists of the components of corporate communication, corporate design, organizational culture, behavior, corporate structure, sectoral identity and organizational strategy (Melewar & Karaosmanoğlu, Seven dimensions of corporate identity: A categorisation from the practitioners' perspectives, 2006). Before getting into corporate design, the is a benefit to emphasize the multidimensionality of corporate identity by shortly defining these components in question:

Corporate Communication

Communication, which already starts in the mother's womb, is a factor that plays a critical role in sustenance of existence for all living beings. Life starts by learning, and all living beings survive with the help of what they have learned. Learning occurs by usage of various forms of communication. Considering this issue in the context of organizations, it may be argued that all organizations have functions of learning and teaching, or in other words, receiving information and providing information. The entirety of all these processes of receiving and providing information is the expression of corporate communication processes. Corporate communication focuses on the organization as a whole and the important task of how an organization is presented to all of its key stakeholders, both internal and external (Cornelissen, 2004). In addition to all these, corporate communication may also be defined as a management function:

Corporate communication is a management function that offers a framework and vocabulary for the effective coordination of all means of communications with the overall purpose of establishing and maintaining favorable reputations with stakeholder groups upon which the organization is dependent (Cornelissen, 2004).

Considering it as a management function, corporate communication that is discussed in the framework of the 4Ps of marketing (Product, Price, Place, Promotion) refers to a much more complicated and planned process that may be considered in the context of the science of public relations and covers various strategies. This is why it covers the concepts of strategic thought, strategic planning, and thus, operational and action plans. In this aspect, there is the question of a controlled communication. How-

Figure 1. The original corporate identity categorization (Melewar and Karaosmanoğlu, 2006)

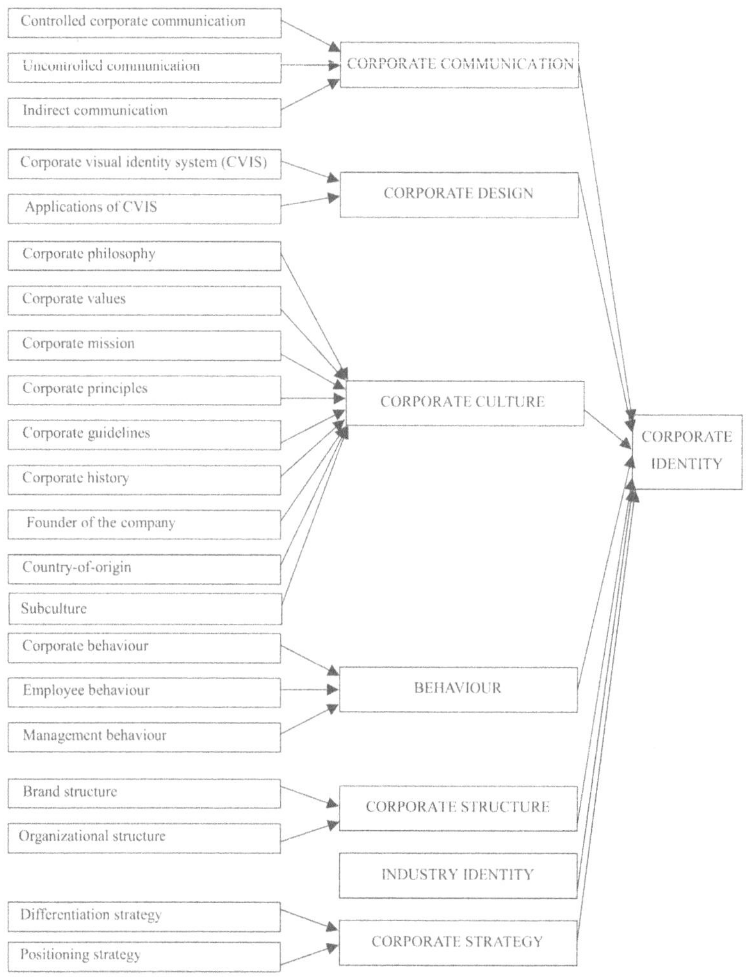

ever, corporate communication has also uncontrollable, spontaneously developing and indirect aspects. Increasingly frequent usage of social media may be considered to be the best example of the uncontrollable dimension of corporate communication. Firms that constantly have to follow social media for this reason are recently spending more and more effort to be able to control social media communication and take action in this field.

Corporate Culture

It is possible to think about the existence of culture as an influential and influenced concept in every community where people are. Culture, which may be defined as a "realized system of making sense, a wholesome lifestyle" (Williams, 1993) refers to the entirety of all types of understanding that affect the identity of an organization directly. In this sense, the concept of corporate communication has been explained with similar definitions. Ayla Okay defined corporate culture as a system of values, beliefs and habits that are shared within an organization and interactively influence the formal structure with the aim of achieving behavioral norms (Okay, 2005). According to another definition, corporate culture is a collection of norms that show how individuals behave and activities that determine interpersonal relationships are carried out with the attitudes, beliefs, assumptions and expectations who work at the same organization even though it is not expressed in written form (Erengül, 1997). Several definitions on the topic of corporate culture state that elements that distinguish all organizations from each other like shared beliefs, values, meanings, myths, ceremonies and symbols form corporate culture (Bakan, 2005). The concepts of corporate identity and corporate culture are a collection of abstract phenomena that form the general structure of the organization, determine the ambiance in the organization and create a living space for employees and a perception or outlook towards the organization for external target audiences (Elden & Yeygel, 2006). Most researchers viewed corporate culture as an effective instrument used to support and reinforce corporate identity to achieve such goals as labor flexibility (Vella & Melewar, 2008). In the light of all these definitions and other definitions in the literature, corporate culture may be explained as an organizations main adopted values, philosophy, mutual views, collective mind, rules, beliefs and assumptions, habits, attitudes, expectations, projections, emotions and history (Akıncı Vural, 2005), (Okay, 2005) (Erengül, 1997) (Balmer & Greyser, Managing the Multiple Identities of the Corporation, 2003).

Corporate Behavior

Corporate behavior is formed out of collection of the behavioral styles of the individuals who work at the organization. In this framework, there are two main approaches that explain corporate behavior. These two approaches may be considered as personnel behavior and managerial behavior styles. It may be stated that managerial behavior style forms personnel behavior style, and the combination of both forms corporate behavior. Considering that organizations cannot be thought of independently of the culture they live in and that behavior is a concept that is completely intertwined with culture, it is clear that cultural factors influence the organization via its employees.

The cultural factors that are related to corporate behavior may be categorized by the cultural dimensions theory of Hofstede. These dimensions proposed by Hofstede define the effects of culture on organizational culture, and therefore, organizational behavior. In the theory of Hofstede, there are six different dimensions of culture, and each community may be categorized based on these six dimensions. These dimensions are as the following (Hofstede, 2011):

1. Power Distance, related to the different solutions to the basic problem of human inequality;
2. Uncertainty Avoidance, related to the level of stress in a society in the face of an unknown future;
3. Individualism versus Collectivism, related to the integration of individuals into primary groups;
4. Masculinity versus Femininity, related to the division of emotional roles between women and men;
5. Long Term versus Short Term Orientation, related to the choice of focus for people's efforts: the future or the present and past.
6. Indulgence versus Restraint, related to the gratification versus control of basic human desires related to enjoying life.

In relation to Hofstede's theory, the behavioral forms of organizations may be classified in the contexts of degree of democracy in decision-making, rate of avoiding uncertainty, level of individualism or collectivism, masculine or feminine character, qualities of short-term or long-term decision-making and whether or not freedom is restricted. For example, employees at firms that are established in individualistic societies are more "me"-centered, while those in collectivist societies are more "us"-oriented. Nevertheless, the issue that should be kept in mind is that each country might not necessarily be at these extreme points that are defined in the cultural dimensions proposed by Hofstede. There may be countries where individualism is intense, and collectivism is less dominant, or vice versa. It is even possible to encounter societies where both approaches are equally dominant. However, the main issue is that the society that is lived in is determining in terms of corporate behavior.

Corporate Structure: Brand structure and Organizational structure

Melewar and Saunders mentioned two separate structural elements in the context of corporate structure which is a key element of corporate identity: organizational structure and visual structure. Organizational structure may also be divided into two foundations: centralized or decentralized. In the centralized structure, the organization accepts a certain location as a center and forms its entire identity based on this center. In the decentralized structure, corporate identity is constructed differently at different centers (Melewar & Saunders, Global corporate visual identity systems: Standardization, control and benefits, 1998). This issue, therefore, affects the visual structure. In the centralized structure, an organization is expressed over a single corporate identity in all markets where it operates. In the centralized structure, it may, for example, develop a unique corporate visual identity system for each different market it operates in. This way, differentiation from market to market in the corporate structure becomes a necessity.

Wally Olins defined corporate identity structures as monolithic identity (usage of a single name and a single visual element by the organization everywhere), endorsed identity (endorsement of an organization for the operational areas it has with its name and identity) and branded identity (ownership of a brand or brands by an organization or its subsidiaries) (Okay, 2005). This classification is a more detailed approach in the context of corporate identity to the centralized and decentralized corporate structures defined by Melewar and Saunders (1998).

Branded structure is among the main factors that both affect and shape the corporate structure. The brand's essence (psychological and physical elements of the brand), brand personality (Who is the brand? What is its identity?), brand value (what it makes the consumer feel), utilities offered by the brand and brand quality (advisability, being a product that provides precise solutions for problems, etc.) may be listed as the components that form brand structure (Aktuğlu, 2004). According to a different view, brand structure is built on three pillars: identity, which includes the sign or group of signs that identify the brand; marketing, which includes the products, and response, which includes the markets (De Lencastre & Côrte-Real, 2010). Therefore, considering brand structure, it is possible to recall all preceding and succeeding components related to the brand. At this point, it is necessary to make a distinction about brand identity and corporate identity which are concepts that are very close to each other, even intertwined. While corporate identity is a blanked concept that defines the entire organization and all elements that are related to the organization, according to Perry and Wisnom, brand identity consists of the controllable elements of a firm, product or service such as an essence, placement, brand name, label, logo, message and experience. The reason for considering these as controllable elements is that they may be changed at any point, while this requires more control in comparison to creating, sustaining and changing corporate identity (Perry & Wisnom III, 2003). Corporate communication refers to rather the internal structure of the organization in the context of some certain elements of it, while corporate visual identity, and thus, brand identity that is an element of corporate design defines the aspect of the organization that faces the outside. As the process of communication with the society has started at the point of exteriorization, a small change in brand identity requires a long-term and detailed procedure, followed by a set of communication activities.

Industrial Identity

The industrial structure where the organization operates and the identity of this industry may be defined as factors that directly influence corporate identity. This is because this structure also determines the environment where the organization belongs to. The competition level in that industry, the size of the industry, its speed of change-transformation and how the organization positions itself in the industry where they operate also affect corporate identity. According to Melewar and Karaosmanoğlu (2006), especially finance firms, banks and petrol companies are a good example for the interaction between industrial identity and corporate identity in the context that they are highly influenced by the industry they are in. Considering them in this framework, the highly frequent exposure of the petrol industry and thus petrol companies to criticisms by environmental protection communities defines the relationship between out-of-control communication and industrial identity (Melewar & Karaosmanoğlu, Seven dimensions of corporate identity: A categorisation from the practitioners' perspectives, 2006). In this sense, it is necessary for such firms to be prepared for a potential activity by environmentalist communities and shape their corporate identity accordingly.

Corporate Strategy

Corporate strategy is the blueprint of the firm's fundamental objectives and strategies for competing in their given market (Melewar & Karaosmanoğlu, Seven dimensions of corporate identity: A categorisation from the practitioners' perspectives, 2006). Melewar and Karaosmanoğlu (2006) categorized corporate strategy as differentiation of corporate strategy and positioning strategy. Nevertheless, as the strategy known as positioning may also include differentiation from competitors, these two concepts may be discussed under a single framework.

The element of competition is at the center of corporate strategy. Thus, competition strategies influence corporate strategies. There are three competition strategies in general (Porter, 2007):

Total Cost Leadership: The objective of this strategy is to bring the firm to a more advantaged position in comparison to its competitors in the market producing substitute products by reducing costs.

Differentiation: The purpose in this strategy is to create something that is accepted as matchless in the sector by differentiating the goods and services offered by the firm.

Segmentation: This involves focusing on a specific group of buyers, a section of the product range or a geographical market.

The choice from among these competition strategies is in parallel with the general structure and objectives of the organization. As the strategy that is selected will also define the strategy of the organization, it will form one of the inputs of corporate identity.

Corporate Design

Corporate design refers to the way of physiological perception of all elements about the identity of an organizations by its target audiences. The scope of corporate design may involve a discussion on several factors including the logo of the organization or the brand, architectural structure of its buildings, decoration and product packaging. The main objective of visual identity is conversion of the philosophy of the organization into a visual language, while visual communication has a substantial meaning and role in this, and this is achieved by corporate design (Okay, 2005).

Corporate design is explained based on three main factors as product, communication and environment (Okay, 2005). These factors may be defined as the following:

Product: Considering that the main objective in the scope of competition strategies is differentiation, rising among competitors in the context of product design is seen as an important factor in terms of gaining strategic advantage. At the first stage, differentiation of the product in terms of its areas of usage is directly proportional to the effort spent by the organization on R&D activities. Designing a different and novel product always plays a key role in the organization's acquisition of strategic advantage. In the second step, the package of the product will not only help in rising among competing products and achieving visual advantage, but it will also get the product to the consumer in an intact and safe way, which will contribute positively to the image of the organization.

Communication: Design of elements of communication defines all visual elements that allow the organization to connect to its target audience. These may be discussed in a very broad spectrum from the logo of the organization and corporate usage of color, and all these elements are shaped within the framework of a corporate visual identity system. Corporate visual identity systems are discussed under a separate section in this study.

Environment: The factor of environment includes several physical design constructs such as the style of decoration, colors, architectural structure and usage of lighting at the working areas and stores of the organization.

All these visual elements that are perceived by target audiences define the outer face of organizations considering the prominence of visuality today, and therefore, these are among the issues that require much care.

Corporate Visual Identity System

The concepts of 'visual, 'visuality' and 'visual culture' have great significance in our period. The most significant evidence for the importance attached to visuality and aesthetics is the growth in the plastic surgery and cosmetics sectors. It was reported that the size of the cosmetics sector in the world is $460 billion, and the medical aesthetics market grows annually by 10% on average (Advertorial, 2019). The most important reason for such large investments that an image is formed in the first second of meeting among individuals and at first sight. In other words, visual image shapes personal image not completely but to a great extent. A similar situation is also applicable to firms. The first image about an organization is perceived by its target audience in the form of visual elements. Such visual elements are discussed in the subject matter of corporate visual identity systems.

A corporate visual identity system is a concept that defines all elements that visually present the organization. Corporate visual identity comprises all the symbols and graphical elements that express the essence of an organization (Bosch, LM, De Jong, & Elving, 2005). "Key elements of a CVI are the corporate name, logo, color palette, font type and a corporate slogan, tagline and/or descriptor, and these may be applied on, for instance, stationery, printed matter (such as brochures and leaflets), advertisements, websites, vehicles, buildings, interiors, and corporate clothing, while sometimes architecture can also be an important element in an organization's visual identity (examples: McDonalds and Ikea) (Bosch, Elving, & de Jong, 2006)." Visual identity is widely regarded as a key dimension of corporate identity: physical aspects of the corporation such as logos and company name are essential features of an organization's identity (Simoes & Dibb, 2008). In the light of all these views, corporate visual identity may be understood to not have the same meaning as corporate identity contrary to popular usage, but it is a key element of corporate identity. That is, the concepts of visuality and aesthetics have a much greater significance today in comparison to the past. Adding the tendency of people to believe what they see more than what they hear onto this significance, people pay attention to visual identity as evidence of the corporate image.

In today's competition-intensive world, the most suitable way of rising among competitors is differentiation. "Like any other identity, visual identity can, in the first instance, be defined in terms of both difference and continuity. Visual identity means difference because it ensures the recognition and proper positioning of a commercial enterprise and because it is an expression of the company's specificity. On the other hand, visual identity means continuity because it testifies to the ongoing industrial, economic and social values of the company (Floch, 2000)." According to this definition by Floch (2000), visual identity is one of the keys of the organization in differentiation among its competitors. Visual identity has four main objectives which may be listed as giving the brand life by providing its positioning and name with a character and a personality, spreading the acceptance and remembrance of the brand, helping the brand differentiate within competition, and connecting different brand elements to each other under the same appearance and emotion (Perry & Wisnom III, 2003). Additionally, visual identity has a highly significant role in terms of protection and sustenance of the organization's image.

Applications of Corporate Visual Identity Systems

The concept of corporate visual identity systems (CVIS) defines how and in what way various visual elements related to the organization will be used. In this context, there are various elements such as organization / brand name, slogan, logotype/symbol, color, typography, pattern (Picture 1), layout, letterhead / template, business card and additional graphic elements (Melewar & Karaosmanoğlu, Seven dimensions of corporate identity: A categorisation from the practitioners' perspectives, 2006) (Okay, 2005) (Bartholmé & Melewar, 2011) (Peng & Hung, 2016). The key elements of a CVI are the corporate

Figure 2. The integral design of the corporate logo, pattern, layout and color plan for TGC (Peng & Hung, 2016)

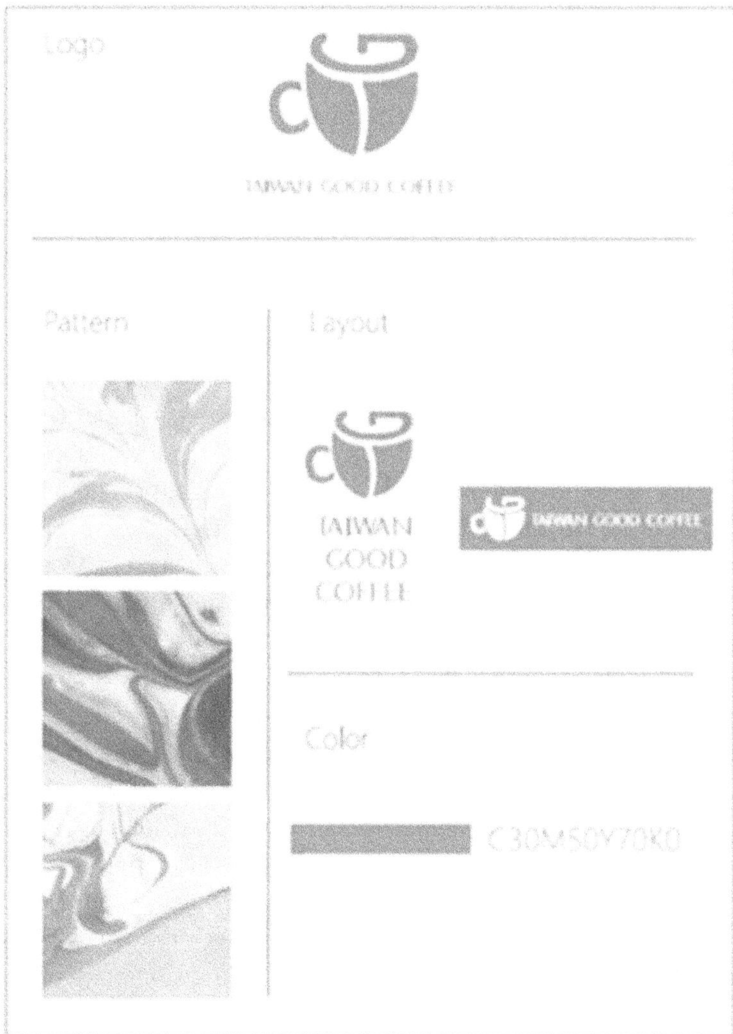

name, logo, color palette, font type and a corporate slogan, tagline and/or descriptor, and these may be applied on, for instance, stationery, printed matter (such as brochures and leaflets), advertisements, websites, vehicles, buildings, interiors and corporate clothing (Bosch, Elving, & de Jong, 2006). As seen here, this corporate visual identity handbook contains several elements. However, this study provides the definitions of the most fundamental elements that are indispensable in a corporate identity handbook and also play a key role in shaping other elements. These elements in question are logos, colors, typography and printed materials.

Logo / Logotype / Symbol

Logo, logotype and symbol are indicators that imaginarily define several messages related to the structure and character of an organization or brand by using visual and symbolic narratives. There are various definitions of a logo in the design literature:

The corporate logo is the official graphical design for a company and the uniqueness of the design requires significant creativity, which must match a firm's strategy and identity: it should be unique and creative in its design (Foroudi, Melewar, & Gupta, 2017).

A logo should be in the form of a printed letter (also known as fond) that will help definition of the character and personality of the brand (Perry & Wisnom III, 2003).

A logo or logotype refers to an organizational signature, symbol or trademark designed for easy and accurate recognition (Rosson & Brooks, 2004).

A logotype is a word (or words) in a determined font, which may be standard, modified, or entirely redrawn (Wheeler, 2009).

Logo / logotype and symbol are the most fundamental and more frequently used visual identity components of an organization. The logo and its variations that are used in almost all printed materials, advertisements and similar visual tools are the most important and indispensable elements of visual identity. In a professional way or not, all entrepreneurs who aim to create a brand think about a logo design right after the brand name. A logo contributes to the recognition of the brand and helps in creating its image in a very short time. In other words, it is like the signature of the brand.

There are certain issues that require attention in designing an effective and appropriate logo. The first and most important of these is the simplicity of the logo. Simplicity, which is also one of the main principles of design, is a factor that increases the effectiveness and message-providing strength of the logo. Another important point is designing the logo in a way that would not diminish the character of the logo on all surfaces it would be applied to. If there are typographical characters in the logo, these characters must be used in a way with high readability and noticeability. This way, the logo, which is one of the most important elements of the visual identity of an organization or brand, will contribute to the image of the organization or brand positively.

Color: Another element whose characteristics are clearly defined in the visual identity handbook is color. A corporate color is an element of visual identity that is used effectively in the image of the organization or brand whose codes are clearly predetermined. The color that is chosen is used in all visual materials of the organization. In this context, it should be chosen carefully. Selection of a color that contradicts the personality of the organization or the brand will lead to building an incorrect image in the eyes of the target audience. As the psychological effects of colors on people cannot be denied, it is a sensitive issue that needs to be planned beforehand what the colors to be used while forming the corporate identity may be associated with and which effects they may have on individuals (Bakan, 2005). The cognitive, emotional and behavioral reactions shown to color vary based on the categories of color. Colors may be categorized as follows (Schmitt & Simonson, 2000):

Saturated Colors: *If a color's rate of saturation is high, it provides a strong impression that it is moving. If the color is shinier, it also looks closer than it actually is. While red, orange and yellow are perceived as warm colors, blue, green and purple are considered to be cool colors. Additionally, different shades provide different impressions of distance. The colors blue and green are perceived to be farther away than red, orange or brown.*

Prestige Colors (Black and White; Gold and Silver): *Black and white reflect the highest levels of saturation and brightness. Metallic colors such as gold and silver have a bright image and carry the visual qualities of these metals.*

Reactions Shown to Color Combinations: *Usage of some color combinations together may attract some associations. For example, using red, brown and other shades of earth colors may be associated with a desert. Using red, white and blue together may be associated with forming the American identity.*

Corporate visual identity handbooks identify the colors of the organization by using certain standard color codes. While color systems that use standardized color codes may be used (e.g. the coding system of the firm Pantone), color codes for four-color process inks (CMYK) may also be preferred (Wheeler, 2009). The important issue is that it is necessary to explain all standards and combinations about color usage in detail in the corporate visual identity handbook.

Typography

Typography is a broad concept that defines and determines the shape, syntax, thickness, size, spaces for letters and similar characteristics and covers the readability and noticeability of texts. It is not possible to argue that a design that is made without considering typographic rules is an effective design. Therefore, typography is one of the most fundamental elements that are defined in the corporate visual identity handbook towards the visual image of an organization. Typographical rules are used to determine and restrict how and by using which typographic characters the brand name and internal correspondence text will be written. All these limitations and typographical rules are found in the corporate visual identity handbook. A unified and coherent company image is not possible without typography that has a unique personality and an inherent legibility, and typography must also support the positioning strategy and information hierarchy (Wheeler, 2009).

While choosing a font in the scope of corporate visual identity, the first criterion that needs to be considered is the readability of the font. At the second stage, considering each font is a separate design element, it should not be forgotten that messages are transmitted to viewers in a visual sense. Based on their form, fonts may provide their viewers with masculine or feminine, stationary or dynamic, entertaining or serious, modern or historical meanings. Hence, for the characteristics of the organization such as its general identity and brand personality, the typographical characters that contain the message to be emphasized should be selected as elements of corporate visual identity. Additionally, which typographical character will be used in which visual material of the organization (correspondence, website, printed materials, etc.) and how it can be used need to be explained in the corporate visual identity handbook in detail.

Printed Materials

After defining the main elements such as the logo, color and typography to be used for all visual materials of the organization in corporate identity handbooks, limitations about the printed materials on which these elements will be used are also identified. Accordingly, corporate visual identity handbooks include templates about designs of the organization's letterheads, envelops (A4, diplomat, A3), identity card, business card, files, CDs, flags, bags, etc.

PERCEIVED DIMENSION OF VISUAL IDENTITY

A perception is a content that is received through the senses, processed by the brain and creates a type of physical or mental reaction after being stored in memory. In other words, it is the process of a person to be aware of and interpret the stimuli or events around them (Mutlu, 2004, p. 17-18). According to another definition, perception, which is a cognitive process, involves making stimuli that reach the eyes, ears and other receptors meaningful in the mind and interpreting these (Çağlayan, Korkmaz, & Öktem, 2014, p. 170). Most human perceptions are visual and obtained through the sense of sight. The sense of sight is possible by capturing of physical energies outside by receptor cells in the eye and transformation of these energies into nervous energy. As a result of the processing of the aforementioned mental energy in the part of the brain responsible for sight, a perceptual product emerges. This process is called 'visual perceiving', while the product that emerges is known as 'visual perception' (Alpan, 2008, p. 83).

People use especially their organs of sight in the process of knowing about, assessing and interpreting what goes around them. Several pieces of information settle in the mind through careful impressions and visual perception. This allow people to more easily and permanently comprehend the events around them (Öncel Taşkıran & Bolat, 2013, p. 54). In this context, it may be stated that visual perception is not just the skill of seeing well, but it is also a visual-based learning process that takes place in the mind. The point that needs to be especially considered here is that perception should occur first for the process of visual sense-making, visual learning or visual communication to start. Considering the process of visual aesthetical production mentioned by Riley (2004), the pioneering role and significance of visual perception may be understood more clearly (Table 1). Making any idea visible requires a starting stage where social concepts and individual perceptions are coded in a material form. The visual aesthetical process that is considered here is an order of visual perceptual relationships the producer finds fit for transforming an aspect of the socio-cultural values of a certain social and cultural context into a visible form (Riley, 2004, p. 299).

Visual perception is 'object-oriented' (allocentric) perception; it provides information about the size, shape and color of an object (Kurt, 2002, p. 121). In visual perception, the individual organizes, classifies and generalizes visual stimuli in a meaningful sense to understand the information that is obtained by the sense of sight. While interpretation of the visual stimulus starts by seeing, the process of making sense of

Table 1. Visual Aesthetical Production Process (Riley, 2004)

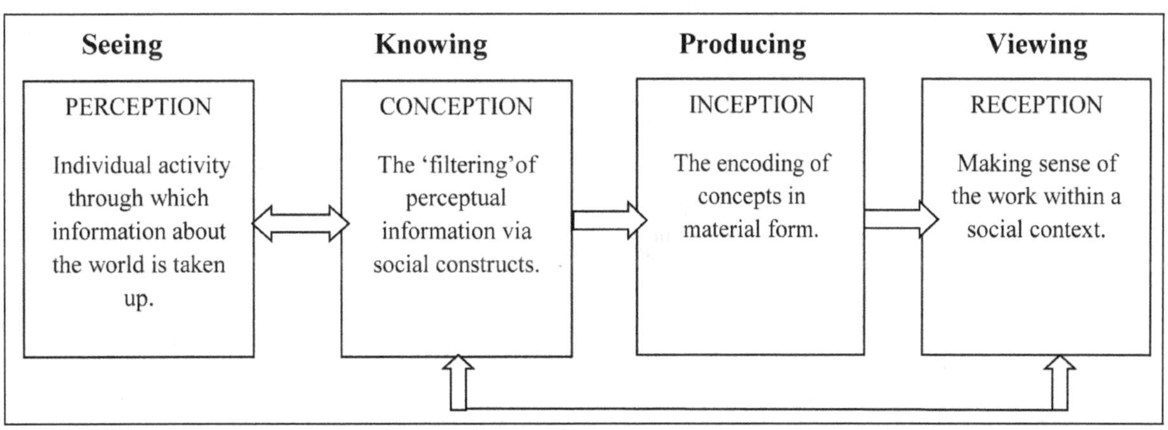

it occurs in the brain. For example, seeing a ball is a sensory process, but recognizing and understanding that it is a ball is an act of thinking, and it takes place as a result of a set of cognitive processes (Çağlayan, Korkmaz, & Öktem, 2014, p. 170). Experimental studies have shown that the perception of visual content often accompanies thinking (Kurt, 2002, p. 122). In this sense, it may be stated that the elements of visual identity for a brand are guiding for people in terms of having an idea about the brand. Strategies on various levels are constructed for the target audience to perceive and make sense of elements of visual identity in the desired way. At this point, the design of the construct that forms the visual content and the combination of design elements are highly important. The colors, lines, textures, fonts and all types of visual images, symbols or marks that are used in design while forming visual identity for a brand are influential on the process of perception. In this process, brands used the aforementioned design elements within a hierarchy to be able to reach the position they want to be perceived by their target audience. With the help of the elements of visual identity or other advertisement-based visual contents that emerge as a result of this process, the process of brands' visual communication with their target audience starts. If careful efforts are spent on all elements of visual identity from the logo to the packing design for an effective visual communication process, the desired level of perception and meaning may be achieved. The following table shows the progression of this process from the parts to the whole in detail.

As the most important priority for a brand is to grab the attention of its target audience, the concepts of attention and visual perception have a critical significance in communication between brands and target audiences. Considering that the human organism receives all messages coming from external stimuli through its sensory organs and perceives and interprets these messages, it is impossible for this organism that is exposed to millions of messages every day to perceive all stimuli that reaches its sensory organs. For this reason, the main goal of a brand is to achieve access to the field of attention of its target audience by leaving all this crow of messages (Elden & Okat, 2015, p. 143). Factors that are effective on perception and interpretation of stimuli are related to both the structure of the stimulus itself and various individual characteristics such as the past experiences and personality traits of individuals. Factors such as the magnitude and size of the stimulus, its similarities and differences to other stimuli and its colors are influential on its perception and in making sense of the perception process (Elden, 2009, p. 399-400).

The strategic communication of brands with their target audiences is highly important for being able to get a permanent place in the mind of the consumer. In this context, the visual dimension of strategic

Table 2. Selection and combination of design elements (Riley, 2004)

SELECT	COMBINE	COMMUNICATE
Elements Of Drawing:	*Combinations Of Elements Produce:*	*Combinations Stand For Physical And Emotional Experiences Of The World:*
Point Line Shape (2d) Texture Tone Colour Plane	Contrast Proportion Scale Pattern Rhythm	Spatial Depth Force Direction Movement Volume,Mass Weight Balance Symmetry Structure Form (3D) Surface Properties Observer'sposition(S)/Mood,Attitude

communication has a very important role. The visual dimension of strategic communication, namely visual communication design, provides benefits for the brand if it is managed effectively in a way suitable for the target audience. While it adds a visual point of view to strategic communication, it is needed to be aware of the meanings and effects of the visuals that are added to the communicational context of the process. In the context of strategic communication, a visual point of view plays an important, effective and functional role in not only the communication activities of a brand but also all levels of communication (Goransson & Fagerholm, 2018, p. 59). So, brands may reach a desirable position in the eyes of their target audiences very fast with the help of strategic communication on the visual level.

As a result of field studies that have been carried out with marketers, it has been known for a long time that perception of brands' logos, packages and other visual elements by the target audience has a critical role in brand awareness. Thus, brands have allocated substantial levels of resources for visual communication. Several successful brands have become recognized immediately with different visual element as a result of the investments and research carried out in the historical process. The swoosh in Nike's logo, the golden belts of McDonald's and the contoured bottle of Coca-Cola may be given as examples of this (Bajaj & Bonda, 2017, p. 77).

Usage of elements of corporate identity within a certain hierarchy makes is easier for the brand to be perceived to be in a desired position by the consumer and differentiate from its competitors. Considering the example of McDonald's mentioned above, it is observed that the style of the brand's visual communication is completely reflected on its elements of corporate identity. Accordingly, in the elements of corporate identity, the symbol, colors, font, lines, textures and sizes were used in harmony in the McDonald's examples. So, it may be argued that the McDonald's brand achieved a desired dimension of perception and sense-making as a result of using its elements of corporate identity in an integrated way in its visual communication process.

There is an integrity in all elements of corporate identity from the mascot of the brand to its product packages. In addition to using the corporate colors, the symbol of the brand, fonts, lines and textures are used in the same hierarchy in all designs. All these design elements that are used in designing elements of corporate identity are effective on the level of the consumer in perceiving the brand visually. This provides the brand with various advantages in the process of visual communication. For example, today, the brand McDonald's is easily and quickly recognized with the help of its corporate identity everywhere from the United States to Japan, from Turkey to Russia. Moreover, these visual contents that provide a significant dimension on the level of corporatization for the brand also increase the value of the brand in the eyes of consumers. Increased brand value directly affects the positive attitudes of consumers towards the products and services of the brand. In summary, visual communication and corporate identity, which is a reflection/form of visual communication, that are highly prioritized by brands for managing perceptions provide brands with significant contributions in this era where the number of competitors and the degree of market competition are increasingly higher. Furthermore, corporate communication, which provides brands with an identity, personality and image, almost provides brands with life force and sustains their survival.

CONCLUSION

Visual communication has been a form of communication that has a principal significance in human life since the ancient times where wall paintings were created. While accepting the fact that written communication is associated with the literacy rates of societies, and its increased rate of usage is related to the intellectual capacity of societies, when it is aimed to influence the general public in societies with people of different intellectual levels, visual communication has always been a method that makes comprehensibility easier. In this sense, it is inevitable for organizations and brands that are operational in the commercial field to prefer forms of communication that are easier to understand for everyone in order to achieve higher profits. In the context of the easiness it provides, its function of shortening the perceptibility and comprehensibility times in the process of communication and its appeal to much larger masses, visual communication is also a form of communication that is preferred the most frequently by organizations and brands today.

In the arena of commerce, where organizations and brands make high amounts of investments, they cannot be expected to leave their communication processes with their target audiences to coincidence. Accordingly, it is inevitable for them to conduct a planned and goal-oriented process. As much as there are controllable factors in this planned process, uncontrollable factors may also be encountered. The main purpose of organizations and brands is to develop systems with maximally controlled and predictable results by minimizing the processes that are uncontrollable. Detailed marketing plans and the corporate identity research based on these plants also indicate the efforts of organizations and brands for achieving the maximum control.

Practices on corporate identity offer approaches that define, protect and shape factors such as corporate communication, corporate design, corporate culture, behavior and structure, sectoral identity and corporate strategy. The visual aspect of these factors is shaped in the context of corporate design. The design aspect of a corporate identity practice is discussed under the name of corporate visual identity, and practices of corporate visual identity define the visual aspect of the communication established with target audiences by an organization or a brand. Accordingly, the corporate colors, logo, symbols, typographical elements, styles of correspondence and all printed materials of an organization or brand are studied in detail by the professionals in this field and identified on their corporate visual identity handbooks. Therefore, in human life, where visual communication is so important, corporate visual identity practices define the planned and systematic aspect of corporate communication and brand communication, provide contribution to the perception of the image of organizations and brands in desirable ways and are prominent as a factor that increases the success of organization in this environment.

REFERENCES

Advertorial. (2019, 02 27). *Bu Yaza Damgasını Vuracak En Yeni Estetik Ve Güzellik Trendleri!* Retrieved from http://www.hurriyet.com.tr/bu-yaza-damgasini-vuracak-en-yeni-estetik-ve-guzellik-trendleri-41131904

Akıncı Vural, B. (2005). *Kurum Kültürü*. İstanbul, Turkey: İletişim Yayınları.

Aktuğlu, I. K. (2004). *Marka Yönetimi*. İstanbul, Turkey: İletişim Yayınları.

Alpan, G. (2008). Görsel Okuryazarlik Ve Öğretim Teknolojisi. *Yüzüncü Yıl Üniversitesi Eğitim Fakültesi Dergisi, 5*(2), 74–102.

Bajaj, A., & Samuel, B. (2018). Beyond beauty: Design symmetry and brand personality. *Journal of Consumer Psychology, 28*(1), 77–98. doi:10.1002/jcpy.1009

Bakan, Ö. (2005). *Kurumsal İmaj*. Konya, Turkey: Tablet Yayınları.

Balmer, J., & Greyser, S. (2003). Managing the Multiple Identities of the Corporation. In J. Balmer & S. Greyser (Eds.), *Revealing The Corporation* (pp. 15–30). London, UK: Routledge. doi:10.4324/9780203422786

Bartholmé, R., & Melewar, T. (2011). Remodelling the corporate visual identity construct: A reference to the sensory and auditory dimension. *Corporate Communications, 16*(1), 53–64. doi:10.1108/13563281111100971

Becer, E. (2006). *İletişim ve Grafik Tasarım*. Ankara, Turkey: Dost Kitabevi.

Bosch, A., Elving, W., & de Jong, M. (2006). The impact of organisational characteristics on corporate visual identity. *European Journal of Marketing, 40*(7/8), 870–885. doi:10.1108/03090560610670034

Bosch, V. (2005). How corporate visual identity supports reputation. *Corporate Communications, 10*(2), 108–116. doi:10.1108/13563280510596925

Çağlayan, S., Korkmaz, M., & Öktem, G. (2014). Sanatta görsel algının literatür açısından değerlendirilmesi. *Eğitim ve Öğretim Araştırmaları Dergisi, 3*(1), 160–173.

Cornelissen, J. (2004). *Corporate Communications Theory and Practice*. London, UK: Sage.

De Lencastre, P., & Côrte-Real, A. (2010). One, two, three: A practical brand anatomy. *Journal of Brand Management, 17*(6), 399–412. doi:10.1057/bm.2010.1

Elden, M. (2009). *Reklam ve Reklamcılık*. İstanbul, Turkey: Say Yayınları.

Elden, M., & Okat Özdem, Ö. (2015). *Reklamda Görsel Tasarım-Yaratıcılık ve Sanat*. İstanbul, Turkey: Say Yayınları.

Elden, M., & Yeygel, S. (2006). *Kurumsal Reklamın Anlattıkları*. İstanbul, Turkey: Beta.

Erengül, B. (1997). *Kültür Sihirbazları - Rekabet Üstünlüğü Sağlayan Yönetim*. İstanbul, Turkey: Evrim Yayınevi.

Floch, J.-M. (2000). *Visual Identities*. London, UK: Continuum.

Foroudi, P., Melewar, T., & Gupta, S. (2017). Corporate Logo: History, Definition, and Components. *International Studies of Management & Organization*, *47*(2), 176–196. doi:10.1080/00208825.2017.1256166

Goransson, K., & Anna-Sara, F. (2018). Towards visual strategic communications: An innovative interdisciplinary perspective on visual dimensions within the strategic communications field. *Journal of Communication Management*, *22*(1), 46–66. doi:10.1108/JCOM-12-2016-0098

Hofstede, G. (2011). Dimensionalizing cultures: The Hofstede model in context. *Online Readings in Psychology and Culture*, *2*(1), 1–26. doi:10.9707/2307-0919.1014

Holsanova, J. (2014). In the eye of the beholder: Visual communication from a recipient perspective. In D. Machin (Ed.), *Visual Communication* (pp. 331–355). Berlin, Germany: Mouton De Gruyter.

Jaenichen, C. (2017). Visual Communication and Cognition in Everyday Decision-Making. *IEEE Computer Graphics and Applications*, *37*(6), 10–18. doi:10.1109/MCG.2017.4031060 PMID:29140778

Kavuran, T., & Özpolat, K. (2016). Görsel İletişim Aracı Olan Dergilerin Tasarlanma Süreci. *Fırat Üniversitesi Sosyal Bilimler Dergisi*, *26*(2), 267–275.

Ketenci, H. F., & Bilgili, C. (2006). *Yongaların 10.000 Yıllık Gizemli Dansı: Görsel İletişim ve Grafik Tasarım*. İstanbul, Turkey: Beta.

Kurt, M. (2002). Görsel-Uzaysal Yeteneklerin Bileşenleri. *Klinik Psikiyatri*, *5*(2), 120–125.

Lester, P. M. (2006). *Syntactic Theory of Visual Communication*. Retrieved from http://paulmartinlester.info/writings/viscomtheory.html

Melewar, T., & Karaosmanoğlu, E. (2006). Seven dimensions of corporate identity: A categorisation from the practitioners' perspectives. *European Journal of Marketing*, *40*(7/8), 846–869. doi:10.1108/03090560610670025

Melewar, T., & Saunders, J. (1998). Global corporate visual identity systems: Standardization, control and benefits. *International Marketing Review*, *15*(4), 291–308. doi:10.1108/02651339810227560

Melewar, T. C. (2001). Measuring visual identity: A multi-construct study. *Corporate Communications*, *6*(1), 36–42. doi:10.1108/13563280110381206

Melewar, T. C. (2003). Determinants of the corporate identity construct: A review of the literature. *Journal of Marketing Communications*, *9*(4), 195–220. doi:10.1080/1352726032000119161

Mercin, L. (2017). Müze Eğitimi, Bilgilendirme Ve Tanitim Açisindan Görsel İletişim Tasarimi Ürünlerinin Önemi. *Milli Eğitim Dergisi*, *46*(214), 209–237.

Mutlu, E. (2004). *İletişim Sözlüğü*. Ankara, Turkey: Bilim ve Sanat Yayınları.

Okay, A. (2005). *Kurum Kimliği*. İstanbul, Turkey: Mediacat.

Öncel Taşkıran, N., & Bolat, N. (2013). Reklam Ve Algı İlişkisi: Reklam Metinlerinin Alımlanmasında Duyu Organlarının İşlevleri Hakkında Bir İnceleme. *Beykent Üniversitesi Sosyal Bilimler Dergisi*, *6*(1), 49–69.

Öncü Yıldız, M. (2012). Görsel Okuryazarlik Üzerine. *Marmara İletişim Dergisi*, *19*, 64–77.

Onursoy, S. (2017). Görsel Kültür ve Görsel Okuryazarlık. *Türk Kütüphaneciliği*, *31*(1), 47–54. doi:10.24146/tkd.2017.4

Peng, L.-H., & Hung, C.-C. (2016). The practice of corporate visual identity—A case study of Yunlin Gukeng Coffee Enterprise Co., Ltd. *2016 International Conference on Applied System Innovation (ICASI)*. Piscataway, NJ: IEEE. 10.1109/ICASI.2016.7539844

Perry, A., & Wisnom, D. III. (2003). *Markanın DNA'sı*. İstanbul, Turkey: Kapital Medya.

Porter, M. (2007). *Rekabet Stratejisi* (Vol. 4). İstanbul, Turkey: Sistem Yayıncılık.

Riley, H. (2004). Perceptual modes, semiotic codes, social mores: A contribution towards a social semiotics of drawing. *Visual Communication*, *3*(3), 294–315. doi:10.1177/1470357204045784

Rosson, P., & Brooks, M. (2004). M&As and Corporate Visual Identity: An Exploratory Study. *Corporate Reputation Review*, *7*(2), 181–194. doi:10.1057/palgrave.crr.1540219

Ruhela, V. S., & Parween, S. (2018). Effect of visual communication in tracking activity schedule among children with autism spectrum disorder. *Indian Journal of Health & Wellbeing*, *9*(5), 748–751.

Rutter, D., Stephenson, G., & Dewey, M. (1981). Visual communication and the content and style of conversation. *British Journal of Social Psychology*, *20*(1), 41–52. doi:10.1111/j.2044-8309.1981.tb00472.x PMID:7237005

Schmitt, B., & Simonson, A. (2000). *Pazarlama Estetiği*. İstanbul, Turkey: Sistem Yayıncılık.

Simoes, C., & Dibb, S. (2008). Illustrations of the internal management of corporate identity. In T. C. Melewar (Ed.), *Facets of Corporate Identity, Communication and Reputation* (pp. 66–80). Oxon, UK: Routledge. doi:10.4324/9780203931943.ch4

Statista. (2019, Nisan 9). Retrieved from https://www.statista.com/statistics/272014/global-social-networks-ranked-by-number-of-users/

Uçar, T. F. (2004). *Görsel İletişim ve Grafik Tasarım*. İstanbul, Turkey: İnkılap.

Uyan Dur, B. İ. (2015). Türk Görsel İletişim Tasarimi Ve Kültürel Değerlerle Bağlari. *Journal of International Social Research*, *8*(37), 443–453. doi:10.17719/jisr.20153710615

Van Riel, C. (1997). Research in corporate communication: An overview of an emerging field. *Management Communication Quarterly*, *11*(2), 288–309. doi:10.1177/0893318997112005

Vella, K., & Melewar, T. (2008). Corporate Identity. In T. C. Melewar (Ed.), *Facets of Corporate Identity, Communication, and Reputation*. Oxon, UK: Routledge.

Wheeler, A. (2009). *Designing Brand Identity* (Vol. 3). Hoboken, NJ: John Wiley & Sons.

Williams, R. (1993). *Kültür* (S. Aydın, Trans.). Ankara, Turkey: İmge Kitabevi Yayınları.

Yaban, N. T. (2012). Sanat Ve Görsel İletişimin Buluşma Noktasi: Ekslibris. *Batman Üniversitesi Yaşam Bilimleri Dergisi*, *1*(1), 973–984.

ADDITIONAL READING

Barry, A. M. (1997). *Visual intelligence: Perception, image, and manipulation in visual communication*. SUNY Press.

Berger, A. A. (1989). Seeing is Believing: An Introduction to Visual Communication. Mayfield Publishing Company, 1240 Villa Street, Mountain View, CA 94041.

Floch, J. M., & Pinson, C. (1990). *Sémiotique, marketing et communication: sous les signes, les stratégies*. Paris: Presses universitaires de France.

Horn, R. E. (1998). *Visual language*. Bainbridge Island, WA: MacroVu, Inc.

Ivins, W. M. (1969). *Prints and visual communication* (Vol. 10). Mit Press.

Lester, P. M. (2013). *Visual communication: Images with messages*. Cengage Learning.

Meech, P. (2006). *Corporate identity and corporate image. Public relations critical debates and contemporary practice* (pp. 389–404). London: Lawrence Erlbaum.

Melewar, T. C., Karaosmanoglu, E., & Paterson, D. (2005). Corporate identity: Concept, components and contribution. *Journal of General Management*, *31*(1), 59–81. doi:10.1177/030630700503100104

Schwarcz, J. H. (1982). *Ways of the illustrator: Visual communication in children's literature* (p. 9). Chicago: American Library Association.

Van den Bosch, A. L., De Jong, M. D., & Elving, W. J. (2005). How corporate visual identity supports reputation. *Corporate Communications*, *10*(2), 108–116. doi:10.1108/13563280510596925

KEY TERMS AND DEFINITIONS

Brand Communication: It is the whole of the strategic communication activities planned for a brand to sustain its market activities without interruption.

Consumer Behavior: It can be defined as a process that includes the decisions about selecting, purchasing, using and disposing of products and services.

Corporate Communication: In the simplest terms, it can be defined as the integrated management of all communication processes in line with the strategic business objectives of the institutions.

Corporate Visual Identity: Corporate visual identity is the visual indicators that enable us to understand who and what the institution is.

Marketing Communication: Marketing communication is a concept that encompasses all the interaction of the products or services that the companies have to sell their products and services to the buyers.

Visual Communication: Visual communication is the transmission of information and ideas using symbols and imagery.

Visual Design: Graphic design is a creative process, which involves organizing a two-dimensional or three-dimensional text and images to develop an image, or to visualize a thought.

Visual Perception: Perception is the ability to interpret the information received from you and your senses.

Chapter 4
How Is a Country Image and Identitiy Construction Reflected via Discourses in Press?

Emel Özdemir
Communication Faculty, Akdeniz University, Turkey

ABSTRACT

This chapter is aimed to put the matter of how is a country image able to be constructed in hand through the medium of the online press, by evaluating The New York Times (USA), The Daily Express (England), Spiegel Online International (Germany), and Le Monde Diplomatique (France) in terms of "Turkish image and identity" throughout four months (January-April) in 2019. The author uses Van Dijk's discourse analysis approach that is based on two main principles, macro and micro discourse analysis, and the content analysis technique. It is possible with this evaluation to determine how Turkish image and identity is established and what kinds of images, expressions, and representations are used by the foreign press, as well as their approach to Turkish identity.

DOI: 10.4018/978-1-7998-1041-4.ch004

INTRODUCTION

In the study of "How is A Country Image and Identitiy Construction Reflected via Discourses in Press?", all the news, images, expressions and representations about Turkey are elaborately studied for making out how the perception and position of Turkey that is affected by the changes with the globalization is reflected via discourses in four online foreign press throughout four months. This study has an objective to assess the Turkish image that is constructed in the globalized world, by analyzing all the news, images, expressions and representations about Turkey in terms of discourse and the content analysis technique, by comparing the online newspapers textes. Accordingly, the Turkish image that is established by the newspapers of "The New York Times (USA), The Daily Express (England), Spiegel Online International (Germany) and Le Monde Diplomatique (France)" and their approach to Turkish identity can be understood with this study. In Addition, this analysis is done, by using various studying areas, such as image and translation studies and it acquires an interdisciplinary qualification.

THE AIM OF THE STUDY

The aim of this study is to demonstrate how Turkey is described and which images, expressions and representations are used for Turkey in the online newspapers of "The New York Times (USA), The Daily Express (England), Spiegel Online International (Germany) and Le Monde Diplomatique (France)". Therefore, we are able to understand how Turkish image is established and what kinds of images and expressions and representations are used by four newspapers. In this analysis, it is aimed see how the position of Turkey is created by the discources of the newspapers in various societies and whether Turkish image has been started to change with the globalization in some ways, or not.

METHODOLOGY

In this study, discourse analysis is one of the methods that is maily used in order to understand how the Turkish image is established in the online newspapers of "The New York Times (USA), The Daily Express (England), Spiegel Online International (Germany) and Le Monde Diplomatique (France)" during four months in 2019. As discourse analysis is a field that is concerned with studying and analyzing written and spoken texts, it is aimed to make out the thoughts and ideologies about Turkey, are not especially expressed in four online newspapers texts from foreign press. It is possible to see inequalities, dominiance, power, bias, prejudices in the texts and how the realities about Turkish image are reproduced within social, political, ideological, historical contexts, by making analysis of the texts with discourse analysis method. By the way, in this analysis, the reality how the discourses, in the newspapers of "The New York Times (USA), The Daily Express (England), Spiegel Online International (Germany) and Le Monde Diplomatique (France) reproduce all kinds of information about Turkey, with some words, phrases, images, photographs is understood and it is clearly seen how the texts about Turkish image reflect a definite point of view with their discourses. All news about Turkey are going to be analyzed with discourse analysis, especially Van Dijk's discourse analysis method, in order to be able to see four newspapers' point of views and thoughts about Turkey and Turkish image that is established in the globalized world. That's to say, it is possible to make out how the ideologies, inequalities, bias, prejudices of all the newspapers are expressed in their discourses with Van Dijk's discourse analysis. Because, Van Dijk's discourse analysis, as one of the most detailed discouses analysis method evaluate all the texts both textual and contextual level and it takes places at two sections: microstructure and macrostructure. At the microstructure level, analysis is focused on the semantic relations between propositions, syntactic, lexical and other rhetorical elements that provide coherence in the text, and other rhetorical elements such quotations, direct or indirect reporting that give factuality to the news reports. Central to Van Dijk's analysis of news reports, however, is the analysis of macrostructure since it pertains to the thematic/topic structure of the news stories and their overall schemata. Themes and topics are realized in the headlines and lead paragraphs (Van Dijk, 1998, p. 31-45). That's to say, it is possible to analyze the discourses, in the online newspapers of "The New York Times (USA), The Daily Express (England), Spiegel Online International (Germany) and Le Monde Diplomatique (France), both structural level and thematic level with Van Dijk's discourse analysis, in order to make out their approach to Turkey and Turkish image and thoughts, ideologies, are not especially espressed in these texts. Because, the realities about Turkey are reproduced in the discourses of these four newspapers, according to their point of view.

The content analysis is the other technique that is maily used in order to understand how the Turkish image is established in the online newspapers. There are different definitions of the content analysis. Walizer and Wienir (1978) defined the content analysis as a systematic procedure, developed to examine the content of recorded information. Content analysis is an analysis technique rather than an observation method. Content analysis results are presented in the form of percentage tables, as in searching researches. The definition of Kerlinger (1973) which defines content analysis as a technique for analyzing and analyzing communication in a quantitative context, in order to measure systematic, objective and variables contains three concepts that require detailed elaboration. First, content analysis is systematic. This means that the content that will be analyzed is chosen according to certain and continuously applied rules. Sample selection should be made according to the appropriate rules and each content should be analyzed in the same way. The evaluation process should also be systematic; each content that is going to be examined must be handled exactly the same way. Encoding and resolution procedures must be associated, too. Systematic evaluation simply means that the evaluation guide is used throughout the entire review (Atabek, 2007). As it is understood, while conducting the content analysis, each process should be done carefully and in accordance with the rules and systematically planned. In general terms, the content analysis can be defined as a set of methodological tools and techniques that apply a wide variety of discourses. These tools and techniques, collected under the name of the content analysis, can be characterized as a controlled interpretation effort. The content analysis is therefore a second reading in the message to identify elements that affect the individuals.

THE UNIVERSE, THE SAMPLING AND LIMITATIONS OF THE STUDY

The universe of this study consists of the online newspapers of "The New York Times (USA), The Daily Express (England), Spiegel Online International (Germany) and Le Monde Diplomatique (France). The sample of the study icludes the online versions of "The New York Times (USA), The Daily Express

(England), Spiegel Online International (Germany) and Le Monde Diplomatique (France)" newspapers in terms of "Turkish image and identity" throughout four months (January, February, March, April) in 2019. Because of the limitations of the study, very significiant examples that are used to reproduce the reality of Turkey and Turkish people is taken place in this analysis, according to Van Dijk's discourse analysis.

THE RESULTS OF THE STUDY

In the study of "How is A Country Image and Identitiy Construction Reflected via Discourses in Press?", it is seen how "The New York Times (USA), The Daily Express (England), Spiegel Online International (Germany) and Le Monde Diplomatique (France) newspapers' points of view about Turkey is reflected in their discourses, by analyzing the words, images, phrases, photographs that are used in their texts, in a detailed way with discourse analysis and the content analysis technique, it is possible to understand the position and place of Turkey has been changed with the globalization, because of the changes of four countries' historical, social, political, ideological structures. It is aimed to see how Turkish image is reporduced by these newspapers' discourses in various societies and whether Turkish image that is maily established, negatively has started to be change and improve, or not. Because, many bad images, phrases about Turkey, such as 'cruel, wild, violent, brutual, bad' are used in different coutries' historical sources, written texts and Turkish people are accepted as an 'other' and a symbol of 'a common enemy' of many society. But, when "Turkish Image" is evaluated during four months (January, February, March, April) in 2019, by using Van Dijk's discourse analysis approach that is based on two main principle, it is understood that 'Turkey and the news about Turkish images' are still reflected negatively in the globalized world, too and many different bad images, expressions are used for Turkey, as it is seen in Table 1.

Acoording to this analysis, there are 139 news about Turkey in four newspapers, as 11 positive, 30 negative, 13 neutral news in The New York Times (USA), 19 positive, 50 negative, 8 neutral news in The Daily Express (England), 2 negative news, 1 neutral news in Spiegel Online International, 3 negative, 3 neutral news in Le Monde Diplomatique. And, as it is seen in the table, there are only 30 positive news, 85 negative, 25 neutral news about Turkey, from 139 news and this number shows how the Turkish image is established in a negative way in a globalized world, too.

In Addition, when the newspapers of "The New York Times (USA), The Daily Express (England), Spiegel Online International (Germany) and Le Monde Diplomatique (France) are analyzed during four months (January, February, March, April) in 2019 according to the news' numbers about Turkey, it is seen that there are 33 news in january, 14 news in february, 30 news in march, 31 news in april. While the most widely spoken news about Turkey are taken place in january, march and april; the least one is

Table 1.The news about Turkey in The New York Times (USA), The Daily Express (England), Spiegel Online International (Germany) and Le Monde Diplomatique (France) during four months in 2019

	Positive	Negative	Neutral	Total
The New York Times (USA)	11	30	13	53
The Daily Express (England)	19	50	8	77
Spiegel Online International (Germany)	-	2	1	3
Le Monde Diplomatique (France)	-	3	3	6

taken place in february. Because, “Syria, Turkey-USA Relations, Turkey’s Economic Conditions, Elections in Turkey”, happened in january, march and april months are very significiant issues that are widely discussed all around the world. It is possible to see news’ numbers according to the months in Table 2.

In the study of “How is A Country Image and Identitiy Construction Reflected via Discourses in Press?”, when the subjects about Turkey that are mentioned about Turkey are analyzed, we see the subjects, such as “Saudi Arabia’s Consuls’ Death, Syria, An N.B.A. Player, Bodrum, Turkey-USA Relations, Working Class Rights in Turkey, Turkish Art, Turkey’s Rights for Religion, Turkey-Greece Relations, Turkey-China Relations, Turkish Courts, Turkey’s Economic Conditions, New Zealand Attacks, Taksim Square, Elections in Turkey, Turkey-Arab World Relations, Turkey Holidays, Turkish Football, Building Collapse in Istanbul, Armenian Genocide, Venezuela Crisis, Earthquake in Turkey, Turkey’s EU Accession, Turkey-Israil Relations, Turkey-Russia Relations, Drug War, Freedom of the Press, Muslim Brotherhood, Turkish Soldier” in four different online versions of the newspapers texts. News’ numbers, according to the subjects can be seen in Table 3.

In order to understand how Turkey and Turkish image is reproduced by the discourses of “The New York Times (USA), The Daily Express (England), Spiegel Online International (Germany) and Le Monde Diplomatique (France), it is very significiant to analyze all the images that are mainly used for Turkey, as they show these newspapers’ point of view about Turkey. When the images about Turkey are evaluated, it is possible to see images, such as “Turkish Millitary, religious Turkish people, uneducated Turkish people, Turkish President, Recep Tayyip Erdoğan, divided Turkish people, Minorities in Turkey, Turkish financial crisis, Turkish elections, Turkish relations with other countries, cheap Turkish holidays” to describe Turkey and Turkish people.

In this analysis, when very significiant examples from this study is evaluated according to Van Dijk’s discourse analysis, it is seen that many words, images, phrases, photographs are used to reproduce the reality of Turkey and Turkish people with their discourses, especially in a negative way. For example, In 25 march 2019, The news with the headline of “Turkey lira REBOUNDS against US dollar as Erdogan issues WARNING to bankers”, it is deliberately emphasized that TURKEY saw its lira rebound today as it entered recovery mode after President Recep Tayyip Erdogan issued a warning to bankers in the discourses of the news to indicate the financial crisis in Turkey. In March 28, 2019, in the news of “Turkey election latest: Turkish lira DIVES as Erdogan RAGES – ‘I’m in charge of economy!”, how Turkish currency falls is told with the expressions of the “TURKEY saw its lira dive by as much as five percent against the US dollar earlier today after the central bank revealed a drop in international reserves ahead of local elections this weekend.” In April 16, 2019, in the news of “Turkey lira CRISIS: Currency WEAKENS near to WORST level against US dollar in SIX MONTHS”, it is possible to see the same

Table 2. News’ numbers according to months

Months	The New York Times	The Daily Express	Spiegel Online International	Le Monde Diplomatique
January, 2019	20	12	-	1
February, 2019	4	8	1	1
March, 2019	8	19	1	2
April, 2019	9	19	1	2
TOTAL	41	58	3	6

Table 3. News' numbers according to the subjects

	The New York Times	The Daily Express	Spiegel Online	Le Monde Dip.	Total
Saudi Arabia's Consuls' Death	5	1	-	-	6
Syria	8	6	4	3	21
An N.B.A. Player	1	-	-	-	1
Bodrum	2	-	-	-	2
Turkey- USA Relations	5	15	-	-	20
Working Class Rights in Turkey	-	-	-	1	1
Turkish Art	2	6	1	-	9
Turkey's Rights for Religion	1	-	-	-	1
Turkey-Greece Relations	1	-	-	-	1
Turkey-China Relations	1	1	-	1	3
Turkish Courts	1	6	-	-	7
Turkey's Economic Conditions	6	9	-	1	16
New Zealand Attacks	1	6	-	-	7
Taksim Square	1	-	-	-	1
Elections in Turkey	7	6	-	-	13
Turkey-Arab World Relations	1	-	-	-	1
Turkey Holidays	-	9	-	-	9
Turkish Football	-	2	-	-	2
Building Collapse in Istanbul	-	1	-	-	1
Armenian Genocide	-	1	-	-	1
Venezuela Crisis	1	2	-	-	3
Earthquake in Turkey	-	1	-	-	1
Turkey's EU Accession	-	2	1	-	3
Turkey- Israil Relations	-	1	-	-	1
Turkey- Russia Relations	2	-	-	2	2
Drug War	-	-	-	1	1
Freedom of the Press	3	-	-	-	3
Muslim Brotherhood	3	-	-	-	3
Turkish Soldier	2	-	-	-	2

kinds of discourses that is planned to establish the image of a country with a deep crisis and chaos, by using "TURKEY saw its lira weaken once more against the US dollar today, floundering near its worst level in six months." Turkish economy is one of the most important rhetorical elements that is used in the news of The Daily Express (England). This image shows how Turkish economy starts to loss its power and be uneffective, when compared its power to before. Besides, when the images that is used in the news of "Erdogan Says Turkey Cannot 'Swallow the Message' From Bolton" in Jan. 08,2019 in New York Times (USA) is examined, it is possible to see some expressions such as "President Recep Tayyip Erdogan of Turkey denounced comments from the American national security adviser, John R. Bolton, that Turkey must agree to protect Kurds if Americans withdraw from Syria." In January, 09, 2019 in the news of "Trump Threatens to 'Devastate Turkey Economically' if It Attacks Kurds", it is said that" In tweets, the president threatened the NATO ally and seemed to offer a blanket of protection for a band of American-backed militias in Syria after U.S. forces leave." When these news discourses are analysed, it is explicitly understood that they are trying to divide Turkish people by emphasizing ethical discourses and expressions, such as Turkish, Kurdish, national and the plan to indicate their reader that Turkey will be refrain from the Trumpt's millitary treaths. It is made out that this words are intentionally prefered to construct the Turkish image negatively. But, this discourse that is deliberately choosed in this news does not totally reflect the reality of Turkey. That's to say, four different newspapers reproduce the reality about Turkey with the images, phrases, photographs and expressions with their discourses.

CONCLUSION

In the study of "How is A Country Image and Identitiy Construction Reflected via Discourses in Press?", the Turkish image, established in the globalized world is analyzed during four months (January, February, March, April) in 2019, by using Van Dijk's discourse analysis approach and the content analysis technique in order to understand how the image and position of Turkey is affected from the changes with the globalization all around the world. It is possible to see how Turkish identity is expressed in four newspapers, from different countries and what kinds of images, phrases, expressions, photographs are used to describe Turkish people.

In this analysis, when "The New York Times (USA), The Daily Express (England), Spiegel Online International (Germany) and Le Monde Diplomatique (France) are evaluated during four months in 2019, "85 negative, 30 positive and 25 neutral" news about Turkey are seen in four different newspapers and these numbers demonstrate how Turkey and Turkish image is mainly expressed, adversely as before. It is clearly made out that the perception and position of Turkey is still desired to be established with many negative expressions in four newspapers' discourse even in the globalized world, so as to maintain bad Turkish image in every area all around the world. But, when four newspapers, used as an example from different countries are analyzed in this study, according to the conditions of the globalized world, it is possible to understand negative Turkish image still continues in many societies, although there are some changes in its qualifiaction, compared to the historical sources about Turkey. That's to say, we see that the Turkish image that is widely desribed as "barber, wild, cruel" in many countries, such as England, Germany, France, Italy has been changed as "religious, racist, aggressive, troublemaker" Turkish people in the globalized world. As it can be understood from this study, "The New York Times (USA), The Daily Express (England), Spiegel Online International (Germany) and Le Monde Diplomatique (France) reproduce the reality about Turkey, negatively with their discourses, by using words, images, phrases, photographs that demonstrate their points of view, inequalities, bias, prejudices about Turkey and their aim is to reflect Turkey as an "undeveloped, unstable" country that restricts freedoms of people in many area and Turkish people as "an uneducated, a non-modern, rude, narrow-minded, conservative". In the study of "An Example of Discourse Analysis: How is A Country Image and Identitiy Construction Reflected via Discourses in Press?", it is possible to understand all the newspapers, analyzed in this study reproduce the reality of Turkey and Turkish image with their discourses, according to their historical, social, political, ideological point of views and they aim to affect the society with the realities about Turkey, established by them.

REFERENCES

Atabek, Gülseren Şendur ve Ümit Atabek. (2007). Medya Metinlerini Çözümlemek, Ankara, Turkey: Siyasal Kitabevi.

Danforth, N. (2008). Ideology and Pragmatism in Turkish Foreign Policy: From Atatürk to The AKP". Turkish Policy Quarterly, 7(3), Washington DC, Spring.

Ertan Keskin, Z. (2004). Türkiye'de Haber İncelemelerinde Van Dijk Yöntemi. In *Dursun, Ç. (Der.), Haber-Hakikat ve İktidar İlişkisi*. Ankara, Turkey: Elips Yayınları.

İnal, A. (1996). *Haberi Okumak*. İstanbul, Turkey: Temuçin Yayınları.

Kerlinger, F. N. (1973). Foundations of Behavioral Research. 2nd edition. New York, NY: Holt, Rinehart, & Winston.

Özarslan, Z. (2003). *Söylem ve İdeoloji Mitoloji, Din, İdeoloji. Çev. Nurcan Ateş, Barış Çoban, Zeynep Özarslan*. İstanbul, Turkey: Su Yayınları.

Sözen, E. (1999). *Söylem, Belirsizlik, Mücadele, Bilgi/ Güç ve Refleksivite*. İstanbul, Turkey: Paradigma Yayınları.

Ülkü, G. (2004). Söylem Çözümlemesinde Yöntem Sorunu ve Van Dijk Yöntemi. In *Dursun, Ç. (Der.), Haber-Hakikat ve İktidar İlişkisi*. Ankara, Turkey: Elips Yayınları.

Van Dijk, T. (1998). Ideology. *Sage (Atlanta, Ga.)*.

Van Dijk, T. A. (1985). (In the Press). News Analysis [Mahwah, NJ: Lawrence Erlbaum Associates Publication.]. *Case Studies of International and National News*.

Van Dijk, T. A. (1999). *Söylemin Yapıları ve İktidarın Yapıları, Medya İktidar İdeoloji*. Mehmet Küçük, Ark.: Der. ve Çev.

Van Dijk, T. A. (2003). Söylem ve İdeoloji, Çokalanlı Bir Yaklaşım. In B. Çoban & Z. Özarslan (Eds.), Söylem ve İdeoloji: Mitoloji, Din, İdeoloji. İstanbul, Turkey: Su Yayınları, (Çev.: N. Ateş).

Van Dijk, T. A. (2008). *Discourse and Power. Basingstoke, UK: Palgrave* Macmillan. doi:10.1007/978-1-137-07299-3

Walizer, M. H., & Wienir, P. L. (1978). *Research Methods and Analysis: Searching for Relationships*. New York, NY: Harper & Row.

Weeks, J. (1998). *Farklılığın Değeri. Kimlik: Topluluk/Kimlik/Farklılık içinde. Der: J. Rutherford, Çev: D.Sağlamer*. İstanbul, Turkey: Sarmal Yayınları.

ADDITIONAL READING

Akdemir, E. (Güz: 2007).Avrupa Aynasında Türk Kimliği, Ankara Avrupa Çalışmaları Dergisi, Cilt: 7, No:l s. I 31-148.

Barker, C., & Galasinski, D. (2001). *Cultural Studies and Discourse Analysis: A Dialogue on Language and Identity*. London, UK: Sage.

Burçoğlu, N. K. (2003). A Glimpse at Various Stages of the Evolution of the Image of the Turk in Europe. In Mustafa Soykut (Ed.), Historical Image of the Turk in Europe. İstanbul, Turkey: İsis Yayınevi.

Burçoğlu, N. K. (2005). Social And Culural Aspects of Turkey's Integration To The EU, Haluk Kabaalioğlu, Muzaffer Dartaıı, M. Sait Akman, Çiğdem Nas (Edit.), Europeanisation of South-Eastern Europe: Domestic Impacts of the Accession Process, Marmara University European Community Institute Publication No: 12, 2005.

Güvenç, B. (1996). *Türk Kimliği*. Istanbul, Turkey: Remzi Kitabevi.

Kula, O. B. (1992). *Alman Kültüründe Türk İmgesi I*. Ankara, Turkey: Gündoğan Yayınları.

Milas, H. (2000). *Türk Romanı ve "Öteki" Ulusal Kimlikte Yunan İmajı*. İstanbul, Turkey: Sabancı Üniversitesi.

KEY TERMS AND DEFINITIONS

Discourse Analysis: It is possible to make out how the position of Turkey is created by the discourses of the newspapers in various societies and whether Turkish image has been started to change with the globalization in some ways, or not, by using discourse analysis. As discourse analysis is a field that is concerned with studying and analyzing written and spoken texts, it is aimed to understand the thoughts and ideologies about Turkey.

Foreign Press: All the news, images, expressions and representations about Turkey in foreign press directly affect the perception and position of Turkey. To be able to assess the Turkish image that is constructed in the globalized world, analyzing all the news in the foreign press is so significant.

Globalization: Today, the concept of globalization is seen as a situation which causes various comments by making many different definitions about it. With the globalization, new trends and changes have taken place in many areas in the world and the world has become a global environment where different cultures live together.

Identity Construction: The Identity is one of our most basic reference point. Any description that is made about a person will refer to his identity. The identity, which is the basic element of our own system of meaning the world enables us to interpret other people.

Otherness: It is understood that all people need "the other", while building their identity and their positioning in the society. People need to "other" to know and describe themselves.

The Content Analysis Technique: The content analysis can be defined as a set of methodological tools and techniques that apply a wide variety of discourses. The content analysis is as a systematic procedure, developed to examine the content of recorded information.

Turkish Image: The significant examples of image from this study is evaluated to be able to see the words, images, phrases, photographs are used to reproduce the reality of Turkey image and Turkish people.

Chapter 5
Journalism and Communication Design in New Media

Rengim Sine Nazlı
Bolu Abant İzzet Baysal University, Turkey

Bahar Akbulak
Bolu Abant İzzet Baysal University, Turkey

Arzu Kalafat Çat
Bolu Abant İzzet Baysal University, Turkey

ABSTRACT

This chapter emphasizes that the changes and developments in information and communication Technologies are reflected in the field of journalism being integrated into new media. The transformation and development in the new media have changed the traditional practices of the perception and communication design of journalism, leaving its place to the elements that include information-communication and communication design. An effective communication design on human beings shows that information-target, mass-message-communication pieces are the interaction of today's consumer's taste and consumption desire.

DOI: 10.4018/978-1-7998-1041-4.ch005

BACKGROUND

Developments in communication technologies and widespread use of the Internet have made it easy, fast and economical for individuals to access information. Today, many individuals see the newmedia as an indispensable part of their lives since the Internet offers a variety of information free of charge as well as giving users a variety of communication possibilities (Balcı & Tiryaki, 2018, p. 11). The new media environments have become important for individuals as new social mediums with the popularity and prevalence of internet technology. For this reason, when talking about the new media, new features added to internet technology become significant. The Internet has enabled individuals to reach information easily and quickly and provides interpersonal communication opportunities independent of time and place (Satar, 2015, p. 55). However, the new media tools are also criticized because many face-to-face and real-life actions are carried to virtual environments through an interface, social relations are weakened, digital addiction has become an issue, the mind is instrumental zed and people are more alienated. Although criticized, "the new media" where the media meets the Internet and creates a new medium, is considered to be a new power that can bring masses together for specific purposes and manipulate masses, beyond being a means of personal and mass communication.

Internet technology, which enables the exchange of information between all computers connected to the network via the global computer networks in the world, was first used for military purposes in the United States Department of Defense in 1962. ARPANET which provided information flow among only four computers in 1969 met with www (World Wide Web) technology in 1991 and has become globally widespread ever since (Kahraman, 2014, p. 17). Only print was shared until 1989, which was then transformed into posts that included visual content by Tim Bernard Lee as a result of the developments in World Wide Web (Sine, 2017, p. 50). However, the widespread use of internet technology took place in the 2000s. In 2004, users' limited activities became interactive and participatory with the help of Web 2.0 technology. Thanks to Web 2.0, usersare able to produce their own content. As a result of participants' active content production, the new media has started to be defined as "social media". Poynter states that there are many definitions to explain the concept of social media. According to Poynter (2012, p. 208) social media defines a new medium broadcasting for a large number of people from a large number of points as opposed to broadcasting by the traditional media from a single point to a large number of people.

Social media channels, called social media tools, are increasing day by day in terms of both the number of users and environment diversity. Allowing users to share content without any limit is the most fundamental feature of social media tools with different technological substructures. In this context, a participatory media process has emerged with the help of Web 2.0 technology and as a result of easier access to social media which extensively increased interaction (Kahraman, 2014, p. 21). In other words, the static structure of Web 1.0 which allows users only to read and do research has turned into a dynamic structure with Web 2.0 application and enabled social interactions among participants. Web 3.0 application has given the web technology a portable feature thanks to the newly developed interface for mobile devices (Sine, 2017, p. 53). Thus, the new media has become an integral part of our lives by allowing everyone to access and share any content, anytime, anywhere without the constraints of time and space.

The development and widespread use of Web 2.0 technologies has provided interactive capabilities for new media tools. Interactivity has enabled users who were passive before to become active and get involved in the media content (Sarı, 2018, p. 277). In this context, the new media is considered as a global communication tool because it offers easy and mobile usage to its users as well as opportunities for content creation, content sharing and interaction. In particular, individuals have changed their positions from being users only download information from the Internet to users who can also upload information to the Internet with the transition from Web 1.0 to Web 2.0 technology. As users have the opportunity to share information, the new media has created a new medium of communication, called cyberpublicspace. This new communication environment has created an alternative environment for the capital-based and holding-based structures of traditional media organizations (Karagöz, 2013, p. 132). In this context, opinions of minorities have started to be voiced with the new media in addition the thoughts of the majority groups in the society generally circulated by the traditional media tools.

In such an environment, the number of tools with which users can produce their own content and deliver to every corner of the world has increased day by day. In fact, the number of personal web pages and blog accounts is increasing day by day as well as social accounts that allow simultaneous information sharing such as Facebook, Instagram and Twitter. Today, the number of users in social networking sites has reached the levels that will surpass the population of many countries (Satar, 2015, p. 62). As a result of combining mobile technologies with internet technology, individuals are now able to reach information anywhere and produce information in any environment. It is clear that all media are intertwined with each other in the age of social media (Sarı, 2018, p. 293). This situation, characterized by the concept of hyper(text), includes texts as well images. Consequently, traditional journalism has also been influenced by the transformation in social media. In some cases, content production in social media has been known to steer traditional media (Tombul, 2018, pp. 141-142). As a result of the changing and developing communication technologies with the introduction of the internet, newspapers now appear on digital platforms. Newspapers now center on technological developments in order to reach more readers and an utterly different period of journalism has been born.

RELATIONSHIP BETWEEN NEW MEDIA AND JOURNALISM

Journalism began in the West two thousand years ago, with the emergence of ActaDiurna, which was handwritten and publicized during the period of Julius Caesar in Ancient Rome. Digitization, the process of converting information into a computer-readable format, was born in the age of electronic computing. Advanced technology is required for distribution and imaging. Taken together, these two forms refer to "an old practice presented in a new context" namely the synthesis of tradition and innovation (Kawamoto, 2003, pp. 3-4). In this process, journalism has digressed from its traditional forms, has been redefined with the World Wide Web and started to be mentioned with the concept of "speed". Moreover, with users starting to create content with the help of Web 2.0, journalism has moved away from its familiar environment to a completely different platform. The news organizations that have started broadcasting their publications by paying attention to the Internet in web environments since 1995 at a time where Web 1.0 was more common, have adapted to the current development in the process following the development of the internet.

The Internet which hosts the World Wide Web and other information technologies has not ignored journalism which is one of the most important impact areas of new media in the context of global culture and trade. Since 1993, more than 4900 newspapers have begun online journalism following the creation of Mosaic, the first graphic browser for the web (Pavlik, 1999, p. 59). Mark Deuze (2003, pp. 208-211), defined online journalism, the fourth type of journalism following print journalism, broadcasting journalism and radio reporting, under four categories:

1. **Mainstream News Sites:** The news on these sites are originally produced for the web. The mainstream news sites that are the most common in online journalism offers editor-controlled news. The newspapers which carried their pages to the Web are included here. Some examples include mainstream sites such as CNN, BBC, MSNBC, Haber Turk and NTV and news sites outside the mainstream media such as www.ensonhaber.com and Alternet.
2. **Index and Category Sites:** The second type of online journalism is often associated with specific search engines (Yahoo, Google, etc.), marketing research companies and sometimes entrepreneurs. Online newspapers often use such sites as a source of news and information. In other words, this index and category sites contain links that share news instead of the original content.
3. **Meta and Comment Sites:** This is the third category of news sites that generally provide news and information about news media and media issues. U.S. examples of these sites are *Mediachannel, Freedomforum* and *Poynter'sMedianews* and examples from Turkey include *Dördüncükuvvetmedya* and *Medyatava*. These kind of news sites, interpreted as crossing the barriers by site content producers, also include alternative news that are not included in the mainstream media. Alternative news sites define themselves as not being a mainstream (corporate, commercial) news organization. These sites - especially Guerrilla News Networks and Independent Media Centers around the world - not only offer their own online news, but also critically interpret the news provided by existing media networks, leading users to the news on the Web outside the mainstream news.
4. **Share and Discussion Sites:** The reason for the success of Internet and new media technologies is that they enable people to connect in real time with other people around the world. In other words, online journalism uses the potential of the internet, which is just a communication infrastructure by itself, to facilitate sharing ideas and develop sharing platforms around specific themes through worldwide anti global activism (Independent Media Centers, often known as Indymedia) or computer news.

During the period when journalism became widespread in the new media, the codes related to journalism were shaped. Thanks to the internet, speed has become a prominent aspect of journalism as well. Free internet newspapers whose sole income source is advertisements have started to shape their content to get more share from advertising. In this process where news and advertising are intertwined, the main factor that determines the newsworthiness of a news site has been the number of clicks that a news item receives. The share allocated to digital advertising is increasing day by day on world basis and it is ahead of other advertising channels. In order to get a good share from this advertising cake, it is necessary to ensure its distinctiveness by using elements such as recognition, content quality, number of visitors and profile, and visual quality (Işık & Koz, 2014, p. 30).

The news presentations that focus on clicks characterize the headlines and the images that accompany these headlines and these presentations practice a form of content-managed formal journalism by seeking the ways of how to get the maximum number of clicks from the news items, rather than seeking the answer to the question of how to deliver the news in the simplest and most accurate way to readers (Özyal, 2016, p. 275). The first criterion that determines the value of the advertising element that stands out in online journalism is related to the user traffic on the site. Newspapers that want to increase the number of visitors on their sites renew them with various methods. Sites are valued by criteria such as how many times the site has been visited and which content has been clicked more. In this context, almost everything about visual content and design is being updated to lure more readers.

ELECTRONIC BROADCASTING AND PAGE DESIGN

Electronic age can be defined as a period with technological developments and innovations/changes. In this respect, we also see the reflections of electronic developments and developments of Internet on the screens. Due to the electronic information system, we are witnessing a period where it is very easy access information and visual information is more prominent than ever before.

Helf said that the rectangular shape of the computer monitor surrounds everything we see on the screen. The attractiveness of the networked interaction indicates the exact opposite conditions. Even if it is limited by a box that is fixed, the virtual environment enthrones intangible movement, substanceless process and unconquerable and time-independent exchange (Armstrong, 2010: 121). When the main pages of online newspapers, i.e. electronic publishing media, are examined, it is seen that communication design which is a part of visual communication design is of great importance because the main pages of online newspapers should reflect the entire content and the viewer should reach all content easily. A quality internet newspaper should interact with the audience.

This interaction is the closest to the definition of "mother tongue" in electronic field. If the viewer is moving through the data, this action, which is a kinetic activity that limits our perception, dominates our senses and turns into a new superior aesthetic highly distinctively (Helfand 2001, cited in: Armstrong, 2010, p. 122).

Page design is of particular importance for newspapers which are important tools of the print media. Page design is one of the most important points in newspapers' production process and increases the power of the content because page design is a process to ensure that news and articles in the newspaper are perceived by the reader with the least possible effort. In addition, page design has the power of creating corporate identity and image with the use of typography and visual material (Teker, 2003, p. 216) because page layout is basically undertaken to perform functional and aesthetic purposes. These functional objectives serve to present the news in the newspaper in an order that can be easily perceived, the establishment of the hierarchy between the news on the page, to establish hierarchy among the news on the page and to focus on the most important news with the help of page design. The aesthetic aim serves to ensure that the newspaper has a different visuality than others, that it can be easily distinguished and the visual habit created in the reader is maintained (Şeker, 2006, p. 31). Online newspaper readers are impatient, preferring to find what they are looking for quacikly instead of reading for long periods of time.

Communication Process Between Visual use and News Content

Undoubtedly, the most important element in journalism is the news. A newspaper has to possess some features to perform its functions. News, photos, graphics and advertising are the basic elements of the newspaper. Design makes it possible to combine all these elements in the form of a newspaper. Each publication published for the purpose of communication and reading is first seen and then read; therefore it attracts attention with its visual effect (Tiryakioğlu & Top, 2010, p. 138).

The newsworthiness of the news included in the newspapers or magazines increases or decreases with the photo used alongside the news. A single photo frame can be more effective and convincing than long lines of information because people see the news first and then read it. Photos have much stronger sensitization effect than texts. The placement of news with or without photos differ in the newspaper (Tiryakioğlu, 2012, p. 69). In this context, visuals that are used should give information about news content, be related to news, be animated and have motion. Again, the visuals used in the news should be persuasive and intelligible and they should support the text. The images should be placed on the left or right side of the news, depending on readers' perspective. When multiple images are used on the main page, they must be set horizontally and vertically at different proportions and they should have different sizes. If the headline is at the top of the image (portrait photos), the image should be facing towards the headline and not the opposite direction.

In general, images and photographs should be taken into consideration in relation with one another because the effect of image and photo together is likely to be stronger. Planning the news item together with the content and visual elements will create an effective communication language. Designing the thoughts and information through text and image will create a mutual interaction and a two-way sharing.

On the page, a news story should not interfere with other news items since this will affect the reading of the news. The image and text in the specified area should not spread out from the boundaries set for the news item. Images outside the area will be in the other news area and will cause readers to become distracted. In this sense, it is important to emphasize the visual elements without distracting viewers from the topic at hand.

Image Use (Stable, Interactive, Moving Images)

The discovery of alternative home pages should provide curiosity and continuity in the viewer. The images used in the main pages are mostly used to support the news. The image is the first thing that attracts the attention of the viewer on the home page. Communication between the message and the audience is effective when it is intelligible and easier to access.

Image detection is different from text perception. The visual element enables the viewer to interact by stimulating the physiological and psychological states of the viewer. Therefore, there are a number of elements to be considered in the use of images: The visual element should be vivid, it should be appropriate to the content of the news and the focal point of the image should be determined to stay away from unwanted details. Positioning the selected image in the correct area on the page will increase the intended effect.

The image selected as the emphasis should be larger than the images used in the other news and should be placed at the top of the page. The use of images should direct the viewer and be descriptive. The images should arouse the interest of the audience. As a matter of fact, in countries with low reading rates, the use of images is rather extensive and pronounced. Therefore, the first thing viewers encounter on the first home page is the image. The use of images should direct the viewer first to the headline then to the text. Excessive and unnecessary use of images will create confusion. Images are effective factors in the appearance and perception of news. Proportional use of the image in a correct page design supports the clear understanding of the news. This allows the viewer to spend more time on the home page. On the other hand, the image used in the pages should be scaled up and down in proportion without distortion. Otherwise, the image will be degraded and credibility will be reduced.

Images are classified as static images (photos, illustrations, graphics), moving images (moving text, videos, animations), interactive images (menus, signs, icons, ads, banners, buttons).

Page Design and Grid

Page design or layout is the arrangement by which design elements ensure communication between the information and the viewer. The main pages of electronic newspapers, designed in accordance with certain rules, are formed by the processes that allow the viewer to perceive easily. The most important point in page design in traditional journalism is the first page because it is directly related to the sale of the newspaper. This page is defined as the newspaper's showcase. First page can be designed in four ways: horizontal, vertical, window and news based (Taş, 1993: 66).

Garcia (1997: 12) explains the purpose of design under three items as "to attract and maintain the attention of the reader, to make the newspaper legible and understandable and to create an identity that can make the newspaper recognizable". Looking at these explanations in general, it is worth noting that page design is a process to ensure that news and articles in the newspaper are perceived by the reader with the least possible effort. Page design also has the power of creating corporate identity and image with the use of typography and visual material (Teker, 2003: 216).

Page design presents the content of the message, the location and importance of the event. Page design is part of the content as well. Any publication is seen before it is read. For this reason, the first step of attracting the attention of readers is to have them take a look. The selection of the text to be read is usually decided after the first eye contact. Design should be attractive so that the newspaper will be read more since the reader will be attracted to the text. However, if the design cannot convey the intended message, it cannot accomplish its goal no matter how attractive the design is; because design is not just a decoration but a communication tool (Tiryakioglu, 2012, p. 7).

Designs of the main pages of the daily newspapers today should follow the order of visual communication design elements such as text, photo, image (moving, static) based on the order of importance. Each item on the main page affects the viewer in a positive or negative way and may cause a different perception of the news. Therefore, the layout of the main pages should be prepared according to the principles of graphic design. The distribution of graphic design elements used on the page should draw the attention of the viewer and ensure that the viewer remains on the main page, the elements should be visible and readable and should be able to create integrity on the page.

Each publication that is printed for communication and reading is seen before it is read and it attracts attention mostly with its visual effect. In other words, the complementary materials (such as texts, photographs, pictures and graphics etc.) on the pages that make up the publication should be used correctly and in place. The layout of the page should be well prepared in order to provide all these on the page in the intended order. In short, without page design, thoughts cannot be explained and the necessary message about the visual image cannot be conveyed (İstek, 2004, p. 56).

The success of the mentioned page design is determined with the help of balance (symmetric-asymmetric), emphasis, unity, color, text and proportion. For this reason, graphic design principles increase the functionality of the page. One of the most important rules of page design is the grid. Grid guides the layout of the design elements in the layout. Grids that provide consistency between the pages also help the designer to work accurately and quickly.

As a word, grid means a line, a path, a mesh. In other words, the grid is the horizontal and vertical lines that help to organize the elements (text, image, shape, graphics, drawing, etc.) that will be used in the page design. These lines are prepared within a system and in a proportional relationship. Designers sometimes create this relationship by means of mathematical or geometrical methods, sometimes they benefit from intuition (Uçar, 2004, p. 147).

According to Ambroseg and Harrise (2010, p. 33); working without a guide provides complete flexibility over the layout of the design elements, but the lack of the structure means that all space relationships between objects must be considered and determined. Working without a grid also makes it more difficult to maintain design consistency across different pages. Grids that form the basic skeleton of the page organize complex information, so both the reader and the designer save time. In addition to this, grids prevent random arrangements and help the reader concentrate on the content and prevent them from losing the emphasis (Taşçıoğlu, 2013, p. 95).

According to Müller and Brocmann; working with the grid system is accepting the universal validation laws. Use of the grid system implies the will that emphasizes and clarifies the importance of education in tasks that value concentration on the principles, improvement of subjectivity instead of objectivity, rationalization of creative and technical production processes, integration of the elements of color, form and material, possession of the architectural domination on the surface and space and adoption of a positive and forward attitude and to have a constructive and creative spirit (cited in: Ambroseg & Harris, 2010, p. 63).

Balance

Balance is quite important for accurate and effective composition in design because misuse of visual elements may disrupt the balance of the composition. Two basic types of balance, symmetrical and asymmetrical, are used in designs. Symmetrical balance is reflected with equal weight on either side of an imaginary center line based on the characteristics the shapes possess. In asymmetric balance, this equation is not important. The symmetry used in design evokes a sense of boredom and inertia after a certain period of time in ensuring visibility and legibility. Dynamic and exciting page designs that are also catchy are often asymmetrical. Many graphic designers try to avoid the simplicity of creating symmetric balance to create original and creative works (Uçar, 2004, p. 66).

Contrast

Contrast means creating opposition, contradiction and difference. Visual elements, colours, fonts, lines are the elements that help create contrast in page designs. The light-coloured text on a dark visual element on the page, for example, can create contrast. The contrasting use of visual elements in a good page design will have a strong impact on the viewer.

Emphasis

Emphasis, which refers to the center, has a very important place in page design. The emphasis on the page in a single location is the primary priority. More than one emphasis causes instability in the viewer and changes in the impact of news on the page. The emphasis determines the most basic point of the page and sets up the viewer's eye tracking system. In this way, the start and end points allow the viewer to reach the point they want to reach by extending their time on the main page.

Studies in the text readings demonstrate that the eye moves on the page in the form of the letter Z. Accordingly, the eye first moves from the top left of the page to the right and then continues down to the bottom left crosswise and the movement of the eye ends in the lower right corner (Tiryakioğlu, 2012, p. 20).

Proportion

The concept of proportion can be defined as the size of elements in relation to one another. This important concept, which determines how photos, fonts, colour and other visual elements will be structured on the page, cannot be used to transfer the accurate message to viewers when the page is incorrectly configured. Therefore, the proportion of the photo to text, the proportion of the font to the headline and the main texts, the proportion between the area covered by the primary news and the other ones are effective in the correct transmission of the message. The area used for the primary news coverage on the page and the use of larger images, serve to bring the primary news to the foreground in a proportional manner as opposed to other news included in the page.

When two or more visual elements are brought together on the design surface, they indicate a proportionality problem. Proportionality for the designer is the relationship between dimensions. There are always proportional relations between the width and the height of the design surface and the widths and heights of the visual elements and the dimensions of the masses formed by these visual elements. The proportional relationships established by a visual element with other elements in the design directly affect perception and communication (Becer, 2009, p. 68).

Unity

Unity in page design is the most important part of the design. Visual language that does not speak the same language causes confusion in regards to the concept. Successful and perceptible design is ensured through the placement of separate parts on the page as a whole. Unity should prevent differences on the page and should maintain the same design concept. However, unity is not simply the use of similar and same size fonts and use of visual elements in the same proportions. This type of focus on unity may cause the page to be stagnant and boring. Balanced use of the colours, fonts and white spaces /spaces on the page will ensure unity.

According to Uçar (2004, p. 156), simple methods such as use of a tasteful colour, sensitive use of different fonts and overlapping the visual axes are good methods to create unity. Unity is not a relationship of sameness or differences.

Colour

Colour is a very strong phenomenon. The intended message is transmitted accurately or inaccurately based on the use of colours. The visual power of the design and the effect of the message can be made more visible with the help of colour. Colour is one of the important elements in creating visual hierarchy. Breaking news on the main pages of electronic newspapers is related to the visibility of the article and the correct use of the colour element. The use of colour in the page design serves the functions of attracting attention, transmitting the message and creating similarity. Colour is one of the most important elements of web pages with its complementarity, distinctiveness, directivity and emphasis and it keeps the interest by influencing the reader (Pektaş, 2001, p. 75).

By adding movement to the design and presenting certain emotion encodings to the viewer and as a powerful communication tool, colour allows differentiation of different types of news by categorizing them. Effective use and communication of colour supports the use of elements in page design. However, there are some setbacks related to the use of colour: colour intensity which is used to draw attention on the page can cause visual pollution and confusion. Therefore, proportional colour distribution is an important design element in composition. Colours are also used to focus on the composition and to emphasize and create a sense of continuity between the visual materials that make up the design (Holtzschue, 2009, p. 4).

In addition to all these, colours have psychological and physiological effects. Warm colours (red, yellow, orange) ensure fast detection and visibility while and cold colours (blue, purple, green) reflects on the viewer a sense of withdrawal. The colour used in the page design to make the content strong must be guide and highlight as well as being distinctive and influential.

Use of Text

Text is a tool created as the continuation of the signs and symbols that people have come to use in order to clarify their feelings and thoughts with forms. Typography, which includes letters, numbers, symbols, lines and punctuation marks used in all fields of writing and written communication, can convey information as well as visual messages. The relation of human beings with visual materials is rather new compared to humanity's relationship with writing. Arrangement of writing and editing texts have created a different branch of design called editing and page design. Especially the inscription culture has necessitated the creation of this arrangement in accordance with aesthetic elements. Although each inscription communicates in a semantic dimension based on the context of its content, it also presents an objective value as an art element (Uçar, 2004, p. 144).

The purpose of text on a Web page is to get in touch with the audience. Correct use of the text with visual elements on the page will allow the transmission of the intended message. Typography is a visual element that guides and provides communication. It establishes the task of communication to make the text easy to read on the page and to make the image understandable. Typography is the arrangement and classification of letters to transmit the emotions and thoughts with signs, symbols and fonts in written communication. According to Bülbül (2017, p. 132), during the process of redefining the text, the reader who communicates with literary texts continues to re-establish the text created with fictional designs by continuously using the creative imagination and imagery process. Sense making is a fictional action fed by the reader's imagination process. In the stages of the formation of the literary message, the readers react to the fictional universe of the text with their own expectations and emotional interactions.

Typographic arrangements play an important role on the home page in ensuring that the appearance of the message is visible to the viewer and remains on the page. Main headlines on the page, which attract the attention first, should support the visual element. In this context, two types of page designs can be created. First, is the use of a wider area on which the visual elements occupy the page and the fonts remain in the background and the second is to increase the effect on the visual element with the larger use of fonts. The viewer usually selects a focal point after browsing the home page. This point is either a visual element or a main headline. In this context, visual elements and news headlines are important items on the pages.

The typographic arrangements used on the page should be simple to understand because using more than one different font will cause confusion. Use of text is the main drive for viewers to read the text by making the news effective for the viewer. However, what should be considered here is legibility of the text since readability is an important element that will determine whether the viewer will remain on the main page.

The relationship between legibility and readability lies in selecting the font and the type size. The concept of time in the electronic age has led to a state of fast consumption. In this age where speed is dominant, competition is experienced in every field including journalism. The persons who prepare and service the news race against time to ensure that the reader stays longer on the home page. The legibility and readability of the news is important in this regard. Font, style, typefaces with or without serif and harmony between uppercase and lowercase letters must be observed to ensure readability. Font is a physical tool used to obtain a character. It consists of a set of letters with a certain shape and a certain typeface. A character consists of a group of characters, symbols and punctuation, each with the same determinant style (Ambrose & Harris, 2010, p. 135).

Legibility

The typeface which is used in the news article is very important, for example, a very thin font will be weak and reduce legibility. If a too-thick font is selected, it will reduce reading time because it will strain the eyes. In order to increase text readability on the page, it may be sufficient to use texts as a single block in a wide area. This depends on the width of the column. It is very difficult to read the block texts presented in more than three narrow columns.

Apart from all these, legibility is defined with a feature which is not so stylistic after all: pleasure. Pleasure is not the same as being trendy, it is rather related to popularity and how commonly it is consumed. We like to believe that what we take pleasure from in regards to culture grow and mature with age, but when it comes to typeface design, something else happens: overuse obsoletes our pleasure. Californian radical font designer Zuzana Licko has a popular theory: “What we read best is what we read the most” (Garfield, 2012, p. 60).

Letter Spacing

The term letter spacing refers to the space pattern between letters. It is possible to improve readability by adjusting these spaces. Letter spacing is scaled in millimetres, units and smaller scales, depending on the sequence (Sarıkavak, 2009, p. 103).

The use of graphic design elements on the page helps transmit the message to the viewer during the communication design process. Using visual elements (photographs, text, shapes, signs, etc.) without allowing for sufficient space makes the content difficult to read. As too much space will cause distortion in reading, allowing for very little space can negatively affect reading as well since it compresses words and deforms text structure. In fact, arranging the text and images very close to each other may create difficulties in both text and image perception. Space between letters, between lines and words as well as spaces between text and visual elements and spaces between visual elements themselves facilitates perception. If there is not sufficient space between visuals and the text, the page design will lose its effectiveness.

Text and Colour

Colour is an important factor in creating visual hierarchy. The visibility of primary and secondary news is related to colour. The violence, excitement or rhythm of the news is again perceived by the use of colour. For example, white lettering on a black background or black text on a white background increases readability. Using more vivid and bright colours on a dark visual element will enhance the appearance of the news. The colour of text used in the news item should be related to the content of the news. Thanks to the colour of the text, pre-information is provided before the content of the news article is read.

The reader usually wants to colour in the text. It is possible to say that colour is effective in information retention in teaching materials and thus facilitates learning (Hartley, 1996, p. 789, Petterson, 1993, p. 285, cited in: Alpan, 2008).

Type faces and Their Features

They are classified as typefaces with or without serif, serif, traditional (penmanship) and decorative typefaces. The selected typeface is very important in terms of clarity of design. The design and the preferred typeface are expected to be related; i.e., the typeface should be associated with the designed product.

While some characters can be serious, noble and reassuring for the reader, the others are found to be sincere, cheerful and enthusiastic (Uçar, 2044, p. 125). For example, according to Garfield (2010, p. 135); "Helvetica manages to give a sense of honesty and awaken confidence. In addition, its decorative character distinguishes it from what exudes an air of extreme authority; even in a company, Helvetica provides a friendly home setting". Thoughts on the meaning of the characters may differ. Ketenci and Bilgili (2006, p. 250) state that the ones who want to give a sharp message should employ "Avangarde" and those who want to give an attractive message should use "American typewriter".

When the selection of typefaces is left to readers, they choose the ornate and aesthetically pleasing ones. Usually, these types which are not so common are not legible. For example, Roman style is more legible than italics. Especially in compact texts, italics, or inclined fonts, should not be used. Italics are always found to be more attractive by the reader. For adult readers, they can sometimes be used for emphasis. However, it is difficult to say that readers are aware of this issue (Alpan, 2008).

Normal, bold, medium, italic, oblique typefaces help increase the emphasis of the news within the text. The chosen style is effective in conveying the positive-negative, important-trivial situations. In common usage, the words for font and typeface are used synonymously, but both terms have different meanings. A typeface is a collection of characters with the same custom design. On the other hand, a font is a physical tool for the production of typeface, whether it is in the form of a lithographic film, a metal or a in computer code to describe the character (Ambrose & Harris, 2010).

Method

This study investigated comparatively the literary codes and the visual elements in terms of design features included on the home pages of the most visited online news sites in Turkey. For this purpose, five of the most visited online news sites in Turkey were identified first. According to data[1] obtained from www.alexa.co, the most visited online news sites in Turkey, are Ensonhaber, Haber7, Haberturk, Haberler and Ntv respectively. After identifying the internet news sites to be analysed, the date ranges to examine their page designs were determined. Page designs and communication design process (page design and grids, visual elements, text usage, integrity between visual elements and the news, use of visuals and texts) of selected news sites' main pages were evaluated and differences and similarities were presented during the period between 8-15 April, 2019.

While the universe of the study was composed of online news sites in Turkey, the study sample consisted of five news sites with the most visitors. Therefore, the research is limited to the main pages of these news sites that were analysed in the framework of the study. In addition, the research was chronologically limited to April, 8-15, 2009.

Findings

This study examined the differences and similarities among the page designs of online news sites that had the highest number of visitors according to www.alexa.com data and the interaction between the main page and the audience was presented. A convenient communication tool must be easy to use and free to access. A means of communication for which users do not have sufficient financial means or information on how to use is regarded to be a bad communication tool; and perhaps it is not even considered to be one by the mass that cannot reach this medium (Poe, 2019, p. 29).

Within the scope of the analysis conducted in this study, these five online newspapers were found to be uniform in their use of graphic design elements and it was determined that each of the newspapers displayed an approach similar to the communication design process in order to reach the audience by using photographs (sliders), moving images, texts and illustrations. In this context, it is possible to say that a simple and open approach is adopted in the newspapers that were examined.

Investigation of the main pages of the newspapers demonstrated that approximately 25 news items were included in each site per day. When these news sites were analysed in terms of page layout, combined use of photos and texts was observed to be very common. The sliders, which represents the proportion of visual unity, were widely used on the main pages. The use of images, headlines or spot news support the visual compatibility with the news on the page. At the same time, while most of the news did not sport a graphic design product, the advertising graphics were observed throughout the news. In this sense, it is possible to say that advertising has an important place in online news sites.

One of the findings point to the fact that online newspapers examined in this study tended to give more priority to texts rather than graphic design elements. However, it was observed that these five online newspapers included photographs and text fields. On home pages, primary news headlines were

Figure 1. Haberler.com web graphic design

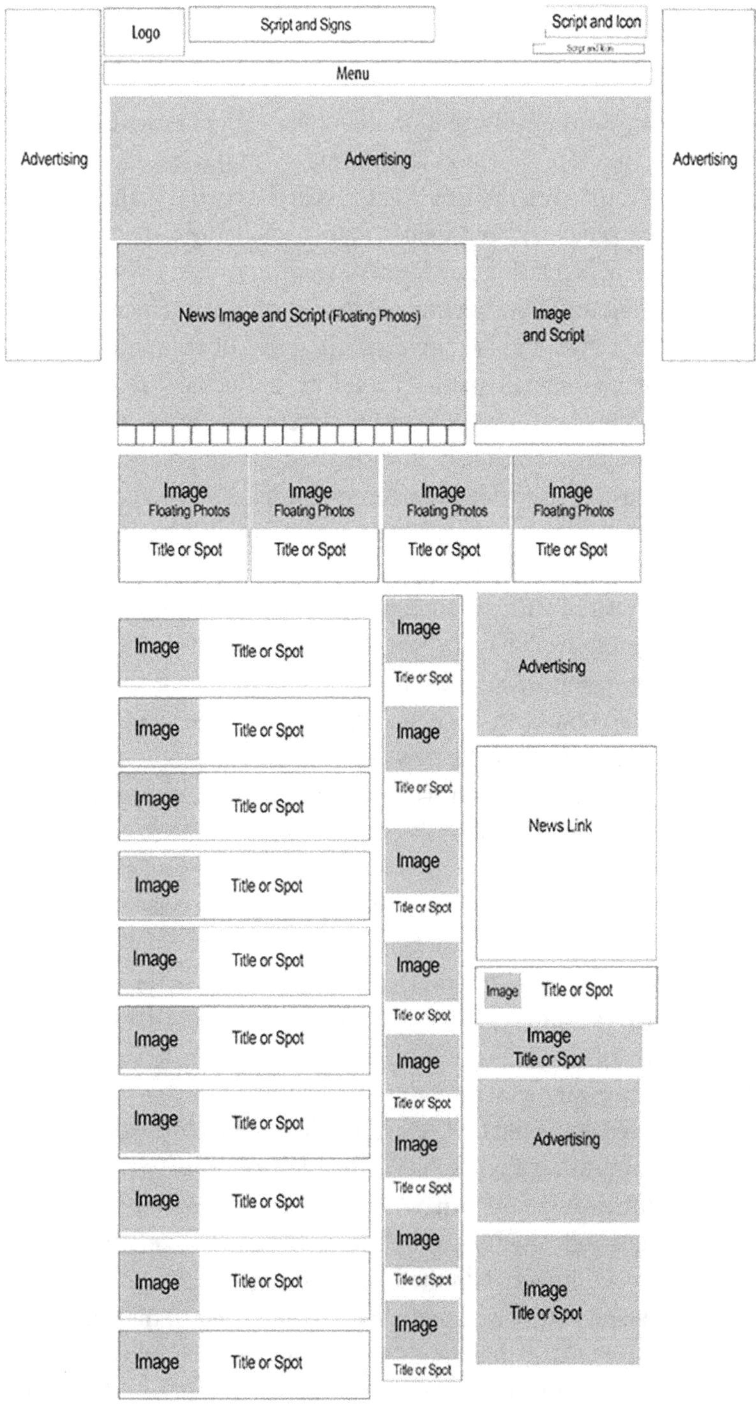

indicated by large fonts and graphics. The primary news on home pages in all online newspapers studied here were presented more distinctly. newspapers. Images in these areas were included large rectangular photographs.

It is thought that the associated news, detailed news and related information are more likely to be accessed by clicking to obtain further information than the original news content. All news content is constructed in hyperlink on these investigated internet pages in the five newspapers. With a click, the viewer can access full news texts from the headlines displayed on the main pages. In order to read the headline news, the viewer may access either the main news texts or the full photo-link texts. This provides the contents of online newspapers. Online newspapers aim to provide easy access to readers by taking advantage of the available technology. Examination of graphic design elements in today's communication design process shows that these elements are important in terms of communication process since use of texts and photos are abundant on the main pages. Therefore, online newspapers should be considered as a whole. All headlines and photos are created to ensure interaction with the audience.

Taking these processes into consideration, it can be argued that graphic design elements should be used in accordance with today's design approach when an online page is designed. In establishing rules for the use of a news site, interaction is a key element to consider and the balance of communication power requires that there is a flow from the broadcaster to the viewer and from the viewer to the broadcaster. An online newspaper page prepared with the audience should in mind be simple and easy to understand. However, the main pages of online newspapers provide direct communication between the editor and the audience. As a symbol of interdependence, the introduction of the news link should support the concept of communication power of the news content developed between the sender and the viewer.

News and photos on the main page of the news section on Haberler.com was divided into sections where priority was given to the news section. Use of multiple images on the same page was observed and large and small fields are placed horizontally and vertically on the page according to news content. The information provided by the news and images directs the viewer and allows going beyond the news content. Based on the idea that repeating the use of news and images created with monolithic geometric forms for a certain period of time would be negative for page attraction, color yellow is distinctively used on the main page to keep viewers' attention vivid. Headlines use the same font and the use of images vary while moving images occupy less space. Advertisement slots are located at the top right of the page. The icons and signs are located at the bottom of the home page. It was observed that the audience could navigate on the page the easily and quickly.

The headline news is located at the top on Habertürk.com home page based on the hierarchy element according to asymmetric composition approach. The whole page consists of two parts. The location of the second and third news stories at the bottom of the highlighted primary news is very complex. In this case, the use of moving images and news types in different areas are likely to have a negative impact on the audience. The use of text and images is quite evident in the headline news. The text and image use are impressive and complementary. Ads are usually located at the top of the page and they are distributed symmetrically. In terms of ease of use, the menus are distinctly prepared to provide easy access by using modern font and color. When the site was evaluated in general, it was determined that it can easily interact with the audience due to clear and easy to understand use of graphic design elements.

The homepage of Haber7.com consists of eight sections. These sections are very important for viewers' easy access to information. The images, selected fonts and colors used in today's design concept express simplicity. As with the other sites, the headline news is effectively located at the top of the page.

Figure 2. Habertürk.com web graphic design

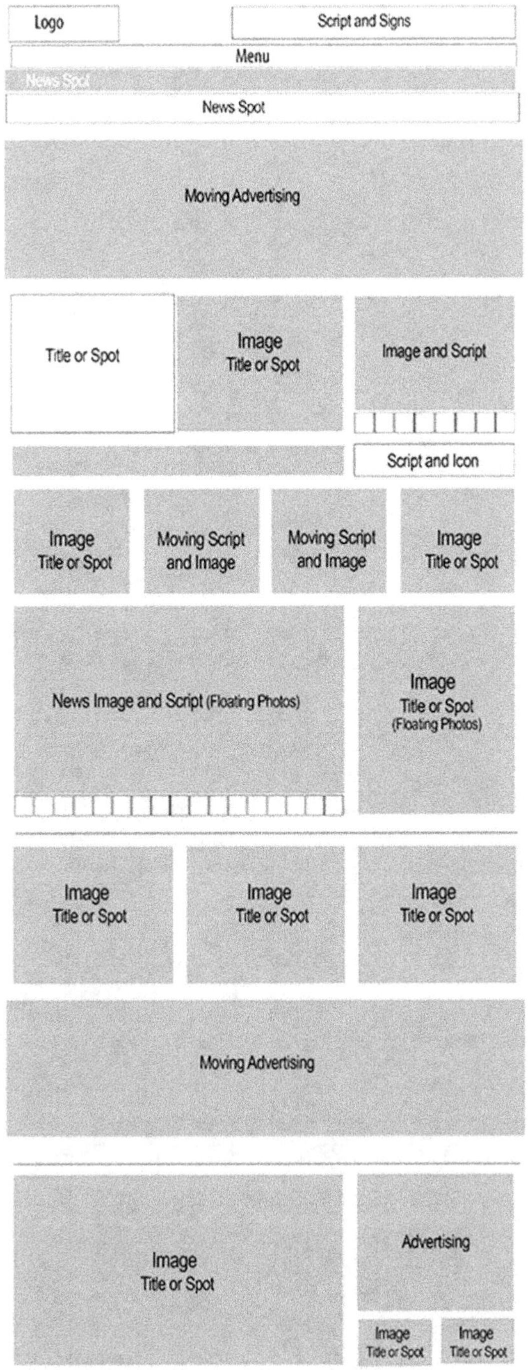

Figure 3. Haber7.com web graphic design

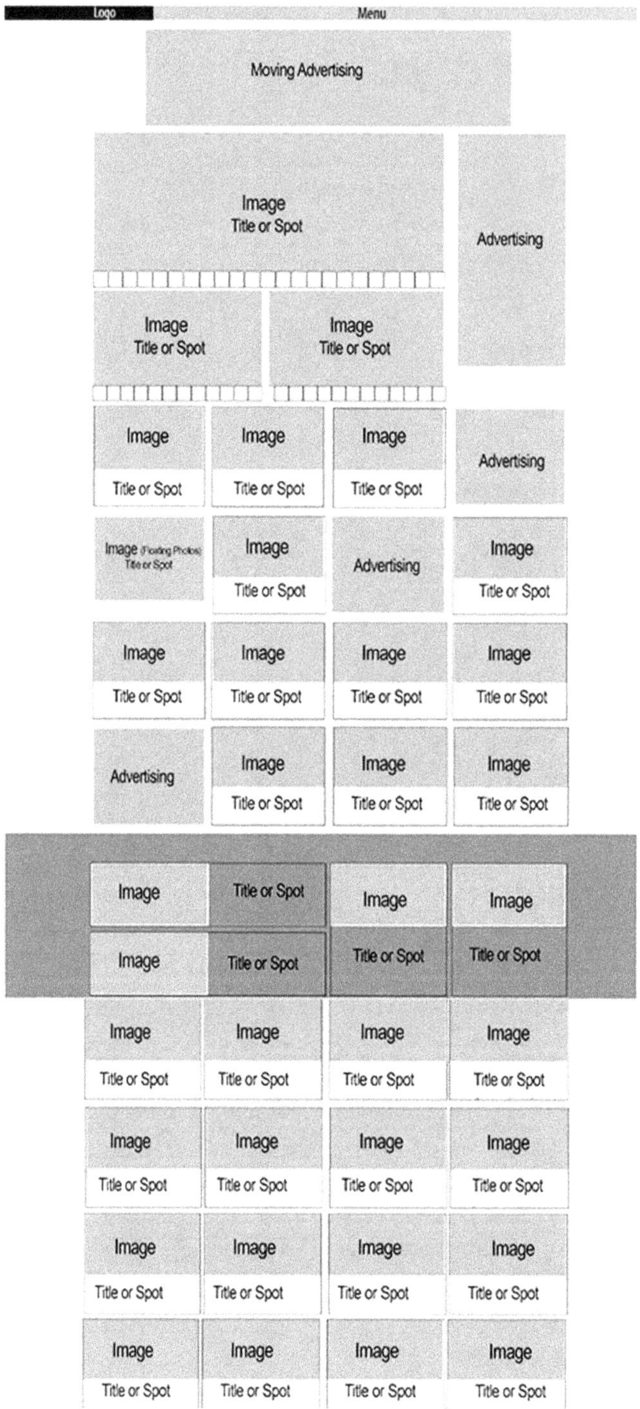

Figure 4. Ensonhaber.com web graphic design

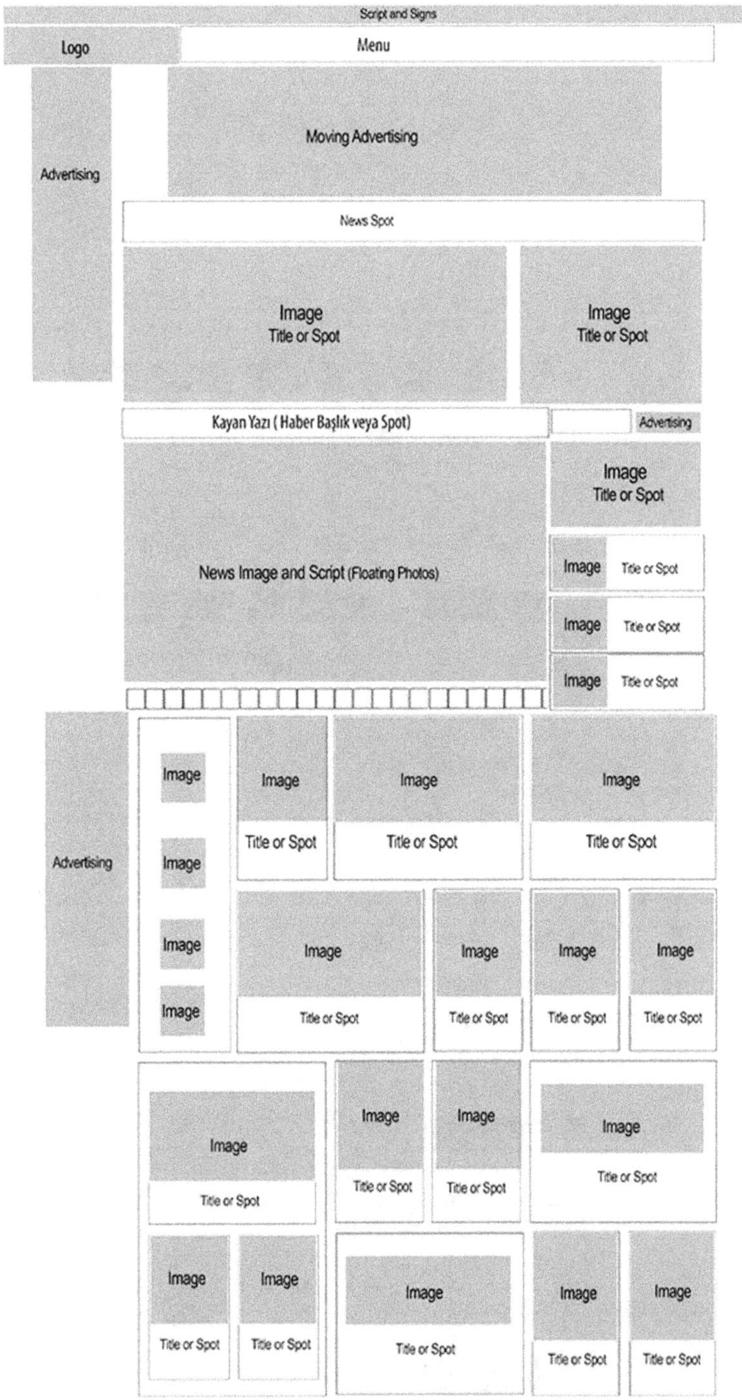

Figure 5. Ntv.com web graphic design

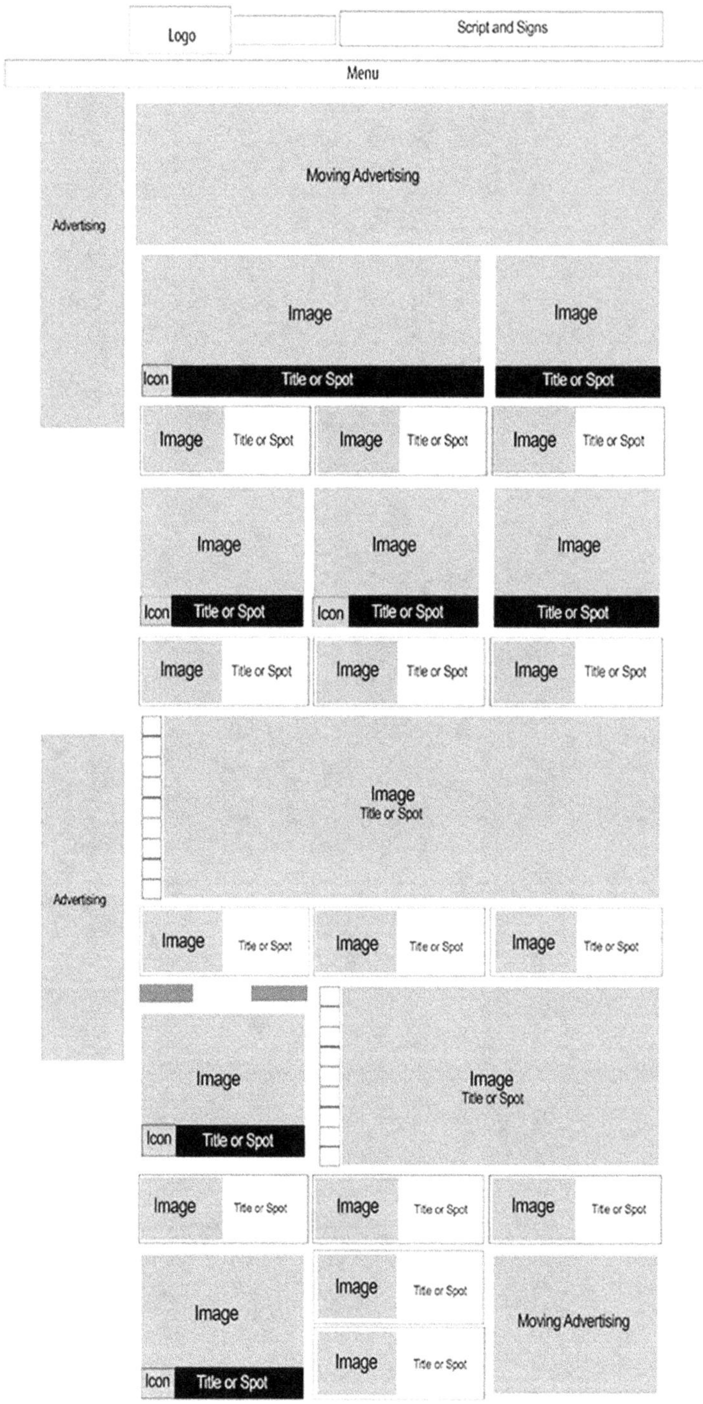

Text and image unity are distinctive. Moving texts and images are abundantly used. Proportion-ratio and the use of color play an active role on the home page of the newspaper prepared in line with the design principles. Advertisement slots are located at the top and left of the page. It was observed that the menu section of the main page for navigation was in the background. Findings show that a simple and clear design concept was adopted with the audience in mind.

It was observed seen that the home page design of ensonhaber.com, another news site examined in the framework of the study, is based on asymmetric composition approach. The headline news is located at the top of the page. While it is intended to emphasize the headline news on the homepage, location of four smaller news items and photos down the homepage may create a negative impact in terms of perceiving the primary news. The main page is divided into sections in general to provide the audience easy access to information. The logo is clear and effective. Various icons and symbols are located at the top of the page next to the logo. The menu for navigation is distinctive. Different sections are marked with symbols and are divided into separate areas so that the layout does not strain the eyes and creates a clear and simple visual effect. The use of typographic elements in the headline news creates interest on the news page and keeps viewers on the main page.

Finally, it was observed that the main page design of ntv.com is based on symmetric composition. The headline news is distinctive; font is colorful and large to accentuate the headline. The logo of the newspaper is located at the top left and it is clearly positioned. This describes the starting point for the viewer on the main page. Signs and icons are located at the top of the page. The advertisement areas are on the left and the top, as they are in the home pages of the other newspapers. Fixed and repetitive image and text fields may create boredom and prevent viewers from staying on the main page for a long time. On the other hand, the headlines are designed as large and colorful. The use of news and images, the use of news, typeface and color support each other.

FUTURE RESEARCH DIRECTIONS

Analyzing and comparing the main pages of the both print and online newspapers will provide assessment from another perspective.

CONCLUSION

As a result, it can be concluded that although the alternative model in the online newspapers investigated in this study carry the characteristics of the general interaction model, it is also significantly different. The newspapers examined in this study include the process of communication called hyperlink, which facilitates many interdependent connections. Different designs of online newspapers offer the opportunity to drastically change the appearance of printed newspapers. In this context, it is possible to say that online newspapers differ from traditional newspapers especially in terms of graphic design elements.

Secondly, this study supports the idea of designing more improved communication process on online newspapers' page designs in the process of interacting with the reader to ensure guidance for viewers, provide solutions and maximize interaction between viewers and written, visual and auditory messages with clear and simple expressions. It was observed use of graphic design elements shape the perception of the audience with production and consumption opportunities and accelerate the communication process. In other words, online news enriched with graphic design elements attracts readers further.

The texts used to highlight news texts and images were supported using typographic elements in each of the five newspapers investigated in this study. It was observed that the headline news items on the main pages of the online newspapers were highlighted and these areas were supported with images and typographic elements. Other news was in comparatively smaller areas. Thus, more news can be presented on the main page. Images and typographic elements on the main page were supported with colors and moving images and advertisements were also placed on the main pages.

In addition to this, the concept of journalism, which has evolved from print media to digital media in the new media axis, has increased the use of visual and moving pictures. The reading habits of the users are strengthened with photographs and sometimes news texts are used as explanations for the photos. It was concluded in this study that a new in-depth journalism and / or multi-faceted journalism has emerged, especially through moving images and hyperlinks, enabling access from one news item to another and even access to various forms of the same news item. Since access to this novel news environment is free of charge, advertising elements have an important place in design.

REFERENCES

Alpan, G. (2008). Ders kitaplarındak imetin tasarımı. *Türk Eğitim Bilimleri Dergisi*, *6*(1), 107–134.

Ambroseg, H., & Harris, P. (2010). *Görsel Grafik Tasarım Sözlüğü*. İstanbul, Turkey: LiteratürYayınları.

Armstrong, H. (2010). *Grafik Tasarım Kuramı*. İstanbul, Turkey: EspasYayınları.

Balcı, E. V. (2018). Sanal Dünyave Dijital Oyun Kültürü, Yeni Medyada Yeni Yaklaşımlar, Ed. Rengim Sine, Gülşah Sarı, Konya: Literatürk Yayınları, pp.11-43.

Becer, E. (2009). *İletişim ve Grafik Tasarım*. Ankara, Turkey: Dost Yayınları.

Bülbül, M. (2017). *İmgesel İletişim*. İstanbul, Turkey: ÇizgiKitabevi.

Deuze, M. (2003). The Web And İts Journalisms: Considering The Consequences Of Different Types Of News media Online. *New Media & Society*, *5*(2), 203–230. doi:10.1177/1461444803005002004

Garcia, M. R. (1997). *Contemporary Newspaper Design*. Upper Saddle River, NJ: Prentice Hall.

Garfield, S. (2012). *Tam Benim Tipim/Bir Font Kitabı, Çev: Sabri Gürses*. İstanbul, Turkey: Domingo Yayınları.

Holtzschue, L. (2009). *Rengi Anlamak, Çev: Fuat Akdenizli*. İzmir, Turkey: Duvar Yayınları.

Işık, U. & Oz, K. A. (2014). Çöp Yığınlarında Haber Aramak: İnternet Gazeteciliği Üzerine Bir Çalışma. *Humanities Science*, *9*(2), 27–43. doi:10.12739/NWSA.2014.9.2.4C0178

Kahraman, M. (2014). *Sosyal Medya 2.0*. İstanbul, Turkey: Media Cat Kitapları.

Karagöz, K. (2013). Yeni Medya Çağında Dönüşen Toplumsal Hareketler ve Dijital Aktivizm Hareketleri. İletişim ve Diplomasi Dergisi, (1),131-157.

Kawamoto, K. (2003). Digital Journalism: Emerging Media and the Changing Horizons of Journalism. In K. Kawamoto (Ed.), Digital Journalism: Emerging Media and the Changing Horizons of Journalism, pp. 1–29. Lanham, MD: Rowman & Littlefield Publishers.

Özyal, B. (2016). Tık Odaklı Habercilik: Tık Odaklı Haberciliğin Türk Dijital Gazetelerindeki Kullanım Biçimleri. *Global Media Journal, TR Edition*, *6*(12), 273–301.

Pavlik, J. (1999). New Media and News: Implications for the Future of Journalism. *New Media & Society*, *1*(1), 54–59. doi:10.1177/14614448990010010009

Pektaş, H. (2001). İnternette Görsel Kirlenme, Ankara, Turkey. *Tübitak ve Bilim Dergisi*, *400*, 72–75.

Poe, T. M. (2019). *İletişim Tarihi: Konuşmanın Evriminden İnternete Medya ve Toplum*. İstanbul, Turkey: Islık Yayınları.

Poynter, R. (2012). *İnternet ve Sosyal Medya Araştırmaları El Kitabı Pazar Araştırmaları İçin Araçlarve Teknikler*. İstanbul, Turkey: Optimist Yayınları.

Sarı, G. (2018) Sosyal Medyanın Yeni Starları: Youtuberlarla Değişen Popülerlik, Yeni Medyada Yeni Yaklaşımlar, Ed. Rengim Sine, Gülşah Sarı, Konya, Turkey: Literatürk Yayınları, pp: 277-296.

Sarıkavak, N. (2009). *Çağdaş Tipografinin Temelleri*. Ankara, Turkey: Seçkin Yayıncılık.

Satar, B. (2015). *Popüler Kültür ve Tekrarlanan İmajlar*. İstanbul, Turkey: Kozmos Yayınları.

Şeker, M. (2006). Sayfa Düzeni Ekollerinin Estetik ve İçeriğe Etkileri. *Selçuk İletişim Dergisi*, *4*(2), 30–40.

Sine, R. (2017). *Alternatif Medya ve Haber Toplumsal Hareketler de Habercilik Pratikleri*. Konya, Turkey: LiteratürkYayınları.

Taş, O. (1993). *Örnekleriyle Çağdaş Gazete Tasarımı*. Ankara, Turkey: Makro Yayın.

Teker, U. (2003). *Grafik Tasarımve Reklam*. İzmir, Turkey: Dokuzeylül Yayınları.

Tiryakioğlu, F. (2012). *Sayfa Tasarımıve Gazeteler*. Ankara, Turkey: Detay Yayıncılık.

Tiryakioğlu, F., & Top, D. (2010). Sayfa Tasarımı ve Kurumsal Kimlik Oluşturma: Türkiye'deki Ulusal Gazetelerin Birinci Sayfaları Üzerine Bir Araştırma. *Selçuk İletişim Dergisi*, *6*(3), 137–146.

Tombul, I. (2018). In R. Sine & G. Sarı (Eds.), *Sosyal Medyada Mahremiyetin Dönüşümü ve Gençlerin Mahremiyete Bakışı. Yeni Medyaya Yeni Yaklaşımlar* (pp. 131–175). Konya: Literatürk Yayınları.

Uçar. Tevfik Fikret. (2004). Görsel İletişimve Grafik Tasarım. İstanbul, Turkey: İnkılâp Yayınları.

ADDITIONAL READING

Bastos, M. T. (2016). Digital Journalism and Tabloid Journalism. In B. Franklin & S. Eldridge (Eds.), *Routledge Companion To Digital Journalism Studies* (pp. 217–225). London, England: Routledge. doi:10.4324/9781315713793-22

Çatal, Ö. (2017). New Technologies Challenging the Practice of Journalism and The Impact of Education: Case of Northern Cyprus. *EURASIA Journal of Mathematics, Science and Technology Education*, *13*(11), 7463–7472.

Deuze, M. (2001). Online journalism: Modelling the first generation of news media on the World Wide Web. *First Monday*, *6*(10). http://journals.uic.edu/ojs/in. doi:10.5210/fm.v6i10.893

Lovlie, A. S. (2016). Designing Communication Design. *The Journal of Media Innovations*, *3*(2), 72–87. doi:10.5617/jmi.v3i2.2486

Sara, G., Rachel, C., Martyn, E. (2013) "The Impact of Design in Social Media Today",2ND Cambridge Academic Design Management Conference, 4-5 September, p.:1-14.

KEY TERMS AND DEFINITIONS

Asymmetrical Composition: In general, the most successful compositions reach balance in one of two ways: symmetrical or asymmetrical. It is relatively easy to understand balance in a three-dimensional object because if the balance is not ensured, the object will be overthrown.

Digital Media: Various platforms where people communicate electronically.

Graphic Design: Organizing the text and images in a visible plane to convey a message. It can be applied in many environments such as print, screen, moving film, animation, interior design and packaging. Its basic principles are alignment, balance, contrast, emphasis, movement, pattern, proportion, hierarchy, repetition, rhythm, and unity.

New Communication Technologies: Communication means for the simultaneous and multi-strata interaction of the communication process based on digital coding system.

New Media: All developments based on the internet and the infrastructure of internet technologies.

Online Journalism: Representatives of traditional journalism regard online journalism as a medium that has evolved from traditional journalism via technology but new generation of journalists do not regard online journalism as a continuation of traditional journalism but as a medium that brings a whole new dimension to news and dissemination. The only case where the two sectors meet at the middle point is the fast news flow feature brought by online journalism.

Web Graphic Design: Unlike the design area, which starts with charcoal in classical terms, web graphic design requires the web designer to have a wide knowledge of computers and web world. In this way, it is separated from the classical concept of art and becomes a technological consumption assistant.

Chapter 6
A Critical Appraisal of Crime Over Social Networking Sites in the Context of India:
Social Networking Sites

Unanza Gulzar
School of Law, NorthCap University, India

ABSTRACT

An inescapable part of our everyday lives, visual communication is a key driver of engagement on social media. These are redesigned their news feed to allow greater emphasis on visual content, resulting in greater interaction. This chapter discusses the current scenario of cyber and social media crime in India and how the government has incorporated the necessities to fight against it. It will also include the types of social media crime enumerating the provisions of Information Technology Act, Indian Penal Code. Through this chapter, the author discusses the various types of cybercrimes, which are cyber defamation, cyber pornography, cyber stalking, fraudulent transaction and misrepresentation, hacking. The author laid emphasis on what legislations are in action to deal with such crimes and how strictly the offenders are punished. The author also discusses the competency of the present legislation and how the loopholes, if any, can be filled to make the virtual world a better place for everyone.

DOI: 10.4018/978-1-7998-1041-4.ch006

INTRODUCTION

The Social Networking Sites like Twitter, Facebook, YouTube, Instagram, Snapchat and LinkedIn in contemporary world, have shattered in popularity and are profoundly entrenched in our day-to-day lives. Moreover, it has changed the mode of communication over the last decade. Further, it has not only transformed the ways we communicate to each other but, at the same time, improved connectedness and interactivity, with access to instantaneous news and information which changed the ways we do business nowadays.

This thriving in social networking platforms has added fuel by the amplified take-up of mobile devices which have replaced the desktop. Today, business entrepreneurs are reaching to their main stakeholders by this inexpensive, valuable communication tool and are thus shrinking the communication gap that once prevailed.

Since their (mobile phones) introduction, the Social Network Sites (SNSs) have grabbed the attention of millions of users and many of users have incorporated these sites into day to day practice. Further, there are various SNSs with numerous technological affordances, assisting a wide spread series of practices and interests. Their main technological characteristics are objectively constant and the cultures that develop nearby SNSs are diverse. Various sites support the conservation of already existing social networks, but others assist outsiders connect based on shared interests, political views, or activities. Different sites provide to various addressees, while other fascinate individuals based on mutual language or religious, sexual or nationality based identities. Also, sites differ in the degree to which they integrate novel information and communication gears, such as blogging, mobile connectivity and Video/photo sharing (Danah Boyd, 2014)

Moreover, with our Social Networking suckle continuously stopped with updates and news, it's easy for the information to get misplaced in the racket and piles of text. Words only will not capture the consideration of our social networking sites audiences. Though, by assimilating and engrossing visual content, this can boost how your audience engrosses and recollects your message.

Now an inescapable part of our everyday lives, visual communication is a key driver of engagement on Social Networking sites. Notably, Facebook and Twitter have redesigned their news feed to allow greater emphasis on visual content, resulting in greater interaction (Blue communication, 2019, Para 1).

Visual Communication in social networking sites is "a powerful medium for delivering the information. In Digital world, Visual communication is a perfect style of conveying your idea details by using the image, video, or minimal texts. Visual communication can be a drawing, an advertisement, a poster or any visual material you can find online on social networking sites platforms. Today, the internet is occupied by visual content and everything is revealed with the help of visual media" (Adqiuckly, 2017). Despite the advantages of visual communication over social networking sites like it connects the people with each other, there is easy and instant communication, we can get real time news and updates, also there is great opportunities for businesses. Simultaneously, it has its bad repercussions also, as the crime rate is increasing day by day over social networking sites. However, the research was carried out with the following questions:

- What are the types of crime happening over Social Networking Sites?
- What is the impact of Visual Communication over Social Networking Sites in the context of Indian judgement Shreya Singhal V. Union of India?
- What are the issues of Privacy over SNS's?

- How the complaint can be filed before Cyber Appellate Tribunal under IT Act, 2008?
- What are the steps taken by government of India to resist cyber-attack?
- What can be the suggestions to come across these issues?

BACKGROUND CRIME OVER SOCIAL NETWORKING SITES

Social networking sites services can only be accessed through a digital platform like laptop, smart phone, desktop and it help to communicate virtually through digital codes. Media purports to spread information, news, story, messages, entertainments, opinions, beliefs etc. between and among the people of the world and the same task is being carried on using the internet in a greater range disguised as social media. Social networking sites has become the indomitable instrument of quick information circulation, a platform of expressing ideas, formation of reference groups, marketing and conducting series of business, educational, cultural and political activities.

Some of the most popular Social Networking Websites all around the globe are like Baidu Tieba, Facebook, Instagram, LinkedIn, Pinterest, Snapchat, Twitter, Viber, WeChat, Weibo, WhatsApp etc (Daxton, 2019). Dependence of technology is so much that we cannot think of educational institutions, banking service providers, commercial sectors cannot in isolation of technology. Also, dependence on technology has led to many challenges like faster obsolescence, complexity in system, and external cyber-crime threats and fraud. In India one cybercrime is recorded in every ten minutes (Kumar Chethan, 2017), and approximate 27,482 cases of cybercrime were reported from January to June. Cyber Crime station have been receiving ample cases and cyber-crime is on the rise with every passing day. In 2017, the station recorded 576 cases compared to 442 in 2016 and 359 in 2015 (Bhattacharya, 2018). Crime over Social networking sites is an intrinsic component of cyber-crime domain today and the same has been criticized because of its trustworthiness, disparity of information, cases of fraud and other grievous offence like profile hacking, photo morphing, Cyber Bullying etc. As "81 percent of Internet-initiated crime involves social networking sites, mainly Facebook and Twitter" (Bhattacharya, 2018). "These platforms are ideal sources for criminals to obtain personal information from unsuspecting people. The vast majority of cyber-crimes consist of identity theft, phishing schemes, fraud, and data mining. One in five adult online users report that they were the target of cyber-crime, while more than a million become victims of cyber-crime every day. Estimates regarding the financial cost of cyber-crime range from $100 billion to an enormous $1 trillion a year in the U.S. This shows how difficult it is to measure the exact financial damage caused by cyber-crime" (Drew Hendricks, 2014).

TYPES OF OFFENCES COMMITTED THROUGH SOCIAL NETWORKING SITES

There are four different kinds of offences committed over social networking sites in India as provided under Indian Criminal Procedure Code, 1973. It is cognizable offence, non-cognizable offence, bailable offence and non-bailable offence. These can be defined as under:

- Cognizable Offences (defined under 2 (c) of CrPC, 1973)-means an offence for which a police officer has the authority to make an arrest without a warrant and to start an investigation with or

Figure 1.

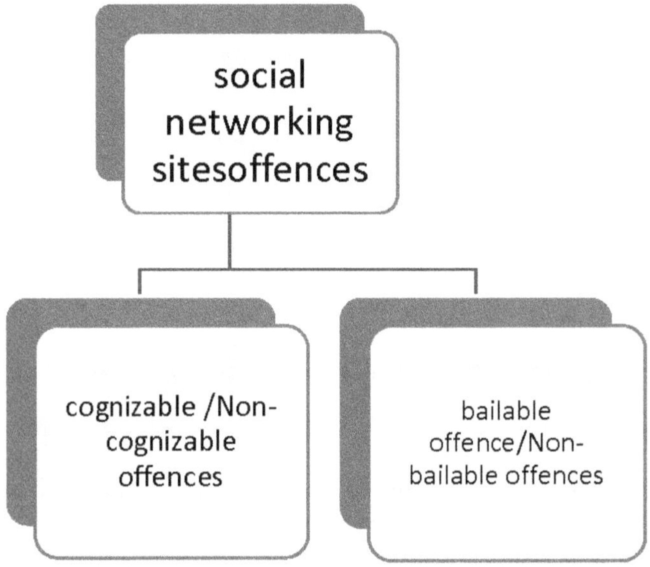

without the permission of a court. Cognizable cases are more serious than the non-cognizable cases as in Non-cognizable cases are of pity nature.

- Non-cognizable Offences- means an offence for which, and "non-cognizable case" means a case in which, a police officer has no authority to arrest without warrant (defined under 2 (1) of CrPC, 1973).
- Bailable offence-means an offence which is shown as bailable in the First Schedule, or which is made bailable by any other law for the time being in force; and "non bailable offence" means any other offence (defined under 2 (a) of CrPC, 1973).
- Non-bailable offence is an offence of serious matter like defrauding a person by copying the personation of some other and making a person believe that he is dealing with such person with whom he wants to deal and in those cases bail cannot be granted.

OUTLINE OF CRIME OVER SOCIAL NETWORKING SITES IN INDIA

There are three different versions of crime over social networking sites like computer related which include cyber stacking, cyber pornography, digital forgery, online gambling, fraud and sale of illegal articles. Secondly, computer network related crime which is hacking and authorized access. Thirdly, data alteration which is destruction or alteration of data, installing malwares, data diddling, salami attack and steganography as highlighted in diagram. These can be discussed as under:

Figure 2.

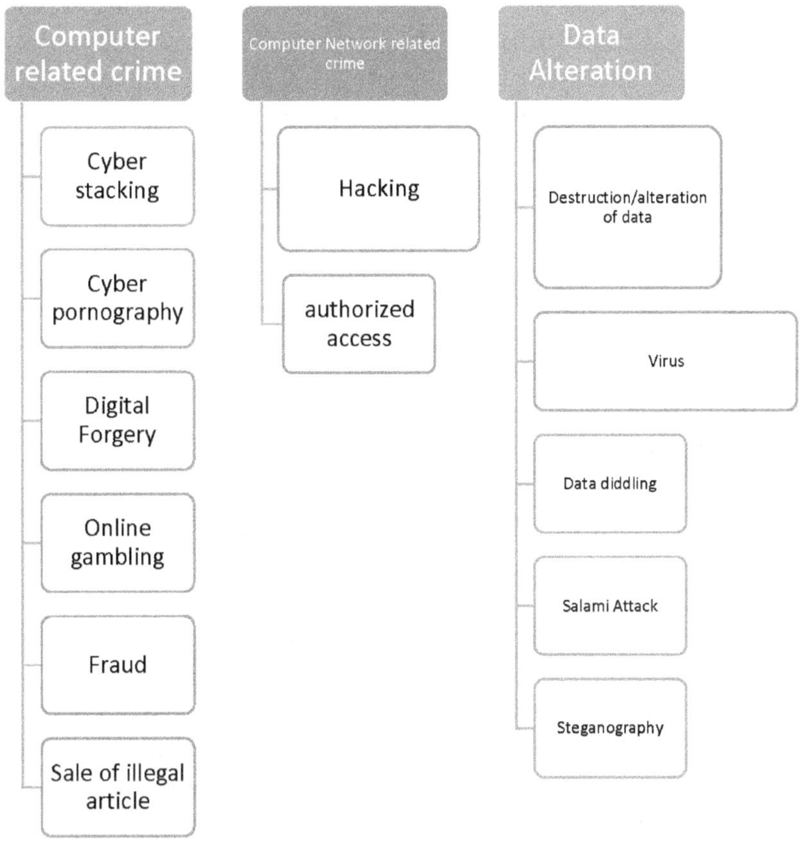

Social Networking Sites Hacking

It can be defined as digital trespassing, without the implied or expressed permission of the owner and misappropriation, alteration, destruction of data. Hacking is included under "computer related offences under section-43 of IT Act. But a mere unauthorized access to the digital profile of the aggrieved party would not amount to the offences unless the complaint has dishonest or fraudulent intention. Hacking of one's profile takes place at the blink of one's eyes that occurs when some other person has total possession and control over another's account by changing all the important credentials for mentally and physically sabotaging the victim. Face book, being the top most hacked amongst all social networking sites.

Social Networking Sites Stalking

Stalk means pursue or approach stealthily. Harass or persecute (someone) with unwanted and obsessive attention. Stalking generally involves harassing or threatening behavior that an individual engages in repeatedly such as following a person, making harassing phone call, leaving written messages or objects etc. When stalking is done through for example over social networking sites platform is called cyber stalking. Section 503 of Indian Penal Code provides criminal intimidation and stalking is criminal intimidation with the help of computers. This type of crime encircles around those acts which are intended for harassing or contacting another with an aim to stalk the individual by putting up an identity that is anonymous by misusing the electronic medium. The Criminal Law (Amendment) Act, 2013 has added a new section known as 354D for penalizing the offence of stalking.

Morphing of Photos

The process of Photo morphing is rather a special effect permitting an individual or morphing or altering one picture or shape into another without an iota of difficulty. It is a child play for a hacker to misuse one's pictures and then morph it.

Romantic Dating Scams

There are individuals who try their best to connect to others through social media, with an intention to communicate and persuade an innocent from moving to an all new different level and form of communication by citing miscellaneous excuses. Further, there are examples where a guy assures to get married to a girl and right before the event of marriage steals and runs away with all her money.

Thus, to conclude these are usual tricks or scams executed and opted by frauds who try to construct and strike a romantic intention for building goodwill and eventually commit fraud and cheating. There are instances where people try taking an offline scam to an advanced online level whereby the explicit videos of women are recorded in compromising postures and later put up on porn sites with a view to blackmail and grab money or to seek revenge.

Cyber Bullying

Posting any sort of unusual and unnecessary humiliating and disgraceful content on the social networking sites or circulating sexual or explicate or vulgar messages through online mode, threats for committing any aggravated acts of violence, cyber stalking through harassing phone calls, messages or threats of child pornography is often termed as cyber bullying. There are multifarious laws in India that permit women, in legally dealing with such menace and harassment and along with the necessary help of the trusted Indian police, one can very efficiently dissolve this specific issue.

Social Networking Sites Defamation

This defamation attributes to the publishing of statements or contents in electronic form that are defamatory in nature. For determining cyber defamation, the court of law takes into view has taken miscellaneous factors such as the time of occurrence, the medium of through which it was published and most importantly the jurisdiction. Due to an absence of any proper border, determining the jurisdiction becomes rather a hard task.

In the case of Joseph Gutnick v. Dow Jones & Company Inc., the High Court of Australia concluded that that the place where the defamatory statement or content has been publication (or the jurisdiction) shall be the place of area where such statements are made and such place shall be the one in which that specific information is downloaded and rather not the place of area where the defamatory statement has been uploaded or the place wherein the publisher's server resides.

Cyber Pornography

Cyber pornography is often called as cyber obscenity which is related to pornography, magazines that provide a medium through online to stimulate sexual attitude. In the case of Ranjeet Udeshi v. State of Maharashtra, the Apex Court, explained the term 'obscene' as something that is outrageously 'offensive to the modesty or decency, and is, impure, verminous, hateful, disgraceful and objectionable in nature. Thus, obscenity which is devoid of a social purpose /gain cannot seek protection under the garb of the free speech that is available to people.

Identity Theft

If any person use or disclose security codes of an user of online medium which is being used for protecting privacy and individuality of each user then he will be liable under section-66C of Indian Information Technology Act, 2008.The offender shall be punished with imprisonment of either description for a team which may extend to three years and shall also be liable to fine which may extend to rupees one lac.

Online Gambling

Gambling is illegal in many countries, thus means that people offer gambling services on the internet from the countries where gambling is permitted and players from countries where gambling is illegal play or bet. The Public Gambling Act, 1867 prohibits gambling.Section-3 of the Act imposes a fine on the person opening a common gaming- house for others

False Statement

Publication or circulation of any statement with an intention to promote feelings of enmity, hatred or ill-will between different religious, racial, language or regional groups or castes or communities on grounds of religion, race, place of birth, residence, language, caste or community or with intent to cause, or which is likely to cause, any officer, soldier, sailor or airman in the Army, Navy or Air Force of India to mutiny or otherwise disregard or fail in his duty as such; "or with intent to cause, or which is likely to cause, fear or alarm to the public, or to any section of the public whereby any person may be induced to commit an offence against the State or against the public tranquility; or with intent to incite, or which is likely to incite, any class or community of persons to commit any offence against any other class or community, shall be punished with imprisonment which may extend to three years, or with fine, or with both." Any person committing the mentioned offence will be punished with imprisonment of 3-5 years and fine (Section 505 IPC).

Cyber-Crime Aggravated by Terrorism through Social Media

ISIS has continued to make screaming headlines around the globe which is due to the insurgent terrorist activities and group which are facing doom by losing its territory. However, ISIS had nefariously misused the social networking sites during the phase of their breakneck expansion all throughout Middle East, while broadcasting and transmitting purely graceless low browed, cruel and inhuman acts of atrocities directly into one's homes, through the misuse of social networking sites streams.

The terrorist groups opt for crimes through social networking sites as the tool of social networking sites and economical and handy facilitating extremely quick and thick and capacious transmission and circulation of the messages, granting communication that is unfettered, free and loose along with a presence of an audience devoid of any filter or rather of any "selectivity" of the predominant and typical news outlets. This further permits the terrorist and other terror groups in engaging with their network efficiently.

Al-Qaeda has been recognized as one of those terror groups which misuse the platform of social media, most extensively and widely. Brian Jenkins, who is a senior advisor (Rand Corporation), rightly opined upon the massive dominance of Al-Qaeda upon the web. Al Qaeda being the first to wholly exploit and misuse the internet.

VISUAL COMMUNICATION OVER SOCIAL NETWORKING SITES: A CASE STUDY OF SHREYA SINGHAL V. UNION OF INDIA

Section 66A infringes the fundamental right to free speech and expression and is not saved by any of the eight subjects covered in Article 19(2). According to Supreme Court, the causing of annoyance, inconvenience, danger, obstruction, insult, injury, criminal intimidation, enmity, hatred or ill- will are all outside the purview of Article 19(2). Sending sexually explicit photographs were covered under this section. Any kind of offensive or menacing message was covered under this section. This section also covered a message which is known to be false, cause annoyance, inconvenience, danger, obstruction, insult, injury, criminal intimidation, enmity, hatred, or ill will.

Social networking sites giants Facebook, Twitter and google in collaboration with the Election Commission of India heading towards conducting a free and fair election in 2019 by maintaining a "Silence period" of 48 hours where no political advertisements can be uploaded, monitoring all posts to prevent circulation of fake news and displaying the notification of sponsored posts (Jain Bharti, 2018).

With the advancement, of social sites being easily available in all styles of communication instruments from smartphones to tablets, it is quite apparent for individuals to share their opinions on every aspect. Whether it's about sports, politics or numerous issues of society, individuals catch this stand as a worldwide gateway to send their opinions to the people. Section 66A is a sub clause of the amended Information Technology Act, 2000 which provided punishment for sending offensive messages through communication services etc.

However, this section got much coverage, after the arrest made by the Mumbai police in November 2012 of two women who had expressed their displeasure at a bandh called in the wake of Shiv Sena chief Bal Thackeray's death, Devu Chodankar in Goa and Syed Waqar in Karnataka were arrested in 2014 for making posts about PM Narendra Modi, Cartoonist Aseem Trivedi and a Puducherry man was arrested for criticizing P. Chidambaram's son over social media. Since then, several arrests have been made by different State police, of various individuals, for the most benign dissemination of online content. Also, in Jammu & Kashmir police arrested two youth for posting abusive comments on social networking sites that forced the Valley's all-girl band to quit and ultimately they got arrested (Times of India, 2014). The arrest caused a huge outrage from every nook and corner over the law (Vishwanathan, 2013). A set of petitions alleged that this section curbs the Fundamental Right to Speech and Expression and so must be declared unconstitutional.

In the wake of this misuse of Section 66-A, in march 24, 2015 the Supreme Court in Shreya Singhal v. Union of India, (AIR 2015 SC) came up with a landmark judgement which redefined the boundaries of freedom of speech on the internet especially over social media. In a writ petition filed in 2012, the law student Shreya Singhal challenged the constitutionality of Section 66A on grounds, inter alia, of vagueness and its chilling effect. More petitions were filed challenging other provisions of the IT Act including Section 69A (website blocking) and Section 79 (intermediary liability), and these were heard jointly by justices Rohinton F. Nariman and G. Chelameshwar. Section 66A, implicating grave issues of freedom of speech on the internet, was at the Centre of the challenge. With the Court's decision in Shreya Singhal & Ors. v. Union of India, were the "The Supreme Court's decision comes at a critical moment for freedom of speech in India. In recent years, the freedom guaranteed under Article 19(1) (a) of the Constitution has suffered unmitigated misery. The tale of free speech on the Internet is similar."

The Supreme Court by quashing Section 66A of the IT Act as unconstitutional, has stepped to the fore with a delightful affirmation of the value of free speech and expression, the Supreme Court judgment shows us that with the right kind of conviction, it is possible to uncover the importance of free speech as a value unto itself within our larger constitutional scheme. It must allow us to believe that we can now challenge the noxious culture of censorship that pervades the Indian state (Parthasarthy Suhrith, 2015).

Section 66A of Indian IT Act: No longer Draconian Law

Section 66A makes it a criminal offence to send any online communication that is "grossly offensive" or "menacing", or false information sent for the purposes of causing "annoyance, inconvenience, insult, injury, obstruction, enmity, hatred, ill will", etc. over Social media. These terms are not defined. What may be offensive for one may not be for another. Neither do they fall within one of the eight subjects for limitation under Article 19(2). It is difficult and impossible, in fact to foresee or predict what speech is permitted or criminalized under Section 66A. As a result, there is a chilling effect on free speech online, resulting in self-censorship.

With this decision, the Supreme Court has struck down Section 66A on grounds of vagueness, excessive range and chilling effects on speech online. What is perhaps most uplifting is the court's affirmation of the value of free speech. In the midst of rising conservatism towards free speech, the Court reminds us that an "informed citizenry" and a "culture of open dialogue" are crucial to our democracy. Article 19(1) (a) shields us from "occasional tyrannies of governing majorities", and its restriction should be within Constitutional bounds enumerated in Article 19(2).

Conflict between Free Speech and Hate Speech under Shreya Singhal Judgment

Previously there was no test which could convert free speech into hate speech that is when Freedom of free speech and expression would get converted into restrictions mentioned in Article 19(2) of Indian Constitution. Freedom of speech is the freedom given to a person to speak the way he likes but of course subject to certain restrictions provided under Article 19 (2) like in case of Sovereignty, integrity of country, public order, morality, incitement of offence etc. Scent contribution of the Shreya Singhal judgement is that it laid down the test were free speech will get converted into hate speech. The Supreme Court of India classified the Speech into discussion, advocacy and incitement. The court further said that Discussion and advocacy are at the heart of Article 19(1) (a), and are unquestionably protected. But when discussion and advocacy reaches the level of incitement than Article 19(2) kicks in, that is, if Free Speech is expected to cause harm, danger or public disorder it can be reasonably restricted for any of these reasons: public order, sovereignty and integrity of India, security of the State and friendly relations with foreign states.

But at the same time Union of India in this case argued that Section 66A is saved by the clauses "public order", "defamation", "incitement to an offence" and "decency, morality". But as the court finds that these are spurious grounds."

Section 66A, however, does not meet the legally prescribed standards set out in the limitation-clauses under Article 19(2), and accordingly was declared unconstitutional. The Union of India in the said case argued that Section 66A is saved by the clauses "public order", "defamation", "incitement to an offence" and "decency, morality". But as the court finds that these are spurious grounds. For instance, Section 66A covers "all information" sent via the Internet, but does not make any reference (express or implied) to public order. Section 66A is not saved by incitement, either. The ingredients of "incitement" are that there must be a "clear tendency to disrupt public order", or an express or implied call to violence or disorder, and Section 66A is remarkably silent on these. By its vague and wide scope, Section 66A may apply to one-on-one online communication or to public posts, and so its applicability is uncertain. For these grounds, Section 66A has been struck down.

The unpredictability and threat of Section 66A has been lifted. Political commentary, criticism and dialogue are clearly protected under Article 19(1) (a). Of course, the government is still keen to regulate online speech, but the bounds within which it may do so have been reasserted and fortified.

Section 66A of the Information Technology Act, 2000 is struck down in its entirety being violative of Article 19(1) (a) and not saved under Article 19(2) by the court.

PRIVACY AND SOCIAL MEDIA: IMPACT OF VISUAL COMMUNICATION

Universal Declaration of Human Rights (UDHR), 1947 was made for the protection and preservation of human rights throughout the nations and it specially protects territorial and communication privacy at international level. Article-12 of UDHR states, "No one should be subjected to arbitrary interference with his privacy, family, home or correspondence, nor to attack on his honor or reputation. Everyone has the right to protection of law against such interference or attacks.

Also Section-66E of Indian Information Technology Act provides punishment of imprisonment which can be extended to 3 years a fine of two lacs rupees or both for violation of privacy in social media.

Violation of Privacy by Visual Communication over Social Media

Privacy means:

- Intentionally captures, publishes or transmits the image of a private area of the person.
- Private areas means-Naked or undergarment clad genitals, pubic areas, buttocks or female breasts.
- Transmit means-Electrically sent a virtual image with an intention to be viewed by public.
- The Supreme Court of India, in a landmark verdict of Justice K. S. Puttaswamy and Anr. v. Union of India and Ors. (2017 10 SCC 1), ruled that the Indian citizens have a fundamental right to privacy. The key highlights of the judgment, which originates from the guarantee of life and personal liberty in Article 21 of the Constitution, are as follows:
    i. Life and personal liberty are inalienable, which are inseparable from a dignified human existence.
    ii. Privacy includes at its core the personal intimacies, the sanctity of family life, marriage, and procreation, the home and sexual orientation.
    iii. Privacy safeguards individual autonomy and recognizes the ability of the individual to control vital aspects of his or her life.
    iv. Those who are governed are entitled to question those who govern, about the discharge of their constitutional duties including in the provision of socio-economic welfare benefits.

Further, obscenity is also an offensive act which is against the moral or generally accepted principle of society. Section 292 of Indian Penal Code says writing, figure, pamphlet, drawings, painting, representation shall deem to be obscene if it

1. is lascivious or
2. appeals to the prurient interest
3. Tends to deprave and corrupt persons who are likely having regard to all relevant circumstances.

The main test of obscenity over social networking sites is whether the tendency of the matter charged as obscene is to deprave and corrupt those whose minds are open to such immoral influences and into whose hands a publication of this sort may fall. Any person committing such an act will be punished with imprisonment of two years and fine of two thousand rupees and on subsequent commission of such an act the imprisonment might extend to 5 years and fine of five thousand rupees.

Bazee Case

A student of Kharagpur IIT was a registered seller in Bazee.com, there he put a lascivious video clip for sale under the filter of e-books and magazines to prevent detection by Bazee.com. Seller was arrested and the issue was raised whether the website will also be liable for selling such illegal pornography or not. The court held that ultimate transmission of the video clip might be through the seller to buyer but the entire process is controlled by an automated system generated by the website so it cannot be stated that the website has no involvement in the process of selling obscenity and liability was charged on Section 67 of IT Act.

Deliberate Outrage of Religious Feelings

If any person deliberately outrage the religious feelings of other by means of any post or writings in social networking sites will be punished under section-295 of IPC.Section-298 purports to prevent any word which deems to hurt religious feelings of any person by convicting with imprisonment or fine or both according to Indian Penal Code.

FORGERY OVER SOCIAL NETWORKING SITES

Making of false document or electronic record or part of a document or electric electronic record is called forgery as defined by section 463 of IPC for example over Social media, such as Facebook.

Fraud

YouTube, Twitter, and LinkedIn, have become key tools for U.S. investors. Whether they are seeking research on particular stocks, background information on a broker-dealer or investment adviser, guidance on an overall investment strategy, up to date news or to simply want to discuss the markets with others, investors turn to social media. Social networking sites also offers a number of features that criminals may find attractive. Fraudsters can use social networking sites in their efforts to appear legitimate, to hide behind anonymity, and to reach many people at low cost. On the one hand legitimate newsletter contain valuable information's while others are the tool for fraud. Fraudsters may use online discussions to pump up a company or pretend to reveal "inside" information about upcoming announcements, new products, or lucrative contracts (US Securities and Exchange Commission, 2019, para. 1).

- **Affinity fraud-**The fraudsters targets different identifiable groups and pretends to be member of such group and provoke the leader of such group to spread false information. Affinity frauds target members of identifiable groups, such as the elderly, or religious or ethnic communities. Many times, those leaders become unwitting victims of the fraud they helped to promote.
- **Advance-fee fraud-** Advance fee frauds ask investors to pay a fee up front – in advance of receiving any proceeds, money, stock, or warrants – in order for the deal to go through. The advance payment may be described as a fee, tax, commission, or incidental expense that will be repaid later. Some advance fee schemes target investors who already purchased underperforming securities and offer to sell those securities if an "advance fee" is paid, or target investors who have already lost money in investment schemes. Fraudsters often direct investors to wire advance fees to escrow agents or lawyers to give investors comfort and to lend an air of legitimacy to their schemes. Fraudsters also may try to fool investors with official-sounding websites and e-mail addresses

- **Binary operation fraud**- A binary option is a type of platform where customer have to register their account or set their profile by depositing a specified amount and the payout depends on the yes/no propositions after expiry of the binary option. US Security and Exchange commission receives numerous complaint of Binary frauds conducted through different social networking sites platform where the customers are not credited with the amount or often their personal information like credit card number, passport, driving license are taken without specified reason. The customers cannot even withdraw the original deposits from this platform.
- **High Yield Investment Program Fraud-**HYIPs are the online platforms which run investment program with a promise of 30%-40% return annually by unregistered individuals and often they are found to be fraud.
- **Spam**-These are false misrepresented information about the investment program.
    - **Virus, worms, Trojan Horses, Logic bombs**-A virus is a program that searches out another programs and infects them by embedding a copy of itself in them. A worm is a program that propagates itself over a network, reproducing itself as it goes. Trojan horse is a malicious, security breaking program that is disguised as something benign such as a directory listed, archiver, game or a program to find and destroy viruses. Mocking bird case is a special case of Trojan Horses where a software intercepts communications between users and hosts and provide system like responses to the users while saving their responses (account id and password). A logical Bomb is a code surreptitiously inserted into an application or operation system that causes it to perform some destructive or security compromising activity.
    - **Salami Attack**-This attack is used for commission of financial attacks. The alternation is made is completely unnoticed in a single case. Salami attacks would be covered by section-477 of IPC relating to falsification of account and section 66 of IT Act.
    - **Data dibbling**-This a kind of attack where the raw data is altered before they are processed by computer. Data can be altered by anybody who is involved in the process of creating, recording, encoding, examining, and checking.
    - **Steganography**- Steganography is the process of hiding one message or file inside another message or file.

CYBER APPELLANT TRIBUNAL UNDER INDIAN IT ACT, 2000

- **Establishment of Cyber Appellate Tribunal-**The central government by notification can set Cyber Tribunal to deal with crime committed through computer networks, computer system or computer resources.
- **Section-**2(1)(j) of IT Act, 2000 defines computer network as inter-connection of one or more computers, computer systems or communication device through the use of satellite, microwave, territorial line, wire or wireless communication media. The definition also includes terminals or complex of two or more computer network.

PROCEDURES TO FILE AN APPLICATION IN CYBER APPELLATE TRIBUNAL

1. The application can be filed by-the applicant in person, by his agent, or by authorized lawyer.
2. The application should be presented in Form No. 14.
3. The application must be presented to the registrar or sent to the registrar through post.
4. Application must be presented in six sets along with one empty envelope mentioning the complete address of the respondent
5. The applicant may attach to and present with his application a receipt slips as in Form No. 1 which shall be signed by the Registrar or the officer receiving the applications on behalf of-the Registrar in acknowledgement of the receipt or the application.
6. Two or more persons having similar prayer can file an application together
7. If the application is found to be in order then it will be registered and given a serial number but if the application is not in proper order then it should be rectified by the applicant within specified time otherwise it will be declined by the registrar.
8. If the application has some formal fault then the register can allow the applicants to rectify the same.
9. If the applicant fails to rectify the application within the specified time period the register can decline the application.
10. Application can only be filed to the registrar.
11. Every application should be accompanied by a paper book containing-a certified copy of the order against which the application is filed and all documents relied upon and referred in the application.
12. A copy of the paper book containing the application will be sent to each respondent by the registrar, the registrar can take into account the number of respondents, their place of residence or work for proper delivery of the paper-book and the applicant need to pay charges when the number of respondent exceeds five.
13. The respondent shall file six complete sets containing the reply to the application along with the documents in a paper-book form with the Registrar within one month of the date of service of the notice of the application on him.
14. The respondent shall also serve a copy of the reply along with copies of documents as mentioned in sub rule (1) to the applicant or his advocate.
15. The Tribunal shall notify to the parties the date and the place of hearing of the applications.
16. Every application shall be heard and decided, as far as possible, within six months of the date of its presentation.
17. If applicant does not appear when the application is called on for hearing, the Tribunal may, in its discretion, either dismiss the application for default or hear and decide it on merit. Where an application has been dismissed for default and the applicant appears afterwards and satisfies the Tribunal that there was sufficient cause for his non-appearance when the application was called on for hearing, the Tribunal shall make an order setting aside the order dismissing the application and restore the same. If the respondent fails to appear the tribunal in its discretion, adjourn or hear and decide the application ex-parte
18. Every order passed on an application shall be communicated to the applicant and to the respondent either in person or by registered post free of cost.

CYBER ATTACKS IN INDIA OF LATE

Momo Challenge-An online game which first started through Facebook and eventually gained popularity among the teenagers in what's App. The players are required to perform a series of task which end up with the final task of killing oneself.

Major Security Breach of Facebook- Facebook reported a major security breach in which 50 million user accounts were accessed by unknown attackers. The stolen data would allow the attackers to "seize control" of those user accounts, Facebook said. Facebook has logged out the 50 million breached users plus another 40 million who were vulnerable to the attack ("The Hindu,"n.a.).

Steps taken by Union and State Government

More than 30 people died this year in a mob violence triggered by vitriolic messages circulated through social media. This prompted the Union government to call on WhatsApp to take immediate action to end this menace. WhatsApp has already taken some steps to prevent spreading of fake news by labeling forwarded messages and launching awareness campaigns. Recently Facebook's WhatsApp is working closely with Reliance Jio to spread awareness of false News and messages ("The Hindu," n.a).

Section-67A of IT Act, 2000 permits the intermediaries to preserve and retain information in digital platform as prescribed by central government. Through the instrumentality of this provision the intermediaries can trace unlawful transaction.

Section-69 of IT Act, 2000: Center, state government or other authorized body can intercept, monitor or decrypt in the interest of sovereignty, integrity of India, defense of India, security of India etc.

Section-69A of IT Act, 2000: Central Government, Designated Office can block access by the public any information generated by a computer system.

Section-69B of the IT Act, 2000: enables Central Government and other competent Authority to provide technical assistance and extend all facilities to such agency to enable online access or to secure and provide online access to the computer resource, generating, transmitting, receiving or storing such traffic data or information.

Section-75 of IT Act, 2000 shall be applicable to any offence or contravention committed outside India by any person irrespective of his nationality provided the offence involves a computer or computer network system located in India.

DIGITAL INDIA COMBAT CYBER CRIME

Department of electronics and information Technology is responsible for the formulation and implementation of national policy in the field of communication, electronics and internet. The mission is to ensure sustainable and inclusive growth of IT and IT industries, promote e-governance and enhance Indian role in global platform of e-governance, ensuring secure cyber –space and enhancing efficient digital services.

Safe and Secure Cyber Space is one of the key elements of "digital India" initiatives. Besides, improving the infrastructure initiatives have been taken to promote awareness about cyber-attacks and to take up proactive responses and mitigations. Some of the initiatives taken to ensure cyber space security are as- 24x7monitaring, malware analysis, NCCC mechanism, application security audit management system, Firewall Access Rule Processing System, Security training on OWASP standard.

- **National Cyber Coordination Centre**-It is a mechanism for situational awareness and early warning. NCCC will scan the internet traffic and incorporate other methodology to asses the threats. NCCC is a multi-stakeholder cyber-security and e-surveillance agency
- It comes under the Indian Computer Emergency Response Team (CERT-In), Union Ministry of Electronics and Information Technology
- It has powers under the Indian constitution with provision of section 69B of the Information Technology Act, 2000
- It will be India's first layer for cyber threat monitoring and all communication with the government as well as the private service providers will be monitored round the clock
- Its mandate is to scan internet traffic and communication metadata (which are little snippets of information hidden inside each communication) coming into the country to detect real-time cyber threats ("India Today," n.a.)

Cyber Swachhta Kendra- (Botnet Cleaning and Malware Analysis Centre) is set up in accordance with the objectives of the "National Cyber Security Policy", which envisages creating a secure cyber eco system in the country. This center operates in close coordination and collaboration with Internet Service Providers and Product/Antivirus companies. This website provides information and tools to users to secure their systems/devices. This center is being operated by the Indian Computer Emergency Response Team (CERT-In) under provisions of Section 70B of the Information Technology Act, 2000 (Cyber Swachhta Kendra, 2017, para.1).

Application Security: The cyber security division of National informatics Centre incorporates various steps to provide application security such as formulation of policies guidelines /advisories. Audit of security services by multi-agency, web application firewall solutions penetration testing for server scanning, malicious codes testing for server scanning.

ONLINE DISPUTE RESOLUTION

Online Dispute Resolution (herein after called ODR) implies a change in dispute resolution medium, both online and offline disputes can be solves through ODR. ODR is not necessarily a platform to solve disputes arise out of online transaction but it's a platform to solve any kind of dispute. For successful working of an ODR an institutional backup is required. The level of technology used in the ODR process should be affirmed. Series of steps starting from Initiation of ODR process, contacting the other party, arbitrator appointment to the final enforcement of agreement through the courts.

CONCLUSION AND SUGGESTIONS

The major drawback of Indian legal system is lack of awareness among the people, citizens are not well aware of the existing provisions that are enacted for the safety and security of people. So, the initiation must be taken to envisage awareness among people about their rights and duties and means to implement them. Legal aid campaign should be more prominently functioning in the area of cyber security. Teenagers who are very much prone to social networking sites must be well aware of the negative sides of it and ways to resolve any kind of disputes without being worried.

Technology and social networking sites falling under it has made our life more comfortable and easier, so we must squeeze the best positive part out of it keeping aside the drawback.

For improving the cyber security, multitudinous precautionary steps are to be taken and kept in mind by the netizens. Some of them are mentioned below:

- One must not send any of his or her pictures to unknown people or strangers for preventing the misusing of the picture.
- The anti -virus software must be updated for guarding against the virus attacks.
- Files must be backed up for enabling the prevention of loss of data due to the virus attack.
- Payments that are made in connection to have access to games and other applications in social networking site sought to be done securely.
- Awareness programmes and classes must be conducted through which education and awareness should be imparted to kids about the growing crimes on social media.
- Security programmes which provide control over the cookies should be preferred.
- Traffic must be monitored and regulated by the owners of websites and such intermediaries for avoiding any deviation or irregularity on the websites.

REFERENCES

Adquickly Blog. (2017, December 21). Importance of Visual communication in Digital Media [Web Blog Post]. Retrieved from https://www.adquicky.com/blogs/importance-visual-communication-digital-media/s

Blue Communications. (2019, January 20). *The role of visual communications in today's digital age.* Retrieved from https://www.blue-comms.com/the-role-of-visual-communications-in-todays-digital-age/

Byod, D. (2014). *The Social Live of Networked Teens*. New Haven, CT: Yale University Press.

Facebook says 50 million user accounts affected by major security breach. (2017, September 28). Retrieved from: https://www.thehindu.com/sci-tech/technology/internet/facebook-reveals-security-breach-affecting-50-million-user-accounts/article25074368.ece

Gutnick v. Dow Jones & Company Inc. [2001] VSC 305

Hendricks, D. (2014, March 4). *Socialnomics.com*. Retrieved from: https://socialnomics.net/2014/03/04/the-shocking-truth-about-social-networking-crime/

Jain, B. (2018 September 6). Google, Facebook, Twitter to self-censor political content. *The Times of India.* Retrieved from: http://timesofindia.indiatimes.com/articleshow/65693822.cms?utm_source=contentofinterest&utm_medium=text&utm_campaign=cppst

Kendra, C. S. (2017, January 20). *Ministry of Electronics and Information Technology, Government of India.* Retrieved from https://www.cyberswachhtakendra.gov.in/

Kumar, C. (2017, June 22). One cybercrime in India every 10 minutes. *The Times of India.* Retrieved from https://timesofindia.indiatimes.com/india/one-cybercrime-in-india-every-10-minutes/articleshow/59707605.cms

Pandit, S. (2013, February 7). Two held in J&K girl Rockband Case. *The Times of India.* Retrieved from http://timesofindia.indiatimes.com/india/Two-held-in-JK-girl-rock-band-case/articleshow/18375067

Parthasarathy, S. (2015, March 26). *The judgment that silenced Section 66-A*. Retrieved from http://www.thehindu.com/opinion/lead/the-judgment-that-silenced-section66a/article7032656.ece

Ranjit D. Udeshi v. State Of Maharashtra (1965) AIR 881, Section 24 of Indian Penal Code- Whoever does anything with the intention of causing wrongful gain to one person or wrongful loss to another person, is said to do that thing dishonestly.

Section 26 of Indian Penal Code- A person is said to do a thing fraudulently if he does that thing with intent to defraud but not otherwise.

Section 294 of Indian Penal Code- punishes obscene act or song with imprisonment of either description for a term which may extend to three months or with fine or both. Section 67 of IT Act provides punishment for such act.

Section 503 of Indian Penal Code defines Criminal Intimidation. Whosoever threatens another with any injury to his person, reputation or property, or to the person or reputation of any one in whom that person is interested, with intent to cause alarm to that person, or to cause that person to do any act which he is not legally bound to do, or to omit to do any act which that person is legally entitled to do, as the means of avoiding the execution of such threat, commits Criminal intimidation.

Section 66A of Information Technology Act, 2000 deals with, Punishment for sending offensive messages through communication service, etc. Any person who sends, by means of a computer resource or a communication device,—any information that is grossly offensive or has menacing character; or any information which he knows to be false, but for the purpose of causing annoyance, inconvenience, danger, obstruction, insult, injury, Criminal Intimidation, enmity, hatred or ill will, persistently by making use of such computer resource or a communication device, any electronic mail or electronic mail message for the purpose of causing annoyance or inconvenience or to deceive or to mislead the addressee or recipient about the origin of such messages, shall be punishable with imprisonment for a term which may extend to three years and with fine. Explanation.— For the purpose of this section, terms "electronic mail" and "electronic mail message" means a message or information created or transmitted or received on a computer, computer system, computer resource or communication device including attachments in text, images, audio, video and any other electronic record, which may be transmitted with the message.

Stewart, D. R. (2019). *Social Media and the Law*. London: Routledge publishing.

Sumit, B. (2018, March 31). Surge in cyber-crime rate. *The Hindu*. Retrieved from. https://www.thehindu.com/news/cities/Visakhapatnam/surge-in-cyber-crimerate/article23395834.ece

Today, I. (2017, August 11). Retrieved from: https://www.indiatoday.in/education-today/gk-current-affairs/story/nccc-cyber-india-1029203-2017-08-11

US Securities and Exchange Commission. (2019, January 20). *Microcap Fraud*. Retrieved from https://www.investor.gov/investing-basics/avoiding-fraud/types-fraud/microcap-fraud

Viswanathan, A. (2013, February 20). Reasonable Restrictions. *The Hindu*. Retrieved from: http://www.thehindu.com/opinion/lead/an-unreasonable-restriction/article4432360.ece

WhatsApp working with Reliance Jio to curb fake news menace. (2016, September 23). Retrieved from: https://www.thehindu.com/business/Industry/whatsapp-working-with-reliance-jio-to-curb-fake-news-menace/article25045384.ece

ADDITIONAL READING

Jeffrey, C. J., & Everett, M. G. (2018). *Analyzing Social Networks*. New York: SAGE Publications.

Marcum, C. D., & Higgins, G. E. (2014). *Social networking as a Criminal Enterprise. Carolina*. USA: Taylor and Francis Group.

Tomayess, I. P., & Isais, P. K. (2015). *Social Networking and Education: Global Perspective*. New York: Springer.

KEY TERMS AND DEFINITIONS

Chilling Effect: Inhibition or discouragement of the legitimate exercise of natural and legal rights by the threat of legal sanction.

Criminal Intimidation: Intentional behavior that would cause a person of ordinary sensibilities to fear injury or harm.

E-Governance: Application of Information and Communication Technology for delivering government services, exchange of information, communication transactions, integration of various stand-alone systems and services between government to citizens (G2C), government to business (G2B).

Firewall Access Rule: the rules that allow or deny traffic to transit an interface. Access rules are processed before other types of firewall rules.

Incitement to an Offence: Encouragement of another person to commit a crime. Depending on the jurisdiction, some or all types of incitement may be illegal. Where illegal, it is known as an inchoate offense, where harm is intended but may or may have actually occurred.

OWASP Standard: Open web application security project is an organization that provides unbiased and practical, cost-effective information.

Social Media: The Websites and applications that enable users to create and share content or to participate in social networking.

Terminals or Complex or Inter-Connected Computer: Is broad enough to incorporate social networking sites in all its forms. It defines both physical and virtual computer systems.

Visual Content: Images, videos, charts, infographics, animations, iconography, and gifs – providing companies with the opportunity to present bite-size information in a compelling manner.

Chapter 7
Mobile-Based Social Media, What Is Cutting?
Mobile-Based Social Media: Extensive Study Findings

Christopher Kipchumba Chepken
University of Nairobi, Kenya

ABSTRACT

In this chapter, results obtained from a longitudinal study on Social Media (SM) use are reported. Previous studies have mostly carried out contextualized research and not a lot of it has been done in Kenya and especially with the emerging mobile application SM platforms. The key objective of the study was to understand the general aspects of emerging SM platforms with a view of mapping out study areas going forward. The study used mixed method approach for an extended period. To effectively carry out the study, seven themes were identified through a preliminary study and literature review. A summary of the results show that mobile app SM platforms are gaining popularity among users. SM uses are majorly socialization, but other uses such as political campaigns, fundraising, and religious uses are taking root. SM groups are dominating; even though SM is reach in functionality, users expect more. There exist various challenges associated with social media use and SM study methodological challenges. Finally, the study established seven key themes which can frame SM studies.

DOI: 10.4018/978-1-7998-1041-4.ch007

INTRODUCTION

Social media (SM) is undoubtedly growing at a very fast rate, probably like its facilitating technologies, such as the Internet and the mobile phones (Morrison, 2014). Research and use of SM as research tools is also growing too fast. The growth has reached an extent the trends, methods, tools and techniques of doing research on or with SM is becoming challenging as their growth escalates (Kaplan & Haenlein, 2010). The difference among the existing and everyday upcoming social media sites, tools and purposes exasperates the complex situation, which among other things is made complex by the various uses of social media, the diversity of users across the globe, the various types of social media, the growing numbers of social media platforms among many other unknowns, which emerge as SM evolves.

The definitions associated with Social Media (SM) can be as diverse as the number of written definitions (Cohen, 2011; Fuchs, 2017; Nations 2018). However, in all these definitions, one theme which comes out is that, social media can be defined from the two words that make it, social and media (Nations, 2018) with Social referring to people interacting with each other by receiving and/or sending information while Media on the other hand referring to an instrument of communication (Fuchs, 2017; Nations, 2018). SM can therefore be defined as any communication tools/instruments which allow people to interact through sharing and use of information. These communication tools are mainly internet-based services, some of which work through the mobile network infrastructure. A major characteristic which differentiates Social media from other communication tools is its two-way communication channel, different from the one way usually provided by the mass media channels (Pan, & Crotts, 2012), such as TV, radio, websites and asynchronous communications (such as emails). This study does not go into deeply defining social media as is defined in sociological and other related theories (Fuches, 2017). A simple definition adopted for this study is that SM, in the context of technology or innovation use could be seen as tools and applications classified as such.

Social Media (SM), as a communication tool allowing users to interact by sending and consuming information can be classified into two major categories, namely Social Network sites (SNS) and Social Media Applications (SMA). Social network sites are web-based services that allow social networking between or among its users (Danah & Nicole, 2007; Miller, 2012). SM applications are mainly computer or mobile device applications (Kaplan, 2012). Under the SM applications, Short message service (SMS) can easily be classified as one of them. Other examples of the current known SM platforms include WhatsApp[1], Telegram[2], Instagram[3], Twitter[4], Facebook[5], Bulksms, Snapchat[6] among many more coming up. These applications can be defined further using social application building blocks as specified by Kietzmann, Hermkens, McCarthy and Silvestre, (2011) and Smith (2007).

These SM flavors have been used in various ways and in different sectors of the social economic fronts. Such uses include, but not limited to trade, family connections, education, news and politics, just to name a few. The level of engagement among the parties using SM varies from friendship (or even enemies or strangers), family ties, personal or organizational contacts, fans, and followers (Danah & Ellison 2007; Miller, 2012) among many other uses which keep coming up every day and are as diverse as the diversity of their users. The use and popularity of these SM platforms is fueled by the low costs of operations and the reach ability (Danah & Ellison, 2007; Miller, 2012). i.e. they are cheaper and reachable compared to other media platforms.

THE PROBLEM STATEMENTS

Several studies have been carried out to try and understand the uses and the general SM arena (Miller et al., 2016). A look at these studies reveals that most of them are contextual in nature in many ways. First, many social media (SM) studies are contextual by way of studying SM use in a particular area of interest such as education (Azam & Dafoulas, 2016); health (Daria & Griffiths, 2011); politics (Young, Giurcanu & Fernandes, 2017); marketing (Pan &Crotts, 2012); news stories and media (Al-rawi, 2017; Jisun, Quercia, Cha, Gummadi & Crowcroft 2013) among others. The second way is by studying a specific type of SM use such as Facebook (Young et al., 2017) and Instagram (Xitong & Luo, 2017) among a specific group of users. The third contextualization method is by studying a specific institution, region or country (Samadi & Gharleghi, 2014). However, it is understood that SM has a mix of many aspects of life and contained in a web of activities, participants, multimedia, and cultural dimensions among many other complexities, which may not be easily seen in a contextualized study. Further, most of these contextualized studies have been one time and are not systematically having follow-up studies to build on previous findings. There is therefore a need to have generalized SM studies encompassing various general aspects during this SM explosion.

In view of the nature and the characteristics of the SM arena, i.e. having many different and diverse platforms, various uses and group dynamics, very fast growth rates, fast two-way communication platform and many other unknown uses and forms, the contextualized studies can only do the much of producing knowledge for the specific context. To be able to understand some of the extraordinaire within the SM, this study was conceptualized in the philosophy of mixed method research (Rohm, Kaltcheva, & Milne, 2013) following an exploratory approach such as those conducted by Wyche, Schoenebeck, & Forte (2013) and David, Crittenden, Keo, & McCarty (2012) with an open-ended view of studying SM with the objective of filling the gaps left by the many contextualized studies. These gaps include the need to identify some of the unknowns in the SM environment and to be able to move towards demarcating study areas which can be followed up by future contextualized studies. This will allow for a development of a long-term plan of SM studies in various disciplines, usage areas, regions or countries, and any other findings which can then generate knowledge on how to effectively leverage SM. The general purpose of this study was therefore to understand the general aspects of SM use and usage variations. More specifically, we aimed at understanding the various uses, use variations of SM within the SM platforms widely used Kenya.

This chapter reports the initial findings of a mixed method research approach and longitudinal study carried out in over one year. In the remaining sections of the chapter, we present the methodology applied, followed by the results of the study and discuss the outcome while presenting suggestions of effectively and efficiently using SM. The key reason for the generalized study is because even though people tend to form social media groups or networks for specifics, e.g. funeral arrangements, such platforms end up being used for many other unrelated things, such as sharing jokes and passing news items. This requires that studies be done in a more general way but using the proposed qualitative and mixed method approach as has been done by other previous studies such as Rohm, Andrew, et al. (2013), and Mao (2014).

As an exploratory study, the initial thoughts were to explore on all the widely used social media platforms. However, as the exploration continued, it was found out that the most widely used platforms within the areas so far explored in Kenya were Facebook and WhatsApp among others such as Twitter and Instagram. Because of the fact that Facebook had been widely studied (e.g. Wyche, Schoenebeck, & Forte, 2013), this study concentrated on WhatsApp as a SM platform without completely leaving out the other platform. The bias was towards WhatsApp because, apart from it being widely used in Kenya, it has recently emerged and hence has not been widely studied. While many of the findings maybe new and interesting, it is important to note that some findings are being confirmed as reported by other studies such by Perrin (2015).

METHODOLOGY

Previous research studies in areas outside SM, such as Chepken (2012) have employed a mixed research approach having qualitative and quantitative data collection and analysis approaches. Data collection methods involve questionnaire, observation and shadowing, structured and unstructured interviews. Even though these study approaches were found to fit into the social media arena, it was noted that some of them needed modifications on how to apply. For example, when observing or shadowing someone, they would be seeing you physically. However, in social media, they are likely to forget that you are shadowing them. Another example is what would be the equivalence of shadowing in a social media environment? In previous studies done by the researchers, they would, for example confirm findings by applying two different approaches, e.g. a questionnaire and chitchat to confirm, say for example age or income. In this study however, it was noted that such combinations could proof challenging. For example, observation to confirm participants age or gender would require a lot of time and logistical challenges. In another example, requesting for permission to study a group of people, say the day labor workers waiting for jobs (Chepken, Blake, & Marsden, 2011) would require you to physically approach them and ask for permission. In the study of social media groups, how one would request for permission to conduct research about or on such platforms is still challenging and has not been documented well.

These complexities associated with SM do not make the social media study/research on or using social media straight and or systematic. This therefore makes the social media research apply the well-known and standard research methods with a bit of variations or techniques. This makes such studies, such as this, to have variations in the application of such methods in ways which may not have been documented and empirically proven to be ideal for generalization.

THE METHODOLOGY APPLIED

In this section, a brief description of the methodology applied is presented. More specifically, the specific data collection and analysis approaches are detailed.

Data Collection and Summary of the Study Population and Sample Size

Table 1. sows the data collection methods and the numbers involved.

Table 1. A summary of the data collection process

Method	Number of Researchers	Number of Participants	Activity/Comments
Observation	5	5	Five researchers observing oneself for an extended period of time.
Online survey	1	68	Randomly selected participants
Social media Group observation	1	15	Observation of social media groups. These were 15 SM groups with several participating members.
Focus group discussions	2	22	The FGDs happened twice with the first one having 12 and the second 10 discussants. The objective was to openly discuss issues related to the study topics
Unstructured interviews	2	20	Person to person conversations of over 10 "friendships" per researcher

Observation

The actual usage and participation were the biggest and the most influential data collection method used. The researchers being users of various social media platforms and members of many different types of groups, the usage patterns in real time were observed. Every detail on how the social media platforms which the researchers were users and groups they were participants were noted. The most widely used platforms were WhatsApp and Facebook and hence most of the observation time was spend on them comparably. Even Telegram is coming up in terms of adoption, it was noted that it is only used mostly among group members and where the group size is larger than 250 members. The actual observation included ten randomly selected WhatsApp groups, over 10 people to person conversations over WhatsApp platform and Facebook. There were five people involved in serious day to day observation of social media use. Out of the five, two were bachelor's students, two were postgraduate and one was the principle investigator. The objective of the observation was to pick out any important unusual and peculiar usage of these social media platforms. This kind of observation was really unique and well thought through because the need to capture every aspect of social media without limiting the emergence of the possible unknowns or unimagined within the realms of social media use.

Observing Oneself

Individually, the researchers observed and reflected on how they used WhatsApp and Facebook. They included: how they deal with an incoming message, how they contribute in a group, how they contribute to individual conversations. The observer paid close attention and in a very conscious mind picked up details of how social media usage was happening. For example, in the morning, I would check how I dealt with messages from the previous night chats. From there, I would also check how I was dealing with them at that time and so on.

Observing Other Users

Observation of other users involved checking how one to one communication occurred between the researcher and the respondent either directly on one to one basis or to the group. This observation occurred mainly on WhatsApp as a social media. However, other observations occurred on Facebook, LinkedIn, Telegram and Google hangout. The duration of the observation was over the 18 months of the study and without any specific item to observe in the beginning. As the study progressed, the observation was based on the categories described on online questionnaire section, which was set out along the way set out along the way. For example, one such parameter was how people behaved when they are together in a social place, resting area or even in a restaurant. These observations were mainly targeting a group of people rather than an individual and the key parameter of observation was the verbal or physical engagement between the participants vis-a-vis their engagement with their devices. The assumption was that every time a participant was not engaging with their colleagues, then they were engaging with their device and on one or more social media platform.

Observing Social Media Groups

This was done mainly through being vigilant within the groups which the researchers were members/participants. This was the biggest observation as it involved over fifteen groups. In the first rounds of observations, it was not clear what to observe because social media groups were known to be diverse, quick in action. Coupled with lack of a theoretical framework for observation, the goal of observation here was general and was aimed at making general observations. The five observers were asked to be open-minded and note anything which looked unusual in any social media groups in which they participated in. They were also requested to identify the kind of groups which existed, for how long they existed, how they ended and the general management of the group, among other issues and characteristics in groups.

The one unique situation was the observation of the electioneering period aftermath. During this activity, the key objective was to find out how the once vibrant groups would maintain the same vibrancy or whether they would end and if they were to end, then how would they wind up. It was also expected that such groups would mutate with the intention of serving other missions. Other observation parameter was whether they would die, how they would die and how long it would take them to die.

SURVEY

Introduction to Survey

In this section, survey was described in many ways according to its purpose. The first was literature survey, also known as literature review (Barbara et al., 2012). This part involved having a critical look at previous attempts to study social media and social networks. The process followed was a systematic literature review approach where relevant articles were identified by first identifying relevant journals in the area of study. The second form of survey was the primary data collection survey. The technique employed was an online questionnaire, to gather information from a sample of respondents on social media use.

Literature Survey

The first objective of the literature survey was to understand the current and past social media studies in order to allow focusing of the study. It involved sampling articles which have social media study models/frameworks and theories. The contribution of the literature survey was a set of factors to consider when studying social media use, which in turn, together with the observations (section xx) were used to generate data collection survey questions.

Data Collection Survey

The main data collection method was a survey involving an online questionnaire. This was used as complimentary to other data collection techniques such as face to face, interviews and focus discussion groups as a. The actual process for each follows:

Online Questionnaires

The request to fill the survey was widely distributed through social media platforms (*Facebook* wall of the researcher and 10 WhatsApp groups) and emails. Emails were sent to 20 randomly selected individuals from the researchers' mailing list. The same request was also sent to group emails for three masters classes. Out of the 130 students in the mailing list, 24 of them filled the questionnaire correctly. The survey was hosted on Lime Survey platform within the University of Nairobi web servers. The questionnaire was divided into seven sections, which were derived from the preliminary observation and the initial literature survey conducted. Apart from the first section carrying the general information, the rest seven sections formed themes associated with potential areas of study in social media. They were: *Social media types and use; usage reasons and frequency; effects and impacts caused by social media; social media groups; social media features; social media and information and politics & the social media.*

Next, the seven themes are described with a view of highlighting what informed their inclusion into the survey.

Social Media Types and Use

The objective was to ask questions which would give insights into what social media platforms are widely used and by who. Since the objective of the study was to curve out study directions, the study needed to get from these questions the widely used social media platforms in order for it to single out which ones to concentrate on. It was not only the most popular but also widely used.

Usage Reasons and Frequency

This theme was meant to collect data which would allow us to understand why users used social media and how frequent they used it. The frequency was measured by asking users how many times they checked their favorite social media platform in a given time interval.

Effects and Impact of Social Media

In many studies, impact and effects of any technology under study is very important. The same was carried out in this study. The effect was described as what would the possible phenomenon be after continues usage of social media. This was done by asking participants to self-report themselves. With regards to impact, the measure of what changed/changes after a prolonged use of any social media. Again, the participants self-reported.

Social Media Groups

From the study preliminary investigations through observation and informal chats with the participants, it was clear that social media group chats are key to social media popularity. In this study therefore, it was important to gain more insights about groups as used in social media through online questionnaires and hence was included as a category.

Social Media Features

Any technology features are very important when it comes to understanding its popularity and use among its users. In this category, the survey questions were biased towards understanding what users like and dislike in specific social media platforms they use. The goal was to get an overall feeling of what features were important to users.

Information

Within the information and data being shared among users of any social media platform category was the kind and the validity/truthfulness of information. Also under scrutiny was the ways in which users authenticated the information they received or shared among themselves. The questions were mainly concentrating on how users confirmed the correctness of information being shared before consuming or sharing the same.

Politics and Social Media

Politics was the odd one out among all the categories of questions. This was because it came out very strongly during preliminary observations as a key use of social media during campaigns and voter registrations. Social media use was seen as a key driving force for politics. Even though the Kenya national politics was the key focus, other political activities, such as University student election politics, were considered during the study.

Survey Data Analysis

The questionnaire results were largely quantitative. However, there were questions relating to participants feeling or own descriptions which were qualitative in nature. As a result, data analysis was majorly quantitative and qualitative for cases were the answers included descriptive result. The analysis spanned three months and a total of 69 responses were analyzed. The findings for each analysis are presented in the results section.

Interviews

Interviews were guided by already obtained results from observation and the questionnaire process. The questions asked to respondents were guided by knowledge from these findings. For example, when finding out the kind of groups one was in and the frequency of activities, the classified categories of groups was used to see if the respondent's answers would fit into the already identified classes/categories. Any other issues outside the identified themes were captured separately and identified as general or special findings. Table 2 shows an illustration of categories of questions obtained from questionnaires and observation.

Focus Group Discussions

Three focus groups discussions were carried out in a span of one year of the study. One was among members of various categories of postgraduate students from outside Kenya. In this focus group, which was conducted among postgraduate students from Mali and Botswana, there were 12 participants. The second focus group was among 10 randomly selected college 2nd year students. The third focus group was among five friends of the lead researcher. The friends were of diverse ages and backgrounds but were all working and staying within Nairobi city, Kenya.

The focus group discussion environment was simple with the lead person introducing himself and describing the purpose of the discussion as getting a general discussion on social media. The discussion was audio recorded and later analyzed for insights of the discussion.

THE STUDY OUTCOMES

Introduction

In this section, the outcomes of the various data collection approaches are highlighted. The reporting framework for the findings is in two sub sections. Section one gives the details of the online survey while section two details the findings of the interviews, focus group discussions and observations.

Table 2. Examples of categories of questions obtained from questionnaires and observation

Question	Group Obtained from Questionnaires	Sample Answer
What is the range of groups you are in?	a) None b) Between 1 and 3 c) Between 3 and 5 d) Above five	- Between 3 and 5
What kind of groups are you in?	a) Family groups b) Alumni group c) Random groups d) Any other group	- Alumni and fundraising groups

Survey Outcome

In this section, results obtained through the survey are presented. The presentation of these results follows the seven themes identified during the survey process, namely the demographics, social media use, effects and impacts, politics, personal gains from social media, social media groups and social media features. The survey responds presented in this section is based on the correctly filled questionnaires of 69 in number.

Demographics

The majority of those who filled the questionnaire were between the ages of 36 and 40 at 27% followed by those between the ages of 26 and 30 at 23% of the respondents. There was 2% who did not say their age. The majority of those who use social media are at the age of between 26 and 30 and 36 to 40 in equal measure. For gender, 66% were men while 20% were female. 14% did not indicate their gender. On the highest level of education, the majority (46%) were reporting that they reached masters level while 28.5% reported to have a bachelor's degree. Another question on demography was on the residence of the respondents. 40% of the respondents said they resided in the urban areas. 26% indicated that they resided in both rural and urban areas and 4% in rural areas.

The Use of Social Media

For this theme, the objective was to find out which platforms are widely used and those that are popular and favorites with users. In summary, WhatsApp was reported to be the most widely used platform (59.42%), followed by Facebook at 56.52%. Table 3 summarizes the outcomes, where the top row shows the question asked and the subsequent rows shows the responses summarized in percentages.

Table 3 shows a summary of the popularity of the various social media platforms as indicated by the respondents. Note that the percentages here exceed 100% because the responds are based on one respondent reported on more than one social medium platform.

In terms of what SM is used for, politics was third after socialization and news. As the name suggests, social media is widely used for family and friend's socialization with 50.72% of the respondents saying

Table 3 Summary of social media popularity findings

		What social media platforms do you use?	According to you, which social media type is widely used in Kenya?	Which is the social media platform you use most?	Which is your favorite social media platform?
1	Facebook	56.52%	52.17	27.54%	24.53%
2	WhatsApp	59.42%	52.17%	49.28%	45.28%
3	YouTube	50.72%	20.29	13.04%	
4	Instagram	23.19%	14.49%	2.90%	0%
5	Twitter	39.13%	26.09%	11.59%	9.43%
6	Google +	34.78%	1.45%	5.80%	1.89%
7	Others	7.25%	0%	N/A	3.77%

so. This is closely followed by news at 36%. Interestingly, politics and education are both at 30%. The other uses include business showcase for marketing, reminders and catching up.

Social media contribution was another factor the study was seeking to establish under the uses. Reading and contributing to discussions topped the uses at 49.28% and was followed by reading a lot and sharing a lot at a distant 8.7%.

Social media has been regarded as one activity/environment which occupies a lot of user's time. In trying to establish how true this was, the participants were asked how frequently they checked their social media. 19% of the respondents (13 out of 69) said that they checked their favorite social media platform twice in an hour. 15% said that they do it twice a day. Only 7.45% said they check it 12 times in an hour. Interestingly, 10% of the respondents could not tell how frequent they checked their social media.

Effects and Impact of Social Media

In this category of questions, the objective was to find out the effects and impacts social media has on its users. The category of questions dealt with SM changing or reinforcing opinions, attractions on social media, perceptions on whether social media is effective, what the respondents liked most, among other things which were left to the respondents to indicate.

On whether social media has changed or reinforced opinions on the user, 72% of the respondents said yes. Table 4 summarizes findings on what usually attracts someone to social media. The responds are listed in the order in which they appeared and not necessarily ranked in according to their popularity.

The question on "Is social media effective and engaging?", 83.72% said yes while 0% said no. The rest of the respondents (about 16%) did not answers the question.

In this study, an effort to find out what makes social media popular or unpopular to some people was made. The questions put forward were 1. What do you like most about social media? And 2. What do you hate most about social media? Table 5 summarizes the outcomes of question one.

Table 4: Things which attract people to social media

#	Themes Associated With Attraction to SM
1	Photos/images – this was mentioned four times
2	Interesting conversations
3	The other attractive factor in social media was about information and knowledge. Participants preferred Informative news, general Information, new facts/details that unknown to them, and ideas. They also preferred to engage people who are Informed and knowledgeable on topics of discussion.
4	Same interests, Shared interests, if i know them (those sharing the information)
5	The presentation, how people discuss issues or framing of information was seen four times in different words but saying a similar thing. The captioning of the discussions also featured.
6	The content of the message or the Message was mentioned appeared eight times. Although this point was presented in various terminologies, such as opinions raised, content of the message was the widely used term.
7	Subject, article topic/title, Good Titles, subject matter
8	Buzz in the circles, topical issues
9	Exchange: This was assumed to be how two or more people in social media exchanged information.

Table 5. What the respondents said they liked most about social media

Characteristic	Descriptions
Cost implications	Cheap usage. Phrases such as cheap to communicate and affordable were used.
Informative nature	Comments such as "*It is very informative and a platform for sharing ideas*" were reported.
Current information and new knowledge	SM allows users to access current information and new knowledge. Phrases such as up to date and instant were used.
Convenience, Effectiveness and efficient	Statements such as availability wherever one is and whatever time, making it a convenient way to communicate, get information and reach out to people indicated this.
Fun, gossip and entertainment	Characterized by Sharing funny content keeping users entertained. Makes a good platform for picking up grapevine.
Education	Phrases used are "*eye opening and education content*".
Connectivity	Ability to help connect one with family and friends.

Table 6 shows the characteristic of SM which make it not liked by the users. The examples of actions that make up the characteristic and the level at which it was mentioned by the respondents are also stated in the table.

Politics

During the preliminary sessions of this study, it was noted that the main use of social media at the time was in politics. As a result, the study sort further clarifications on the same. The first thing was to find out the proportion of people using the social media for politics. The response was that for 69 respondents, only 44% answered in the affirmative. 34% said no and 20% did not answer this question. The next question was a confirmatory one, which was trying to find out the source of political information. Television was rated as the top source of political information at 39.13%. This was followed by social media at 31.88% and print media which scored 20.29%. Radio, internet and other people as a source of political information scored very low.

Personal Gain from Social Media

This item of the study was seeking to find out what drive's users to social media. One question towards this was about what one gains from social media. On top of the list was opportunities at 31.88%, conformity at 13.4% followed by anticipated reciprocity from others at 10.14%. prestige, acceptance, monetary gain, altruism (unselfish concern for the welfare of others) did not rank high here.

With the intention of finding out how users perceive social media, choices were given as ***divisive, cohesive, divisive and cohesive, productive, unproductive, both productive and unproductive and not sure,*** to choose from as what they saw social media to be. Both divisive and cohesive was ranked highly at 40.58%., both productive and unproductive was at 42.03% while productive was ranked 42.03%. Interestingly cohesiveness divisiveness and unproductive were both ranked below 6% of the respondents.

Table 6. What the respondents said they hated most about social media

Characteristic	Examples	Rank in the list
Unsolicited content	Sharing of unsolicited content such as nude images or pornographic videos. Posting of Junk/spam messages and use of bad language	highly ranked
valueless information	Sharing of information that does not add any value to the reader.	Moderately ranked
Personal security	Invasion of privacy, unintended recipients getting access to information and sharing private and sensitive content. Unknown friends, cyber bullying, Insecurity (both personal and cyber security) were also mentioned.	Moderately cited
Social media abuse	Characterized by sharing of messages carrying negative Publicity, Rumors, conman ship, gossip, hate messages, unsubstantiated rumors, abuse on media, incitement, bullying, misinformation (wrong information), false stories, lies, propaganda and vague posts.	highly ranked
Content challenge	SM does not solve personal challenges even with people to talk to through the medium; Long messages with threats e.g. if you don't forward to a certain number of people, something will happen or not happen.	Highly ranked
Spread of believes	One interesting finding was the fact that SM fuels the ever-growing dogma culture or exuberates (believes being spread without any proof)	Low ranking
Usage control	The lack of filters (Posting anything and everything) on what to post or not was mentioned. Others included lack of ways to control users, unchecked mannerisms, posting disturbing messages e.g. sudden death of close relatives or friends, posting horrors photos and shocking news.	Highly ranked
Unappealing usage	An example is people using the media to advertise themselves. Personal advertising or advertising their lifestyle and show-offs were some of the phrases used.	Moderately ranked

Social Media Groups

The question was how many groups one belongs to. The objective was to find out the importance of SM groups such as WhatsApp groups and also understand how participants used the groups. The findings are summarized in Table 7.

Table 7. Social media groups responds

Question	Responds
How many groups are you a member of in social media	A majority is between 1 to 5 (34%) groups followed closely by between 6 and 10 (29%). There were 26.38% who were in groups of above 10.
What is the percentage of groups, if any which you have put on mute form	There is one responded who indicated that are in mute mode. However, those who had muted below 25% of their groups were the majority at 39% over 75% and between 25 and 50% were both at 15%. Irrelevant postings was said to be ones being muted the most.
What is the percentage of groups which you have left compared to those you are in	The highest was below 25% at 52.62%. between 25 and 50% was next at 13%.
How many groups have you been removed from?	On average, the respondents said that they had been removed from about 3.6 groups.
Which types of groups are you in?	Development groups topped at 36% followed by wedding and political at 27.54% and 23.19% respectively. Medical bills and funeral arrangements were at 15% and 20% in that order.
Are social media groups effective?	Surprisingly, no respondent believes that social media is not useful. 47.35% believe that it is useful while 38.84 said that it is sometimes important.

To complete this finding of social media groups, a quick observation, involving physical counting of the SM followers/friends, active members, group contacts and the number of people in a SM group was carried out. The objective of the follow-up observation at this point was to be able to answer the questions which the researchers could not ask the respondents on the surrey because they were thought to be difficult as they would be time consuming. Results of this observation on groups indicate that a typical social media user has over 230 followers/friends or contacts in their social media. Out of these, only about 20 (only about 8.7%) were active. Of all these contacts/friends/followers, about 44 (about 19%) were groups. In each of these groups, the average number of people in the group was 100 or thereabout.

Social Media Features

Under this section, the objective was to find out the most liked and hated features of social media. The questions posed were what features do you like most in your favorite social media platform? And what features do you hate most in your favorite social media platform? Features that stood out included the ability to share files and communicate cheaply, messaging, videos and voice calls, read confirmations (e.g. in WhatsApp) the ability to have a shared environment (e.g. groups), share buttons, and the speed of communication. Those that were said to be most hated were, some social media platform not being able to support a variety of file formats, the poking feature, lack of filtering features, too much storage space occupied by the applications and the availability of voicemail.

On the question: "*What would you like to see in your favorite social media platform?*", the features which respondents would like to see in their social media platforms are shown in Table 8. These features were categorized into two main characteristics, namely functionality and content.

Functionality: This mainly has to do with what the user wants to see the application do. For example, some of the things mentioned included the ability of the SM platform to be able to support different types of file formats. Another example is the ability of the social media platform to do a better categorization of content.

Content: In this category, respondents were talking about the kind of content that should be allowed by the social media platform.

Table 8. Features which respondents would like to see in their social media platforms

Expected feature	How it was described by respondents
Extra functionality	- Ability to use it with Bluetooth for localized communication - More collaboration - Nice interaction - Dislike button - More functions - Effective live transmission - Fact checking sections - Screening of unnecessary uploads. - limits in giving some sensitive information - To support many more types of file format such as GIF Massage bookmarking search option - Better categorization, for Tech stuff vs normal stuff. Get a window or page that is focused on IT work and shows only that Constructive
Content	- Development and investment, constructive ideas, religious and family content, respecting privacy, mature participation and sober respectful discussions and more fun.

Some of the terms/phrases used were not clear on the exact functionality or feature needed. Future studies will endeavor to be specific on some of these terms/phrases.

Content and Information: In this category, the objective was to find out the kind of information shared through SM, how people authenticate the same and the type of media frequently send. The findings are summarized next in Figure 1, Table 9 and in text form.

What is the percentage of information you receive through social media which is false, propaganda or otherwise not clear? Figure 1 shows the responds for this question.

Do social media perpetrate rumors? A big percentage of 63.33% said yes, to this question. Note that the question was not specific on the kind of rumors.

Authenticating Information on Social Media: In this part, the objective was to find out how users authenticate information received or being sent through social media. The question asked was: How do you authenticate information received through social media. The response was mainly touching on confirming from other credible sources such as searching on the web e.g. Google, checking from the mainstream media such as print media, TV and radio. Another way mentioned was crosschecking from other social media platforms. For example, if one is using WhatsApp, they could confirm the same using Facebook or twitter.

Forms of Information Exchanged: The next item the study sought to investigate was the form of information being shared in social media. The findings are summarized in table 5.

The multimedia aspect of WhatsApp was seen to be very heavy. From Table 9, the percentages of text, images, audio and video were seen to be very competitive with video ranking number three at an

Figure 1. Percentage of false, propaganda or otherwise unclear information received through SM

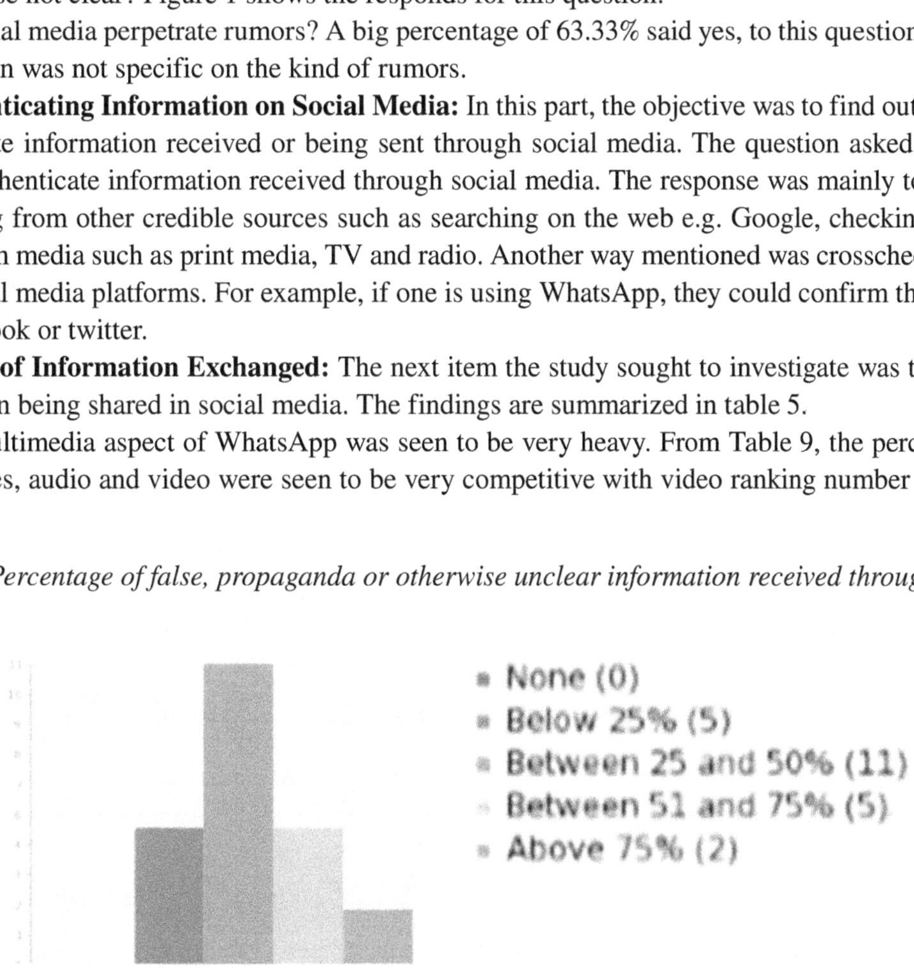

Table 9 Forms of information shared through SM

Form	Percentage	
	Received messages	Send messages
Text	30.43%	30.43%
Images/photos	28.99%	
Video	24.64%	18.84%
Audio	14.49%	17.39%
Multimedia (text, video images combined)	11.95%	18.84%

average of about 21.74% of the messages send or received. Text and images were on top of the list and this can be attributed to their cost and ease of exchanging them. These results were contrasting claims by Yoo & Alavi (2001) which indicated that the current Social media groups are largely multimedia but excludes group audio and video.

Importance of Information Shared on Social Media: One of the arguments which have been propagated about SM is that most of the information shared is not important or just sideshows (as they can be referred to). To find out how true this is, respondents were asked about the percentage of information which they receive through social media and are sideshows. The results are shown in Figure 2.

Politics and Campaign*:* Even though politics was among the uses of SM, it was noted as an interesting area of study during the 2017 electioneering period in Kenya. It was therefore decided in this study that there was need to solicit opinions and perceptions of respondents on whether and how social media was being used in the Kenyan politics. Because there was no any theoretical framework for studying social media and politics found in literature search, the aim of this study was to generate, based on the responds from participants, areas of interest to study as far as *How SM is used in politics; The effectiveness of SM in politics; How to effectively use SM in politics; and What can and cannot be done using SM in politics*. The outcome to this is expected to guide future studies.

To collect information about each of these areas, a series of questions, which were meant to add into the researchers' ideas and then later on form a formidable study area, were formulated. The first question was finding out if the respondents agreed on whether social media is effective in campaigns of any nature. The outcome was that over 70% of the respondents agreed that Social media is effective in campaigns of any nature. This means that they believe that social media can be used to do campaigns of any nature.

During the time of the study, there was a mass voter registration campaign in Kenya. And because of the thought that the study respondents would associate with the exercise, the study sort to use this situation to firm up the question of effectiveness. The question was: ***How effective was social media in mass voter registration which has just been completed in Kenya?*** The outcome saw over 50% of the respondents saying that social media was effective (42.3% effective and 11.52% very effective).

The next point was to find out what respondents thought as ways of making social media effective. The question was what would interest one in a politician using social media for campaign. The following is a summary of what the respondents said about this:

Figure 2. Responds from the question on the percentage of sideshows information

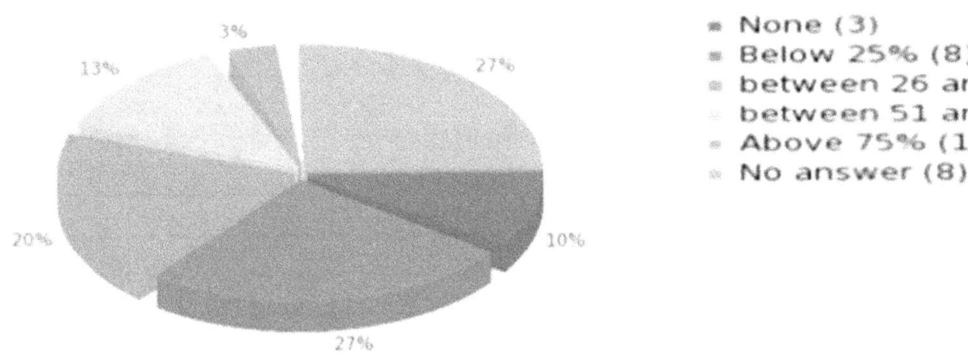

1. **Handling Argument:** For example, how they handle negative comments from critics or those opposing them, whether they create peace or chaos, the kind of strategies employed in dealing with arguments, being able to manage new information emerging about them etc.
2. **Engaging the Electorates:** Through ideas raised, putting forward good policies, the willingness to interact with electorate; availability to his supporters to answer questions, availing their plans online and engaging supporters online so that they can be on the knowhow without attending a physical meeting.
3. **Embracing Technology.** One such example was a politician using social media and technology in general to showcase their development. For example, one would use graphs and statistics to show their development record.
4. **Content Share:** It was reported that the metrics of measuring how good a politician was a included the kind of content shared, the choice of words used and the timing of the messages. The ideas and ideologies, the campaign strategies and the kind of people following/supporting the politician were also mentioned as key metrics.

Findings from Observations, Interviews and Focus Groups

The observation process on the social media use was narrowed down to WhatsApp and Telegram. The reasons behind this were due to their popularity fueled by the fact that they are installable applications on almost any mobile device. Mobile phones, for example, are the dominant devices both offline and online (Underwood, 2016) and literally every mobile device can have WhatsApp or Telegram installed (Nafaâ, Zeadally, & Sayed, 2013). Other reasons include the finding that these platforms were seen as the widely used social media apps and that there was need to narrow down in order to be able to handle the study scope.

After about 20 days of observation, it was discovered that a majority of the activities being carried out in WhatsApp and Telegram related to groups. Even though people communicated on one to one basis, a bulk of the exchanges happened in group chats. This therefore changed the direction of observation to concentrate on WhatsApp and Telegram groups. In this section, therefore, a lot of the outcome relates to WhatsApp groups use. But first the general observations about the social media groups use are reported.

In the focus group discussion, the findings largely reflected findings from observations, interviews and the questionnaire survey. Because of the reason that the findings reflect those seen earlier, only those which are different and unique are highlighted.

Social media use. The widely used social media platforms are Facebook, WhatsApp and LinkedIn professional in that order. It was noted that Facebook is dominant because it has been around for some time compared to WhatsApp which has been in existence for about 2 to 3 years. It was reported that Telegram was yet to pick up.

The main reason for social media popularity was said the cost of operation which was said to be cheaper compared to other alternatives such as SMS. Other benefits included its effectiveness as compared to other means such as email. In fact, it was seen to be almost being more effective than the mainstream media. Some of the major announcements on the traditional media, e.g. print or television were being forwarded and broadcasted through social media. The popularity of social media was so profound to the extent that it was said people always knew about its negative impact but never stopped using it. The negative impact never bothered them.

The social media uses were classified as mainly education, religion and political. The use of social media was said to be influenced by the kind of friends one had. It is likely that one would start using social media if most of his/her friends use.

For effective use and safety of using social media, the focus groups suggested that one has to be careful with the kind of information/data being shared, think before forwarding any received message which may affect you negatively in future; avoid posting own pictures/images, and not posting other people's personal data such as pictures before seeking for permission from them. This discussion brought about the issue of data justice which relate to how personal information should be used for example by marketers (for example a mobile service provider collecting information on the top up of airtime amount and then given loans based on that was said to be data/information injustice). It was noted that information use must be preceded by permission to avoid breaching the data justice protocol.

Social Media Groups- WhatsApp and Telegram

The group chat feature in WhatsApp and Telegrams allows users to chat (send and receive messages) with up to a maximum of 256 and 2,000 people at once respectively. With groups, WhatsApp becomes as a mass action platform. Members of a group form the masses and any action in a group becomes a mass action. In the observation, it was seen that this characteristic which makes WhatsApp a mass action tool featured often. When the bullets come, the majority will disperse. This means that when key issues came in, it was observed that the minority remained with "*the police and their teargas*"- i.e. the issues. The remaining ones are those who stand by the truth and believe in the cause they are championing regardless of whether the cause is correct or not. Such examples emanate from fundraisings (*harambees*), mainly for hospital bills, weddings or raising money for campaigns. The challenge associated with this is the amount of money one pledges to contribute as it may not be matching the "voice" they have on the platform. It was observed that a lot of times, the voice is indirectly proportional to the contribution.

In the same groups, it was observed that members posted content they did not believe in, they have never thought of or are just forwards from other groups or users. These forwards were unsubstantiated, unconfirmed or even unknown to the sender. In such cases, when the need for accountability arises, only the believers of the message will remain to defend their cause. The rest of the mass action team would go underground and, in some cases, even deny having forwarded or even seen the messages.

The average number of groups one was in at a time was three. These groups were classified as mainly for communication and information sharing. Usually politics of any nature was unavoidable in such groups. It was further mentioned that leaving a group was difficult among members. Once one was in a group, they were likely to stay in and even if they left, they were likely to come back in few days.

Following is a summary of SMG findings. The first finding is that SM groups came in two major types, namely short term and long-term groups. Long term groups are created for long term agendas or for basic socialization without a specific objective. Short term groups are usually for very specific objectives with timelines such as hospital bill fundraising. Whether a group is long term or short term, various groups get started different ways such as by consensus, by an individual or by a few people. These groups then grow in different ways either positively or negatively. Once groups have been created and are meant to grow, there are various ways in which members get added or joined SM groups, for example, some people join voluntarily while other just find themselves having been added. These have brought challenges of running SM groups. When it comes to dying (end of a SM group life), SM groups die in different ways, for instance, while some die a natural death, others are killed by their administrators. It was also found out that there was a myriad of challenges, also reflecting in human physical groups. Such challenges included for example, members posting irrelevant content or dominating members running the group single handedly.

CONCLUSION

The chapter has reported the findings of a longitudinal study on general aspects of social media. In summary, it was noted that even though several social media studies have been done, there is still a lot to be studied about SM owing to its dynamism associated with the social nature of humans and the mobile phone pervasiveness which might be fueling social media adoption and use. Further, SM groups enhance socialization, which is a key human characteristic and the pervasiveness of mobile phones allows for SM mobile applications to follow the growth rates of internet and the mobile phones. As such, SM studies should be continuous and done from different perspectives.

A summary of the results show that mobile app social media platforms such as WhatsApp are gaining popularity among users; social media uses are mainly for socialization but other uses such as political campaigns, fundraising among others are taking root; social media groups are dominating the SM arena; even though SM is reach in functionality, there is still no shortage in what users expect them to do; Perceptions about social media may be different from the actual facts. For example, it was found out that the use perceptions and even empirical literature on social media puts Facebook ahead of WhatsApp while the actual self-reporting has their popularity reversed; there exist various challenges associated with social media use and SM study methodological challenges such as ways in which permission is obtained to study a group in SM exists and are inherently different from studying physical groups. Even though these findings may not be very different from other studies such as Robin, McCoy & Yáñez (2017), their importance is contributes knowledge on what people do with the modern social media which then gives directions of what to concentrate in future studies of social media.

REFERENCES

Abdulahi, A., Samadi, B., & Gharleghi, B. (2014). A study on the negative effects of social networking sites such as Facebook among Asia Pacific university scholars in Malaysia. *International Journal of Business and Social Science*, *5*(10).

Adler, A. (2013). *Understanding Human Nature (Psychology Revivals)*. Abingdon, UK: Routledge. doi:10.4324/9780203438831

Al-Rawi, A. (2017). Audience Preferences of News Stories on Social Media. *The Journal of Social Media in Society*, *6*(2), 343–367.

An, J., Quercia, D., Cha, M., Gummadi, K., & Crowcroft, J. (2013). Traditional media seen from social media. In *Proceedings of the 5th Annual ACM Web Science Conference* (pp. 11-14). New York, NY: ACM. 10.1145/2464464.2464492

Appel, H., Gerlach, A. L., & Crusius, J. (2016). The interplay between Facebook use, social comparison, envy, and depression. *Current Opinion in Psychology*, *9*, 44–49. doi:10.1016/j.copsyc.2015.10.006

Boyd, D. M., & Ellison, N. B. (2007). Social network sites: Definition, history, and scholarship. *Journal of Computer-Mediated Communication*, *13*(1), 210–230. doi:10.1111/j.1083-6101.2007.00393.x

Chepken, C. (2012). *Telecommuting in the developing world: a case of the day-labour market* (Doctoral dissertation, University of Cape Town).

Chepken, C. K., Blake, E. H., & Marsden, G. (2011, September). Software design for informal setups: Centring the benefits. In *proceedings of the 14th Southern Africa Telecommunication Networks and Applications Conference*.

Cohen, H. (2019). Social media definitions. Available at https://heidicohen.com/social-media-definition/

Fox, M. (2018). Fake News: Lies spread faster on social media than truth does. Available at https://www.nbcnews.com/health/health-news/fake-news-lies-spread-faster-social-media-truth-does-n854896

Fuchs, C. (2017). Social media: A critical introduction. Thousand Oaks, CA: Sage.

Jabeur, N., Zeadally, S., & Sayed, B. (2013). Mobile social networking applications. *Communications of the ACM*, *56*(3), 71–79. doi:10.1145/2428556.2428573

Kaplan, A. M. (2012). If you love something, let it go mobile: Mobile marketing and mobile social media 4x4. *Business Horizons*, *55*(2), 129–139. doi:10.1016/j.bushor.2011.10.009

Kaplan, A. M., & Haenlein, M. (2010). Users of the world, unite! The challenges and opportunities of Social Media. *Business Horizons*, *53*(1), 59–68. doi:10.1016/j.bushor.2009.09.003

Kietzmann, J. H., Hermkens, K., McCarthy, I. P., & Silvestre, B. S. (2011). Social media? Get serious! Understanding the functional building blocks of social media. *Business Horizons*, *54*(3), 241–251. doi:10.1016/j.bushor.2011.01.005

Kim, J. Y., Giurcanu, M., & Fernandes, J. (2017). Documenting the Emergence of Grassroots Politics on Facebook: The Florida Case. *The Journal of Social Media in Society*, *6*(1), 5–41.

Kitchenham, B., Brereton, O. P., Budgen, D., Turner, M., Bailey, J., & Linkman, S. (2009). Systematic literature reviews in software engineering–a systematic literature review. *Information and Software Technology*, *51*(1), 7–15. doi:10.1016/j.infsof.2008.09.009

Kumar, S. & Shah, N. (2018). False information on web and social media: A survey. *arXiv preprint arXiv:1804.08559*.

Kuss, D. J., & Griffiths, M. D. (2011). Online social networking and addiction—A review of the psychological literature. *International Journal of Environmental Research and Public Health*, *8*(9), 3528–3552. doi:10.3390/ijerph8093528 PMID:22016701

Leslie Becker-Phelps. (2016). Social Media Fosters Insecurity: How to Overcome It. Available at https://www.psychologytoday.com/us/blog/making-change/201603/social-media-fosters-insecurity-how-overcome-it

Madden, M., Lenhart, A., Cortesi, S., Gasser, U., Duggan, M., Smith, A., & Beaton, M. (2013). Teens, social media, and privacy. *Pew Research Center*, *21*, 2–86.

Mao, J. (2014). Social media for learning: A mixed methods study on high school students' technology affordances and perspectives. *Computers in Human Behavior*, *33*, 213–223. doi:10.1016/j.chb.2014.01.002

Miller, D. (2012). Social networking sites. Digital anthropology, 156-161.

Miller, D., Costa, E., Haynes, N., McDonald, T., Nicolescu, R., Sinanan, J., & Wang, X. (2016). *How the world changed social media*. UK: UCL Press. doi:10.2307/j.ctt1g69z35

Moreno, M. A., Jelenchick, L. A., Egan, K. G., Cox, E., Young, H., Gannon, K. E., & Becker, T. (2011). Feeling bad on Facebook: Depression disclosures by college students on a social networking site. *Depression and Anxiety*, *28*(6), 447–455. doi:10.1002/da.20805 PMID:21400639

Morrison, K. (2014). The growth of social media: from passing trend to international obsession. *Social Times 2014*. Available at https://www.adweek.com/digital/the-growth-of-social-media-from-trend-to-obsession-infographic/

Nations, D. (2018). What is social media. *What are Social Media*. Available at https://www.lifewire.com/what-is-social-media-explaining-the-big-trend-3486616

North Central University. (2017). Available at https://www.ncu.edu/blog/dangers-social-media-marriage-and-family#gref

Pan, B., & Crotts, J. (2012). Theoretical models of social media, marketing implications, and future research directions. In M. Sigala, E. Christou, & U. Gretzel (Eds.), *Social Media in Travel, Tourism and Hospitality: Theory, Practice and Cases* (pp. 73–86). Surrey, UK: Ashgate.

Pan, B. & Crotts, J. C. (2012). Theoretical models of social media, marketing implications, and future research directions. Social media in travel, Tourism and hospitality: Theory, practice and cases, 73-85.

Patton, M. Q. (2005). Qualitative research. Encyclopedia of statistics in behavioral science.

Perrin, A. (2015). Social media usage. Pew Research Center, 52-68.

Primack, B. A., Shensa, A., Sidani, J. E., Whaite, E. O., Lin, L., Rosen, D., ... Miller, E. (2017). Social media use and perceived social isolation among young adults in the US. *American Journal of Preventive Medicine*, *53*(1), 1–8. doi:10.1016/j.amepre.2017.01.010 PMID:28279545

Riddle, J. (2017). All Too Easy: Spreading Information Through Social Media. Available at https://ualr.edu/socialchange/2017/03/01/blog-riddle-social-media/

Robin, C., McCoy, S., & Yáñez, D. (2017). WhatsApp. In G. Meiselwitz (Ed.), Lecture Notes in Computer Science: Vol. 10283. Social Computing and Social Media. Applications and Analytics. SCSM 2017. Berlin, Germany: Springer. doi:10.1007/978-3-319-58562-8_7

Rohm, A., & Kaltcheva, V. D., & R. Milne, G. (2013). A mixed-method approach to examining brand-consumer interactions driven by social media. *Journal of Research in Interactive Marketing*, *7*(4), 295–311. doi:10.1108/JRIM-01-2013-0009

Sanders, C. E., Field, T. M., Miguel, D., & Kaplan, M. (2000). The relationship of Internet use to depression and social isolation among adolescents. *Adolescence*, *35*(138), 237. PMID:11019768

Shokri, A. & Dafoulas, G. (2016). A Quantitative analysis of the role of social networks in educational contexts.

Smith, G. (2007). Social software building blocks. Available at http://nform.com/ideas/social-software-building-blocks/

Underwood, L. (2016). Mobile to become dominant device by 2019. Available at https://wearesocial.com/uk/blog/2016/04/mobile-to-become-dominant-device-by-2019

Valenzuela, S., Halpern, D., & Katz, J. E. (2014). Social network sites, marriage well-being and divorce: Survey and state-level evidence from the United States. *Computers in Human Behavior*, *36*, 94–101. doi:10.1016/j.chb.2014.03.034

Williams, D. L., Crittenden, V. L., Keo, T., & McCarty, P. (2012). The use of social media: An exploratory study of usage among digital natives. *Journal of Public Affairs*, *12*(2), 127–136. doi:10.1002/pa.1414

Wyche, S. P., Schoenebeck, S. Y., & Forte, A. (2013, February). Facebook is a luxury: An exploratory study of social media use in rural Kenya. In *Proceedings of the 2013 conference on Computer supported cooperative work* (pp. 33-44). New York, NY: ACM. 10.1145/2441776.2441783

Yang, X., & Luo, J. (2017). Tracking illicit drug dealing and abuse on Instagram using multimodal analysis. *ACM Transactions on Intelligent Systems and Technology*, *8*(4), 58. doi:10.1145/3011871

Yoo, Y., & Alavi, M. (2001). Media and group cohesion: Relative influences on social presence, task participation, and group consensus. *Management Information Systems Quarterly*, *25*(3), 371–390. doi:10.2307/3250922

ADDITIONAL READING

Tomayess, I. P., & Isais, P. K. (2015). *Social Networking and Education: Global Perspective*. New York: Springer.

KEY TERMS AND DEFINITIONS

Mobile Social Networks: Social networks, web-based, mobile application or mobile browser through the web site to enable them to integrate with them through a number of plug-ins.

Social Media: Web sites and applications that allow users to create and share content or join social networks.

Social Media Group: Generally, they are groups that people form through social networks for special occasions or occasions.

Social Networks: information, comments, messages, pictures and so on. a private website or other application that allows users to communicate with each other by sending.

ENDNOTES

1 WhatsApp - www.whatsapp.com/
2 Telegram – www.Telegram.com/
3 Instagram - www.instagram.com/?hl=en
4 Twitter – www.twitter.com/?lang=en
5 Facebook - www.facebook.com
6 Snapchat - www.snapchat.com/

Chapter 8
Erdogan vs. Erdogan: A Polarized Post-Truth Case in Social Media Reality

Savaş Keskin
Bayburt University, Turkey

ABSTRACT

This chapter examines social media relations, which build virtual Erdogans as two opposite realities, with netnography method because of community composition and cultural sharing contents. It will be analyzed visual 'Erdogan' productions in Anti-Tayyip (Opponent) and Erdoğan Sevdalıları-Lovers of Erdogan (Fan/Supporter) communities and it will be drawn post-truth biography of a leader in visual culture of social media. Two different/opposite virtual realities of Erdogan, which are reproduced in social media sociality every day, lead to expansion of polarized political climate in the context of organic society and absorb the political identity of Erdogan.

DOI: 10.4018/978-1-7998-1041-4.ch008

INTRODUCTION

The close relationship between politics and media is a quite common tradition both in Turkey and in the world conjuncture. However, an expansion towards social media and the distribution of agenda-building initiative to political poles represent the next level in this relationship. Even though the hereditary codes are absorbed to a certain degree by the fact that agenda-setting user interests are distributed to multiple social forms in the public sphere of social media, it is clear that politics expand to digital media. Some scientific authorities refer to social media as a kind of public arena since it encourages participation. But it turns into a cyberspace with political topics or individual demonstrations penetrating further by the day, thus it can be argued that this platform is under a clear invasion (Fuchs, 2014). A correlation can be found between the increase in political content of digital action on many platforms, especially on Twitter and the creation of interest related to politicization, as well as motivation on social media and traditional habits. In this context, the agenda that shapes the reality of everyday life has a facade that creates intense attention on the social media. McComb and Shaw (1972) argue that the traditional mass media can form a picture in the minds of the audience and determine "not how they will view something, but what they will view." On the other hand, this view focusing on social media's 'prosumer' culture (Fuchs, 2017, p. 72) seeks the possibility of political truth in "what the audience talks about, as well as how they talk about it." After all, users are the actual actors of a system of truths produced by the political group they adhere to, therefore, they have the merit for determining the agenda together with centralized sources of information. Moreover, the discourse is not based solely on the verbal structure of language, but it creates an arithmetic of visual symbols, as well. The sequence of post-truth, which dominates feelings and opinions increasingly more and becomes more public every day is contained within a network of interactions and identifications with no clear-cut boundaries.

The Oxford English Dictionary (OED) declared 'post-truth' as the word of the year in 2016 (en.oxforddictionaries.com) based on the Brexit Referendum in the UK and Donald Trump's political campaign for Presidency. In the related article about this 'awarded' concept, the situationality was defined with reference to the difference between 'being' and 'appearing.' This concept finds its power in a truth which is only there through the qualities and the feel of this phenomenon. Furthermore, there is no need for the plain truth. Inspired by politics, post-truth penetrates into each and every cell of society with its expansionist oscillation to have supremacy over all convictions. By 2019, post-truth is not only a practice in the political arena, but has its place in the ranks of advertisement and corporate universes. It thus dominates social convictions and resists to touch base with the truth. It turns into a cliché in the face of objective data and the extremes of emotional discourses and convictions (Zarzalejos, 2017, p. 13). The political attitude accompanying politicized feelings and convictions cannot be distinguished from the sensitivities and opinions within the community in general. The post-truth politicization is like a heavy odor descending on social life, it is like a fisherman betrayed by his smell on a first date.

Yanık (2017) emphasizes mass sentiments with the reference that "in forming public opinion on certain issues, objective truth is far less important than personal beliefs" in describing post-truth. After all, the truth does not mean anything on its own and the reality as we know it up until this era is more often than not the modern reality. This rational perspective does not touch the specially structured practices, discourses, politics, and institutionalization. Post-truth, on the contrary, includes all of them and is included by all of them. Truth manufacturing, guided by the politicization of processes is based on the new needs of an age that wears out and consumes up the codes of the past. In the post-truth universe with abstract institutions, images, simulations, and motivation of affections are trending, there is no place for the static and fragile catalysts of the past. As Descartes said, when the future catches up with the speed of humanity, the past will lose all validity (Palma, 2017, p. 17). And this is the case now! For this reason, the post-truth is a concept intrinsic to the present. Its history cannot be deliberately established, but is renewed in the form of current periods. Post-truth politics, which uses a set of maneuvers including the traditional values of the masses can advocate one truth today and deny it tomorrow. What truly matters here is not that what is clearly black is put under control through the created convictions that it is white. The real problem is the universalization of the mass convictions about white beyond its all other qualities (Yanık, 2017). In the opposing views represented by the groups called Anti-Tayyip and Erdoğan Lovers [Erdoğan Sevdalıları], not only the truths at the opposing end of the antagonism, but many truths within their own ranks as well are ignored and reduced to their own subjective truths. In the post-truth, there is no pure white or pure black. The sides clustering on the poles of white and black must fight in a gray universe. The truths scattered in the shades of gray are perceived from the position of black or white. Various Erdoğans as products of social media can only make sense in the direction of the meta-meanings and hybrid codes of the post-truth universe.

This study examines those aspects of post-truth manufacturing in virtual communities that cover Recep Tayyip Erdoğan and reviews the eternal spiral of truth between the poles of demonization and canonization. The fact that the study focuses on the relationship between virtual communities and Erdoğan is strongly connected to the deterministic social media influence of on the truth. After all, the virtual communities that modeled a 'social mediatic Erdoğan' out of Erdoğan assign a certain identity to their own Erdoğan. This interactive biography enhanced through the virtual community performance of the 'assigned Erdoğan' reflects some aspects of the organic identity, but is essentially virtual and synthesized in its entirety. Furthermore, the convergence of organic and virtual lives, the blurring of the borderlines in the post-truth universe causes the creation of a synthetic Erdoğan to interfere with the emotions, opinions, and perceptions in the organic life. This study uses the oft-mentioned 'politician with a mediatic personality' concept in the form of a 'politician with a social mediatic personality' and discusses this personality's ties to new media's culture of participation. The culture of participation and convergence, celebrated by Jenkins (2017) with an optimistic outlook, breaks the resistance of a firm stance of truth even though it offers individuals relative possibilities of creativity and productivity. Therefore, digital reality as a work of the partisans of a polarity is as fragile as a snowman exposed to sunshine.

Besides his political influence and popularity in Turkey, Erdoğan was elected the '2018 World's Most Influential Personality' in a poll by the international news site Rassdnews, gaining 77% of the votes of 300 thousand participants (www.sputniknews.com). As a political or mediatic/social mediatic subject with influence in Turkey and around the world, Erdoğan is a leader whose image was designed in a highly effective way and to whom masses assign meaning according to their own ideological and religious convictions. The answers to polls, in the social media, and in the news discussions titled "Why Erdoğan?" usually point to an ideological attitude or polarized convictions with unidentified authenticity. The polarity of opinions about Erdoğan and the truth wars between two opposing sides may be defined under many artificial categories: the December 17-25 events, Gezi Park Movement, patriotism, neoliberalism, piety, the coup attempt of July 15, Presidential Residence, nepotism, crony capitalism, dictatorship, and political asylum, among others. The general aspects of the polarity of opinion related to Erdoğan and their reflections on virtual communities are also discussed under these categories.

By their very definition, virtual communities attract like-minded people of similar interests, which in turn leads to partisanship, identification, and polarization. Polarization in virtual communities is a result of group opinions reinforced and radicalized through social media performance (Şimşek, 2018, p. 6). This polarization based on partisanship is characterized by a loss of rationality and drifting away from a logic of analogy as individuals take sides not only in terms of social interests, but also of convictions about truths, public consensus, feelings, and opinions. The Facebook groups Anti-Tayyip and Erdoğan Lovers used as the samples in this study reflect a polarization in the form of conflicting sides as clearly seen in their names. The respective outputs of these communities, each with a large number of followers, are identified with the post-truth in character. The partisan polarization created by post-truth cannot be considered outside the context of politician attitudes. Politicians distorting the truth with mediatic discourse is not an invention of our age, but actually was seen in all ages and places. The characteristic of our age, however, is the supporters repeating, expanding, defending this partisan distortion and turning it into a belief, no less. A good example of this was seen in the eve of US presidential elections: according to polls, Republican Trump supporters who penetrated Twitter had a higher vote of confidence than the news media (Maxwell, 2016).

Autonomous relations in virtual societies do not only reinterpret the traditional codes and legacy, but pressure perceptions to change, as well (Keskin & Baltacı, 2018). Transformational movements on a line of low tension may uproot most traditional perceptions with reference to digital coded convictions. In an era when centralized sources of information and info-capitalists experience a loss of power, the ability to construct a truth is transferred to the 'pronetarian' (Atikkan & Tunç, 2011, pp. 103-104) online generation/classes, thus social media becomes the place to build public consciousness. Social media platforms, especially Twitter and Facebook are now important places to follow the news and access information. This gave rise to maneuvering through social media accounts to manipulate national and international public opinion (Özcan, 2018, p. 3). The new generation opinion leaders referred to as 'influencer' perform a key task either as individuals or communities. Public opinion and the convictions needed as a basis for the truth are all manufactured and manipulated at this level of actors, which turned into a measure of personal and collective recognition.

The focus of this study is the confrontational nature of the Erdoğan productions controlling the public perceptions of the two virtual communities. The fundamental problematic of the study is the idea that the fragile social manifestations infiltrated by the post-truth to the last cell and formed by a cultural roughness and negligence are based on doubt, thereby eroding the truth the society needs (Modreanu, 2017, p. 8). Erdoğan is undoubtedly an actor who has a determining influence in social life and a central role in political processes. Creating dubious truths about such an actor and universalizing them through public interaction leads to a degree of polarization hard to reverse. Indeed, one of the basic premises of this study is that one of the most important reasons of the current polarization in Turkey is the confrontational nature of relationships in virtual communities. The Open Society Foundation's index report (2018) on the relationship of post-truth and new media shows that the spreading of fake news under the pressure of post-truth is a cause of mistrust of institutions and experts, the rise of polarization, hate speech, and escalation of criminal action in the real world. This extremism is nourished by the culture of new media participation which also shakes the public foundations of journalism. And the fact that individuals tend to choose belief over fact enhances it all the more. This situation shows that even though post-truth is a sociological phenomenon, it overlaps with mass psychology (Glaveanu, 2017, p. 376), thus weakening lending more weight to beliefs and convictions at the expense of rational cognitive awareness. Moreover, the new era is not only characterized by politicians managing the perceptions of the public, but also by social media groups defining the discourse of politicians. The truths manufactured in social media is a form of pressure on politicians forcing them to comply in time. The Brexit Referendum where voters stood by their own agenda precluding the politicians to implement a new policy is a case in point (Imai, 2017, p. 4). Erdoğan is the passive code of representation in virtual communities as opposed to his active presence in the traditional media. Therefore, commentary about him can lead to a fluid and uncontrolled recognition redirected according to the empathetic actions and the intensity of emotions of the masses. A politician relying on image management and his supporters are not likely to prefer such an outcome.

This study is an attempt to map the truth of Erdoğan as a subject of an indefinite number of images created in the Facebook groups Anti-Tayyip and Erdoğan Lovers. This map is the product of an attempted theoretical explanation of the multi-layered and complex representations occupying the space between the negative and positive poles.

THEORETICAL BACKGROUND: POLITICAL TEST AND DEMISE OF THE ABSOLUTE OR MODERN TRUTH

Baudrillard's (1998) proposition of simulation heralds the end of the absolute truth defended by modernity and idealism in different fronts. Political balancing that surrounds civil societies point to a fantastic transformation in the nature of truth. The post-truth universe resembles a form of hyper-reality constructed by the simulacra and transfers the sensual relationship with the truth to a level of fantasy. After all the post-truth is revealed through the symbolic interaction between the simulacra manufactured in politics and the symptoms and diagnoses. The reality TV, news, and social media practices which include the symptoms of factual reality, but use them to construct new and 'fake' realities are an agent in the establishment of a 'balance of truth' in politics (Suiter, 2016, pp. 25-26). The fact that the truth is broken indefinitely through the transitional layers between traditional and new media instruments increases the need for experts to enlighten the public. The political truths that exist in the realm of humans need support, arguments, and advocates. Otherwise, it is not possible to maintain the political value of the factual evidence upon which the truth is built.

While the concept of post-truth is mainly discussed by the examples of Brexit and Trump's election victory, Forstenzer (2018) dates its history back to cover the Watergate scandal. After all, truth is not today's fashion, it is not even peculiar to the political sphere. The idea that truth cannot be defined precisely and therefore is an adventure full of uncertainties makes reality questionable and tells us that post-truth is no novelty. The reason to this is that post-modern era is characterized by uncertainty, a feeling of disintegration, an increasing dependence on the present time, and a failure to predict the future (Bauman, 2016). In such an age it is quite difficult to understand history, to connect with it or be a part of it. Nevertheless, the post-truth era, representing the end of truth has a specific history. This history is based on the spreading of 'lie' through populist propaganda in the aftermath of nationalism and xenophobia (Glaveanu, 2017, p. 375). It is quite a valid idea that a typology of lies targeting desires, independent of truth, and deserving consideration on its own right (Aytaç & Demirkent, 2017) is the founding catalyst in the history of post-truth. After all, what desiring an object entails is not the satisfaction, but the re-creation of desire. This means that in the relationship with reality, objective truth is shaped by desires as much as it shapes them. When the subject is unable to access the truth that is part of a realm of fantasies or to shape objects according to desires, a feeling of void develops and gives rise to anxiety. The essence of anxiety is not an inability to resist the truth, but the threat that the truth which cannot be constructed as desired may destroy desires (Žižek, 2005, pp. 73-75). In the age of post-truth, there is no doubt that lie is a functional mortar used to fill in the fallacy of truths enveloping the subjects.

According to Modreanu (2017, p. 8), the newly-acquired flexibility of ethic codes under the influence of new public figures contributed to the post-truth as much as relativism, the decline of society, or Facebook narcissism. People of this age when lies became natural and legitimate give up in time on their search for a belief after the truth and turn towards proofs of truth after their beliefs. Any evidence inconvenient for the faith is either ignored or distorted. With the audience laying hands on media and visual design programs which were formerly the realm of the professional (Manovich, 2012), the visual culture evolved into a culture of manipulation. Now, the signs of factual reality can be re-designed in order to construct any truth. As one of the environments where this is seen most frequently, virtual community pages are constructed through compositions that reverse the known truth (Özmen & Keskin, 2018).

Harsin (2018, p. 3) states that the post-truth narrators of truth point to a historical and special concern in their accounts of authority and public facts. This concept, which is of interest to critical theories in the context of power relations can be explained in two major characteristics. The first one includes the two distinct but interrelated forms of truth: honesty/reality and knowledge/legitimate beliefs. The second is that it brings up descriptive problems similar to all other major historical periods (modern/post-modern, industrial/post-industrial, or traditional/post-traditional). The first of these two characteristics contains the unity of objective reality and faith as foundations of the truth, whereas the other refers to a historical differentiation. Because the age of post-truth raises a generation where the concepts of the age of truth are interwoven and become invalid.

As one of the first people to discuss the concept of post-truth, Keyes explores its outlines and roots in his book *The Post-Truth Era* (2004). Keyes discusses post-truth through the key concepts of lie-truth, dishonesty-deception, ethics-politics-media, morality, and sex. He expands the academic, techno-political, media, identity, vital and perceptual dimensions of this discussion to the US and European contexts. Within the collective relations he describes as dishonesty, politicians, lawyers, and therapists are mentors and role models. According to Keyes (2004, pp. 63-71), the rise of lie is closely related to factors such as internal security, recreation, indulgence, adventure, and tax collection. The pinnacle of this relationship is the cooperation between politics and media. On the other hand, the position of the viewer is constantly re-established on the axis of the sense of pride for the confidence based on repeating mediatic lies, for having fun, indulgence, adventure, and paying taxes.

In terms of community building on social media, fake news and trolling are the two important reflections of communicative and informational activities in the context of post-truth. In communities where a visual culture prevails, it is rather easy to mask or distort information through pictures. Communities often make use of pictures to highlight or suppress certain aspects of the effect a truth is supposed to create. The new generation of internet images called caps and memes (Kırık & Saltık, 2017) form a nurturing soil for fake news and trolling practices.

According to Karagöz (2018, p. 681), fake news is a complex form of information created by a haphazard mixture of satire, parody, desktop manufacturing, manipulation, advertisement, and propaganda as an effective part of post-truth. The news practice as an expression of the informational dimension of truth is transformed due to the continuous flow of uncorroborative fake news on social media. This practical fake news shaping the truth for the masses is often strongly believed to be true. More often than not, fake news circulates to support the truths of an opinion and is a risk for the reputation and image of opposing view. This deceptive journalism does not cite a source, it is timeless and lacks confirmation, yet it moves faster than mainstream news agencies and its effect spreads to a much wider area. Kaya (2018), researching this new-generation journalism through the Anadolu Agency, found that fake news influenced agency journalism in many ways and often had negative consequences. However, it is also a remarkable finding that the confusion of information due to fake news increased confidence in news agencies. Fake news also highlighted the importance of global channels of verification and corroboration. Thus, mechanisms as information literacy, fast-checking, and verification are more necessary than ever in the face of post-truth policies. A public sharing portal where statements by politicians and their supporters can be checked can be possible only through these mechanisms (Taniguchi, 2017, p. 3). A review of social media usage figures in Turkey makes it clear that a similar mechanism is needed. The collectives of teyit.org and dogrulukpayi.com founded for this purpose (Ahi, 2017) investigate the accuracy of information patterns in the agenda of the social media agenda and share their results with the public. But it is clear that the results of these verification efforts are barely as effective on the public as fake news or trolling.

Trolls are among the most important carriers and routers of post-truth on the social media universe and they represent the online resources that misuse and abuse information to exploit public opinion (Karataş & Binark, 2016, p. 434). Trolls are usually not 'real' persons, they use design programs to change the indicators of reality as they wish and create parody accounts to misguide the public. Trolling practice has a viral base and it is on the rise thanks to the newly flexible visual culture, availability of means to interfere with digital codes and the vulnerability of masses to digital deception. According to Karataş and Binark (2016, p. 439) trolls are often interested in caps/meme content. The transitional language that visual and written discourse can build together is one of the most important reasons. After all, the situational signs in a picture often require a descriptive statement. And describing situations is the essence of representation (Hall, 1999) and every case represented conveys the truth of its describer.

AIM AND METHOD

The study was designed according to the principles of qualitative data configuration of the critical paradigm and modeled as a cultural analysis. In the acquisition and analysis of data, a netnography approach called a social network ethnography was adopted. Anti-Tayyip and Erdoğan Lovers Visual and written shares in Facebook groups were evaluated as community feelings and insights linking the culture with the ruins and the parts.

The possibilities of creating culture and identity social media offers its users as part of its participatory culture can be used to assign an identity to others as much as for self-identification. The idea that virtual communities give individuals and groups an identity (Donath, 1998, p. 3), as accepted in the related

While the surname Erdoğan is the symbol of a positive statement, the name Tayyip is used as the basis for negative ones. The struggle observed in the macro structure of the name reveals itself more clearly in its micro extensions. What these people actually perform is an act of catharsis and self-consecration by way of creating their own saints and demons. Both group admins and members design the wisdom for their own images through the sanctity they attribute to their leaders and the demonization of the other. The power of Erdoğan Lovers comes from their sanctified leaders, while the Anti-Tayyip supporters find their strength in the other Erdoğan they demonize.

Figure 1 shows a selection of direct visual representations of Erdoğan. Since visual representations from both sides may contain hate speech, only the satirical ones are included here.

The first thing the images show is that the power that determines the truth about Erdoğan is nurtured by convictions. The Erdoğan personality of the Erdoğan Lovers is the personality of group members. Likewise, the Erdoğan personality of the Anti-Tayyip group is the personality of the other, i.e. of the

Figure 1. Visual representations with direct reference to Erdoğan

<table>
<tr><th>Group of Anti-Tayyip</th><th>Group of Erdoğan Lovers [Sevdalıları]</th></tr>
<tr><td>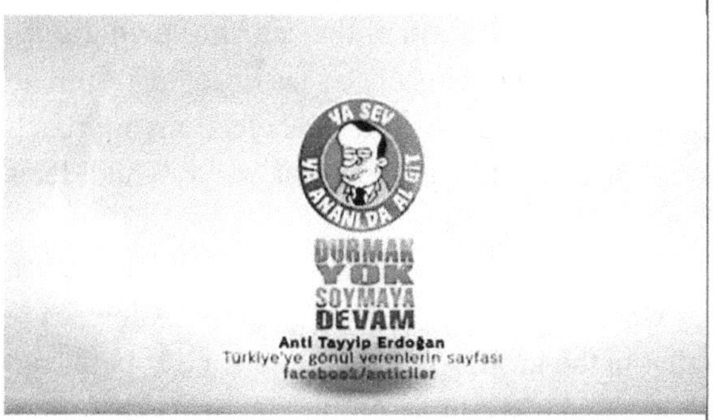
</td><td>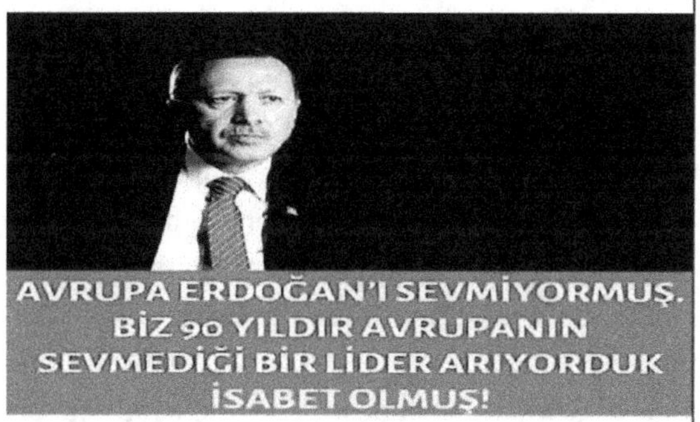
</td></tr>
<tr><td>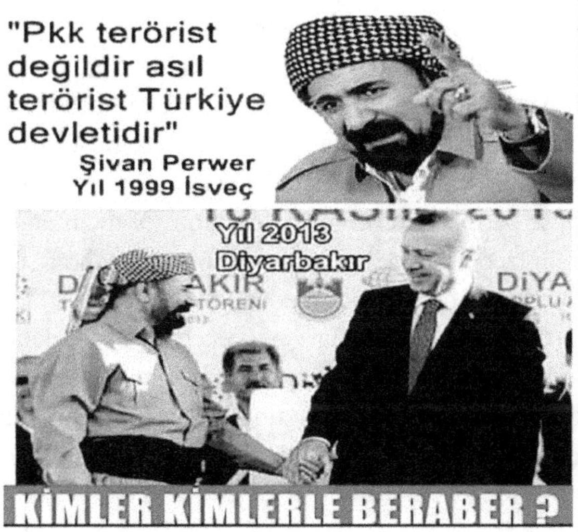
</td><td>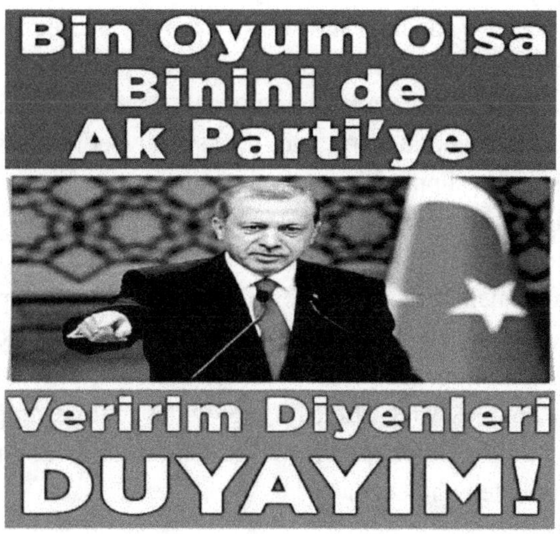
</td></tr>
</table>

enemy. A comparison of the two groups points to a paradox. Because in this comparison, Erdoğan is his own other. All convictions meant by the good and the evil are collected in one person, but from different routes. Erdoğan's figurative presence in these communities basically serves as a means of irrational purification.

There are strong bonds between canonizing and demonizing. After all, canonizing someone is a similar process to demonizing their opposite. One of the basic similarities between the groups is that the subject they attempt to construct appears to be singular, but in fact is plural and asymmetrically covers all opponents or supporters of the power. The breakthroughs of alliance seen in the political conjuncture spread to supporters' virtual life. For this reason, these groups bring up parts of the political scheme Erdoğan or the opposition represents more often than Erdoğan himself. However, in groups where opposition is taken to extremes, differentiating reasons are generally based by political convictions, whereas similarity is measured by work. In the culture of visual representation prevailing in these groups, Erdoğan occupies a far smaller place than others. The sides in the polarization intensifying around the person of Erdoğan judge and condemn each other. In doing so, defense of the truth is not their main concern. Figure 2 offers a selection of opposing identities each group associates with Erdoğan.

The essence of social action in the groups is the instrumentalization of Erdoğan for political tendencies and convictions. This is evident in the fact that most group postings a share is given to all representatives of the party or pole in question and all positivity or negativity is arranged around Erdoğan. As seen in the visuals, the identities of other actors are defined with reference to Erdoğan. While the demonization of opposing actors is a routine practice, sometimes the canonization of the actors of own group are based on the political pole that Erdoğan represents. Erdoğan's opponents or 'Us' as Erdoğan's opponents confront and clash at every front of this war and as the agenda allows. As the local elections approached, the actors covered in these representations were mostly selected from among the mayor candidates of major cities, as well as the leaders of political parties.

Almost all postings are shared without citing any source and with a claim to truth. Reference to opinion leaders, supporting the claim with visuals or resorting to fake news are common devices used to intensify the effect of truth. It is possible to argue that in the groups where rational and concrete arguments of truth are not questioned, the irrational effects typical to a culture of fandom or sports supporter bake are evident. According to Jenkins (2006), the rising wave of admirers in the participation culture of the new media is a result of ordinary people flowing towards each other. In a fan culture, the channels between the fandom and the stars are always active and open to participation in order to glorify a certain interest and conviction. In group participation, text and emojis are widely used in accordance with the general cultural logic of social media. However, some users in the Anti-Tayyip group upload the visuals to support their comments. The admiration is channelized and directed by page admins, who are anonymous persons. The visual content shared by these people determine the opinions of the members and their relationship to truth. In a social platform organized around the theme of Erdoğan, others speak on his behalf and activities of mass participation provide all the support needed for such speech. Thus, the legitimacy and universalization of a case of post-truth is secured. Fans are generally dependent on centrally provided cultural codes and their productivity is usually shaped by an information flow from the source. Fan participation provides routine support for the content. Opposing voices are stigmatized as being manipulated from the outside and serving 'evil.' Most of the time, social considerations are mobilized and excluded.

Figure 2. Representations with no direct reference to Erdoğan

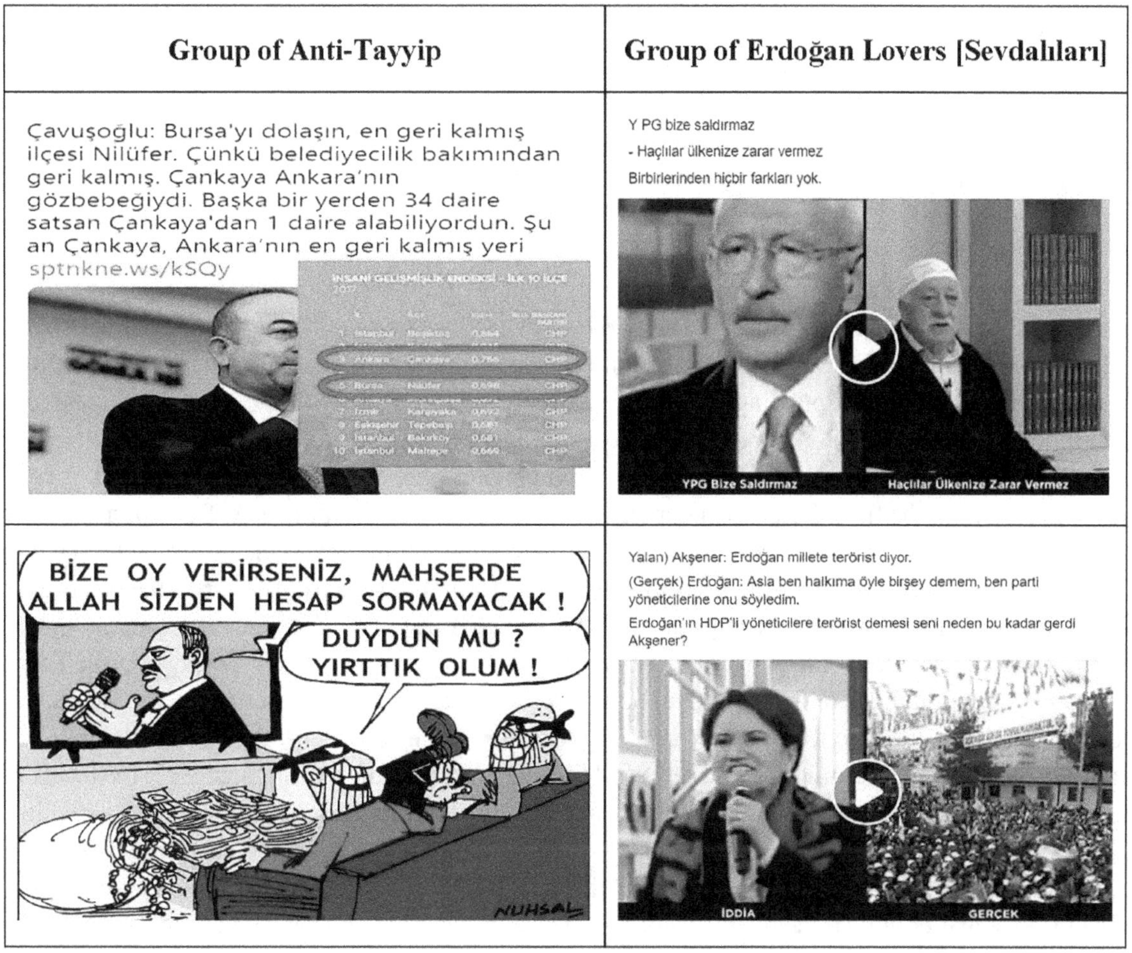

Baker (2017, p. 326) finds it problematic to describe the supporter culture that nourishes admiration as an identity. After all the supporter is both an individual and not. Most of the time they are considered unreliable because their loose ties. And with the backing of crowds, they can turn into a monster all of a sudden. Therefore, group struggles easily produce hate speech that escalates to the point of lynching the opposite. Moreover, it is uncertain who populate these groups. Taking a personality as their reference is not descriptive or explanatory enough about their identity. Who the group members are or who they are not becomes increasingly blurry behind the façade of collective supporter culture. A very clear sense of good and evil prevails. But it is extremely difficult to tell who is good and who is evil.

The Power of Visual Manipulation and the Cultural State of 'Fallacy'

In traditional media representations, Recep Tayyip Erdoğan is the political actor who generally establishes the agenda. He defines the course of political speech with his discourses. And in virtual communities, he is the political actor on whom the agenda is based. Even though the agenda on the social media has strong ties with the fragments of the same on the traditional media, here the structure of the discourse is defined by virtual interactions. Moreover, the cultural production as the basis of this deterministic relationship is a process whereby visual materials representing the same or similar events are continuously manipulated.

Visual production in the groups includes the use of photographs, videos, cartoons, caps, photoshopping, and infographics. When a group wants to build a truth, they usually use images, namely photographs. It is possible to distort and misguide the perception of reality photography creates in memory by way of retouching, manipulating, or subtitling (Sontag, 1993, p. 100). In political photographs, the intervention is more evident. After all, in political photography, it is possible to create the exact opposite message with very small changes (Freund, 2016, p. 142) or a symbolism enhancing the effect of verbal deception through the use of a photoshopping application (Berger, 2018, p. 44). The cultural logic of Facebook presence dictates the use of short texts to describe the situation in the visual. Thus, a culture is developed out of the discourse patterns that distort the truth. In the sharing culture where an event or action depicted in a photograph or video is verbally defined, it is possible to argue that the mechanisms of defining a situation take sides according to convictions. An example is the controversy over the Taksim rally on March 8 on the occasion of International Women's Day. Attempts to establish the truth by opposing poles through video footage shows that photography alone is not capable of telling the truth. The truth produced by highlighting some aspects and ignoring others in the existing images is the result of a deception similar to adopting a certain hue of white and ignoring all other hues. Where the visual itself falls short of depicting the image people seek to build in their minds, verbal practices are used. Image serves a symbolic function, it is an extension of the verbal depiction even if it was sought as an evidence of truth. As a matter of fact, members do not have a clear request about what happens in the image. The general collective movement supports a mechanism that develops the verbal status identifier into a truth.

Figure 3 shows the images of the said events during the March 8 rally.

The visuals include amateur and live footings of the same event. These do not have a political worth on their own right. As for the political worth of the existing truth, it is established by the virtual actions of social media practitioners. The rally was shown as an innocent action in the Anti-Tayyip group with selected parts of it presented, as well as the intervention of the police, described as an oppressive apparatus of the government. In the group of Erdoğan Lovers, the same footage was cut in a way convenient to produce a political meaning and opposing convictions were targeted. Naturally, the images were not given a chance to create their own independent language. The truth established by highlighting certain elements of an image and eliminating others formed the synergy of conflict between the two sides of an argument. The fact that the arguments on the contents of the video were expanded to the whole political context through the International Women's Day bears witness to the impossibility of civil action. Thus, civil movements are somehow subjected to political definitions that parties can turn to income. Furthermore, the prevailing public attitude is to develop a certain opinion to maturity instead of exploring the reasons behind a rally.

Figure 4 illustrates the cultural activity transformed by visual productions in groups.

Figure 3. Both sides constructing their agenda on the Rally of March 8

<table>
<tr><th>Group of Anti-Tayyip</th><th>Group of Erdoğan Lovers [Sevdalıları]</th></tr>
<tr><td>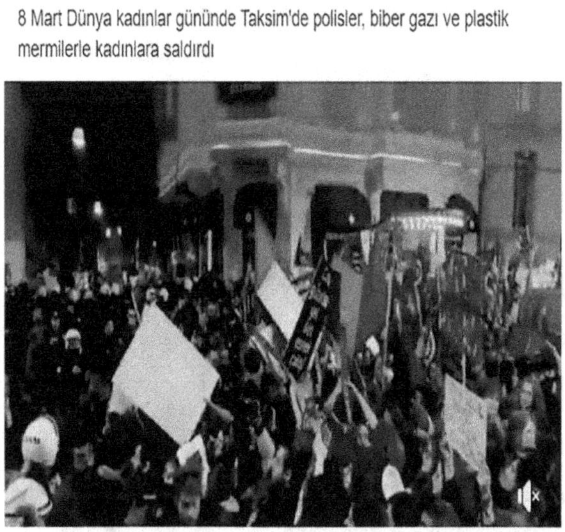
</td><td>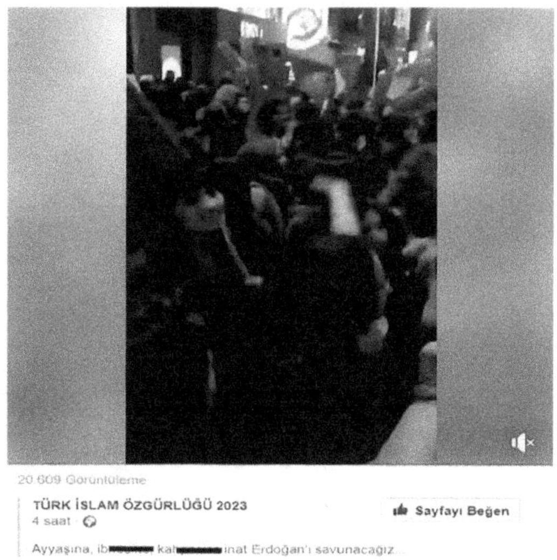
</td></tr>
</table>

The state of reality stretched and skewed by the antagonistic ends of truth can always be reconstructed with the power of visual manipulation. Oppositional convictions produce visual designs as an evidence to an accusation where they have no material evidence – expecting otherwise would be an immense mistake. Similarly, Erdoğan supporters propagate convictions in the visual context in order to sanctify their leaders without any success in proving their concrete connection to reality. The mass consciousness these images carry to the layer of truth is often made up of ideals, attributions, expectations, and fantasies. As in the imagination of an eschatological world, questioning the truth is left for later, to the 'other world,' i.e. the social media. However, this is not the questioning by fair and objective minds in search of the absolute truth. The biased nature of this questioning determines the volume of the mass as the soil that grows the seeds of conviction cultivated by each side on their ranks. The sequence of convictions achieving majority in numbers will facilitate the manipulation of the mass perceptions of truth and the control of intellectual connections with reality. After this stage, virtual consciousness weakens in its resistance against the truth and becomes more dependent on the flow of information. After all, the realities served by the resources of the virtual community are supported with visuals, which in turn are prepared according to convictions. It becomes difficult to tolerate any truth other than own convictions and the diffusion of virtual manipulation expands towards traditional values.

Figure 4. Examples of visual manipulation in groups

Group of Anti-Tayyip	Group of Erdoğan Lovers [Sevdalıları]

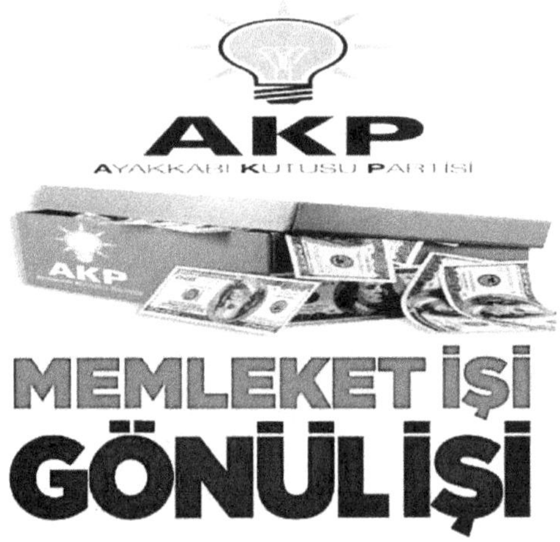

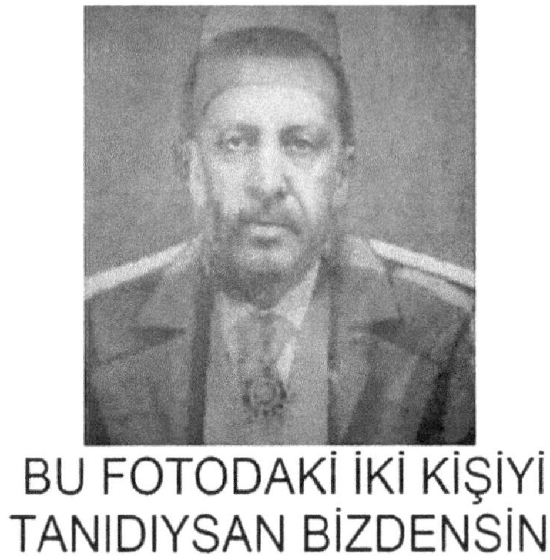

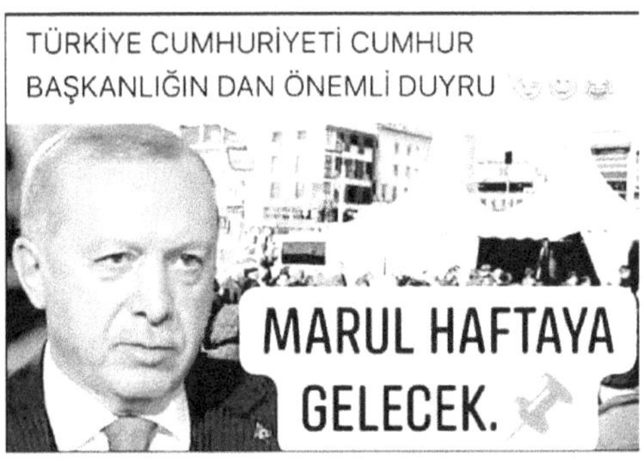

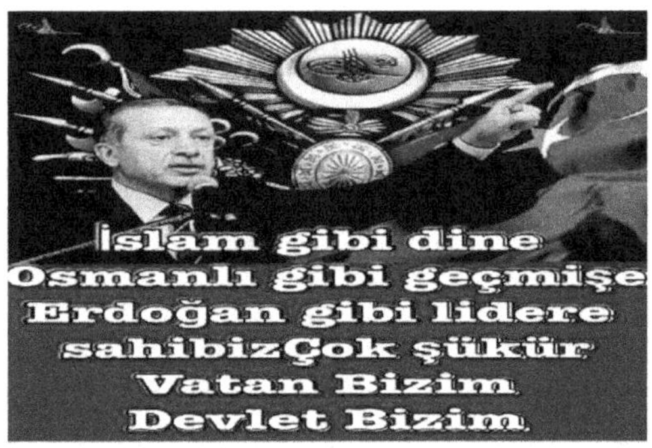

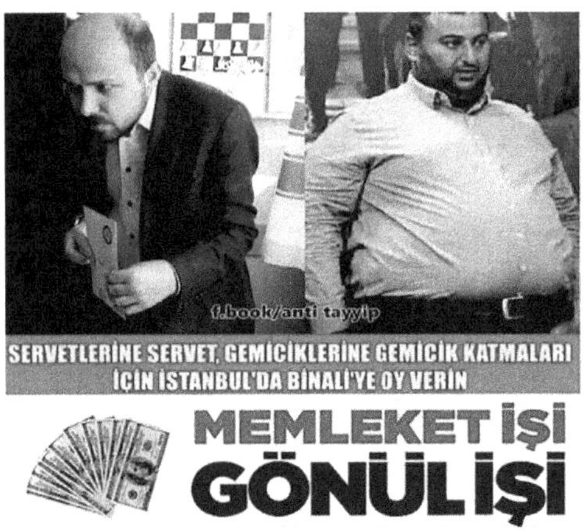

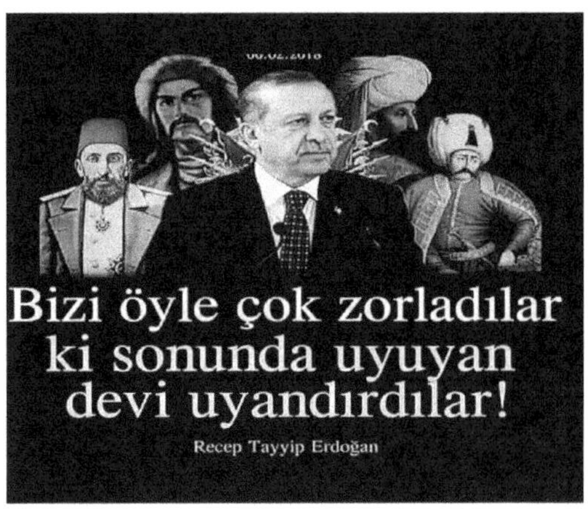

CONCLUSION: THE ILLUSION OF TRUTH VERSUS THE TRUTH OF ILLUSION

An explanation of the post-truth discussed in this study can be found in the movie *Prestige,* based on a story around magicians: "You don't really want to know. You want to be fooled." Volunteering to be fooled took one step further with the advent of social media: now masses lend voluntary support to a political system they believe to be a deception and are loyal to the truth of their own deceptions. The magic show is no longer on the stage, it is everywhere and the audience is no longer a passive captive, but actively takes part in the tricks of illusion. So, it is not possible to tell who is real, who is not and a whole social life goes under the influence of magic. Thus begins the story of the post-truth! Therefore, the Erdoğan of the Anti-Tayyip group and the in the group called Erdoğan Lovers are not the same person. But then, these social media images are not the real Erdoğan anyway. Social media Erdoğans are visions emanating from the insight created in the worlds of belief and conviction of those who described them.

Baker (2018) argues that the society we are forced to live in is a society of conviction like the ancient Greeks who advocated *doxa* against *episteme.* In his view, sociology shifted to doxology and is interested in the production of emotions. Now, way beyond the objective truth, relations with self-existence take place within a range of convictions. In such a society, life is not experienced in a universe of objects, but it is experienced, and at the same time stuck, in a universe of images. Otherwise, it will take a long time to face the 'truth' that life is but an illusion and it will burden the future with an eschatological episteme.

The most important anguish and tension in this age is to defense of the truth. The fact that social reality games controlled by images avoid the touch of truth explains how come magic and magicians stopped being criminal in the eyes of the masses who are engulfed in magic themselves. Indeed, in this age, shouting that the "King is naked!" is the greatest risk, because the image visualized in the eyes of the society indicates that each eye puts another dress on that nakedness. The problem is not missing the nakedness, noticing is the problem. As the Matrix character who betrays the true hero Neo says, everyone knows that the steak in their plate does not really exist. But no one cares as long as the taste is real. Similarly, the groups of opponents or supporters who seem to be interested in who Erdoğan really is in fact only indulge in titillating their own desires and emotions. It is understandable that the reflections of truth on the senses and emotions are misguiding among the suspicious attitudes of episteme. Therefore, the course of philosophy is defined by the idea that a single and mighty truth in the universe of illusions continuously generates and re-generates itself and that *Telos* means to reach the truth breaking through misconceptions. However, it may seem impossible to philosophize on a society where illusion turned into truth and a sequence of sacred rituals settled at the core of the societal.

As noted from the very beginning, the main concern or the subject of investigation of the study is not Erdoğan or any other particular leader for that matter. The main concern is to open an academic window to a cosmology of illusions full of Erdoğan's virtual representations. The polarized distortions seen in the example of Erdoğan undeniably includes the most essential truth of everyday politics in Turkey. However, the relation of Virtual Erdoğan with the outcome of this study points to a strong field of attraction covering the whole society.

Erdoğan as a person is not the question here. Erdoğan is an arena of representation where opposing views and convictions collide. A war of convictions ongoing at different levels between two extremes, those (like somebody from USA) referring to Erdoğan as the Dajjal (Yeni Akit, 2018) on the one end and those (like members from AKP –Justice and Development Party-) seeing him as a Prophet (NTV, 2010) on the other serves to mask their own essence and truth turned into a representation of his person. Consequently, what groups really manufacture is not Erdoğan, but themselves. In their polarized struggle to gain an identity for themselves, they benefit from Erdoğan's political identity that feeds his favorable popular personality and convictions. For this reason, what is more seen even more frequently than Erdoğan in these groups is the arguments the parties choose to legitimize themselves. These arguments are indirectly attributed to Recep Tayyip Erdoğan or his counterpart in the opposition and multi-content hybrid identities are constructed. In these groups, the identity of Recep Tayyip Erdoğan and that of his opponents are portrayed as responsible for all the represented actors and their actions, enhanced and remembered with the positive or negative images of others. This is because the Angels and Demons in these groups copycat each other and throw the exact same accusations against each other. In this way, Erdoğan can fill in the simultaneous parts of a sacrifice offered for the sins of society and a savior to meet divine expectations.

REFERENCES

Ahi, G. (2017). Gerçeklik ötesi (post-truth) ve popülizm. Retrieved from https://digitalage.com.tr/makale/gerceklik-otesi-post-truth-populizm/

Atikkan, Z., & Tunç, A. (2011). *Blog'dan Al Haberi: Haber Blogları, Demokrasi ve Gazeteciliğin Geleceği Üzerine*. İstanbul, Turkey: Yapı Kredi Publications.

Aytaç, A. M. & Demirkent, D. (2017). Hakikatin kendinden menkul bir enerjisi yoktur, hakikati savunmak gerekir. *Ayrıntı Dergi*, Retrieved from http://ayrintidergi.com.tr/ahmet-murat-aytac-hakikatin-kendinden-menkul-bir-enerjisi-yoktur-hakikati-savunmak-gerekir/

Baker, U. (2017). *Beyin Ekran*. İstanbul, Turkey: İletişim Publications.

Baker, U. (2018). *Kanaatlerden İmajlara: Duygular Sosyolojisine Doğru*. İstanbul, Turkey: Birikim Publications.

Baudrillard, J. (1998). *Simulakrlar ve Simülasyon. Oğuz Adanır (Trans.)*. İzmir, Turkey: Dokuz Eylül Publications.

Bauman, Z. (2016). *Postmodern Etik. Alev Türker (Trans.)*. İstanbul, Turkey: Ayrıntı Publications.

Berger, J. (2018). *Bir Fotoğrafı Anlamak. Geoff Dyer (Prep.), Beril Eyüboğlu (Trans.)*. İstanbul, Turkey: Metis Publications.

Donath, J. S. (1998). Identity and deception in the virtual community. In P. Kollock & M. Smith (Eds.), *Community in Cyber Spaces*. London, UK: Routledge.

Forstenzer, J. (2018). Something has cracked: post-truth politics and Richard Rorty's postmodernist Bourgeois liberalism. In Tony Saich (Ed.), *Ash Center Occasional Papers Series*, Cambridge, MA: Harvard Kennedy School Publication.

Freund, G. (2016). *Fotoğraf ve Toplum. Şule Demirkol (Trans.).* İstanbul, Turkey: Sel Puplications.

Fuchs, C. (2014). Social media and the public sphere. *TripleC, 12*(1), 57–101. doi:10.31269/triplec.v12i1.552

Fuchs, C. (2017). Google kapitalizmi. In F. Aydoğan (Ed.), *Yeni Medya Kuramları* (pp. 71–83). İstanbul, Turkey: Der Publications.

Glaveanu, V. P. (2017). Psychology in the post-truth era. *Europe's Journal of Psychology, 13*(3), 375–377. doi:10.5964/ejop.v13i3.1509 PMID:28904590

Hall, S. (1999). İdeolojinin yeniden keşfi: medya çalışmalarında baskı altında tutulanın geri dönüşü. In M. Küçük (Ed.), *Medya, İktidar, ideoloji* (pp. 77–126). Ankara, Turkey: Bilim ve Sanat Publications.

Harsin, J. (2018). Post-truth and critical communication. In *Oxford Research Encyclopedia of Communication* (pp. 1–36). Oxford, UK: Oxford University Press. doi:10.1093/acrefore/9780190228613.013.757

https://www.facebook.com/anticiler/

https://www.facebook.com/ErdoganaSevdalilar/

https://www.aa.com.tr/tr/gunun-basliklari/cumhurbaskani-erdogan-eger-racon-kesilecekse-bu-raconu-bizzat-kendim-keserim/888565

https://en.oxforddictionaries.com/word-of-the-year/word-of-the-year-2016

https://tr.sputniknews.com/ortadogu/201810231035793591-dunyanin-en-etkili-musluman-lideri-erdogan/

Imai, T. (2017). The collapse of faith in policy and the establishment. My Vision, 31. Retrieved from http://www.nira.or.jp/pdf/e_myvision31_A.pdf

Jenkins, H. (2006). *Convergence Culture: Where Old and New Media Collide.* New York, NY: New York University Press.

Jenkins, H. (2017). Medya yöndeşmesinin kültürel mantığı. In F. Aydoğan (Ed.), *Yeni Medya Kuramları* (pp. 33–45). İstanbul, Turkey: Der Yayınları.

Karagöz, K. (2018). Post-truth çağında yayıncılığın geleceği. *TRT Akademi, 3*(6), 678–708.

Karataş, Ş., & Binark, M. (2016). Yeni medyada yaratıcı kültür: Troller ve ürünleri caps'ler. *TRT Akademi, 1*(2), 427–448.

Kaya, Z. (2018). Yeni medyaya geçiş sürecinde ajans haberlerinin yapısal dönüşümü (Anadolu Ajansı Örneğiyle). (Unpublished doctoral thesis). Atatürk University İnstitute of Social Science, The Department of Basic Communication Science, Erzurum.

Keskin, S., & Baltacı, F. (2018). Sinemanın Toplumsallaşması ve Sosyal Temsili: Facebook'un 'Kadife Olmayan' Perdesi. In M. G. Genel (Ed.), *İletişim Çağında Dijital Kültür* (pp. 163–196). İstanbul, Turkey: Eğitim Yayınevi.

Keyes, R. (2004). *The Post-Truth Era: Dishonesty and Deception in Contemporary Life*. New York, NY: St. Martin's Press.

Kırık, A. M., & Saltık, R. (2017). Sosyal medyanın dijital mizahı: İnternet meme/caps. *Atatürk Journal of Communication*, *12*, 99–118.

Manovich, L. (2012). Media after software. Retrieved from http://manovich.net/index.php/projects/article-2012

Maxwell, L. (2016). Donald Trump's Campaign of feelings. Retrieved from http://contemporarycondition.blogspot.com/2016/07/donald-trumps-campaign-of-feeling.html

McCombs, M. E., & Shaw, D. L. (1972). The agenda setting function of mass media. *Public Opinion Quarterly*, *36*(2), 176–187. doi:10.1086/267990

Modreanu, S. (2017). The post-truth era? *HSS*, *3*, 7–9.

NTV. (2010). Peygamber diyen Ak Partili'ye ihraç. Retrieved from https://www.ntv.com.tr/turkiye/peygamber-diyen-ak-partiliye-ihrac,tcyQEPpYpEKHylMB1Qixfw

Open Society Institute. (2018). Common sense wanted resilience to 'post-truth' and its predictors in the new media literacy index 2018. Retrieved from http://osi.bg/downloads/File/2018/MediaLiteracyIndex2018_publishENG.pdf

Özcan, M. (2018). Öznenin Ölümü: Post-Truth Çağında Güvenlik ve Türkiye. INSAMER, (January), 1-11.

Özmen, S., & Keskin, S. (2018). Sosyal medyada öz-temsil ve 'ötekiliğin' öteki boyutu: Karikateist toplumsalı üzerine inceleme. *IntJCSS*, *4*(2), 533–558.

Palma, A. (2017). 'When the future catches up with us, the past will no longer be valid' Descartes could be a yardstick. *Uno Magazine*, *27*, 17–19.

Şimşek, V. (2018). Post-truth ve yeni medya: sosyal medya grupları üzerinden bir inceleme. Global Media Journal Turkish Edition, 8 (16), 1-14.

Sontag, S. (1993). *Fotoğraf Üzerine. Reha Akçakaya (Trans.)*. İstanbul, Turkey: Altıkırkbeş Yayınları.

Suiter, J. (2016). Post-truth politics. *Sage Journals*, *7*. doi:10.1177/2041905816680417

Taniguchi, M. (2017). Confronting to 'Post-truth Era'. My Vision, 31. Retrieved from http://www.nira.or.jp/pdf/e_myvision31.pdf

Yanık, A. (2017). Popülizm (V): Post-Truth. Retrieved from http://www.birikimdergisi.com/haftalik/8463/populizm-v-post-truth#.XM4S6I4zZPY

Yeni Akit. (2018). *ABD'lilerin uykuları kaçtı! 'Erdoğan'ı Deccal olarak görüyorlar!'*. Retrieved from https://www.yeniakit.com.tr/haber/abdlilerin-uykulari-kacti-erdogani-deccal-olarak-goruyorlar-465167.html

Zarzalejos, J. A. (2017). Communication, journalism and fast-checking. *Uno Magazine*, *27*, 11–13.

Žižek, S. (2005). *Yamuk Bakmak. Tuncay Birkan (Trans.)*. İstanbul, Turkey: Metis Publications.

ADDITIONAL READING

Brannon, L., & Brock, T. (1994). The subliminal persuasion controversy: reality, enduring fable, and Polonius's weasel. In Psychological Insights and Perspectives. S. Shavitt ve T. C. Brock. (Eds.). Persuasion: Allyn & Bacon.

Messaris, P. (1994). *Visual literacy*. Boulder, CO: Westview Press.

Mirzoeff, N. (2013). *The visual culture reader* (3rd ed.). New York: Routledge.

Mitchell, W. J. T. (1994). *Picture theory*. Chicago: University of Chicago.

KEY TERMS AND DEFINITIONS

Agenda-Setting: It is the founding mechanism that determines the common issues that people or groups talk about and share and is the process of creating a reality in which users are involved in social media.

Post-Modern: It is an age of phenomena where images replace reality, and the difference between existence and appearance becomes ambiguous.

Post-Truth: It includes a collective culture which is personal convictions and emotions are more efecctive than objective reality. Participants/partisans reproduce numerous truths by participating in political discourses.

Recep Tayyip Erdogan: President of Turkey Republic and popular politician who appeals to a wide audience. In this study, he is the people that is built on the post-truths of polar virtual groups.

Social Media: It is a digital communication and socialization environment that gives to individuals the power to restore reality.

Social Polarization: Social groups to be parties to around different convictions and to take sides against each other with sharp attitudes.

Turkey's Politic Culture: It is structure which includes today's political tradition and relationships in Turkey as a result of accumulation of thousands of years.

Virtual Community: It is a form of community where digital interactors come together.

ENDNOTE

1 The respective figures of membership were recorded on 09.03.2019.

Chapter 9
Is Somebody Spying on Us?
Social Media Users' Privacy Awareness

Şadiye Deniz
Ege University, Turkey

ABSTRACT

One of the concepts that have a strong and dominant effect in transforming the culture, individual, and society of social media has been privacy. Everything that belongs to our domestic space in modern times, which should not be known/seen by others, is made public by ourselves in the postmodern age with new media tools. In social networks focusing on vision and surveillance, privacy is restricted, eliminated, or stretched by individuals themselves for the creation of ideal profiles. The privacy settings that a person thinks are under his control seriously affect the way he uses social media. This chapter will try to determine which subject/situation/images are perceived as intimate among university students, and how the boundaries of social media and privacy are drawn and transformed. The study is based on the assumption that the level of privacy awareness and the level of knowledge control influence the quality and frequency of social media sharing of users.

DOI: 10.4018/978-1-7998-1041-4.ch009

INTRODUCTION

According to social-media usage statistics released in 2018, there are 51 millions of active social media users in Turkey (https://digitalreport.wearesocial.com/). Such a ubiquitous usage of social media has emerged as an inevitable result of the ease/cheapness of tools that make one's appearance on these domains comfortable as well as the increasing domination of attitudes such as self-formation, visibility, watching/being watched by others. Social media and tools that can connect us with such domains have almost become our bodily organs with a similar familiarity and practicality of usage in real life. Aside from this fun aspect social media also offers another salient feature by allowing us to be watched by other third parties– which we may volunteer at times - Hence the aim of this study is to identify which issues/cases/images are considered private among university students and how the boundaries of privacy are drawn and transformed by social media. It is thus another objective of this study to exhibit whether or not sampling group had anxiety for privacy when using social networks and to detect if a correlation existed between anxiety for privacy and social-media usage. Via conducting random sampling technique in this study, a questionnaire was administered among Faculty of Communication students in Ege University. This study is based on the assumption that privacy awareness and information control level influenced frequency of users' social media sharing and quality of the news they share.

RELEVANT LITERATURE ON PRIVACY CONCEPT AND ITS RELATION WITH SOCIAL MEDIA

Privacy is a concept that varies from person to person and in which individuals are not to specify its borders. Rather it is determined by socio-cultural sphere as well. As a result of being a domain in which all human perceptions on privacy intersect with the notion of privacy originating from the culture of one's member society, it is viable to argue that there is a corresponding variety and subjectivity in privacy concept equal to the number of members in a community. Yüksel, while emphasizing this ambiguity, stated that "it is remarkably difficult to define privacy and set its boundaries" (Yüksel, 2003, p. 78). Actions that we hide from others, or we want others not to view but we practice in our private life people we know, issues and information can engulf our privacy circle. That being the case it seems unviable to reach a consensus on this concept. Nevertheless Şener (2013) reported that there are certain common grounds and he continued such;

What constitutes privacy? The answer to this question may differ with respect to the variables such person, gender, age, social class, member culture, social environment, social status hence it seems impossible to provide only one definition of privacy. Still there are also certain consensus points; in societies that experienced modernity human body, specific organs in body, sexuality, family, romantic relations, home and inner-home objects, many goods that we personalize and would rather hide from others or reveal to others in a controlled way may constitute the framework for privacy.

This concept is the derivative of Arabic word 'haram' (unlawful) and refers to "sincere, intimate, hidden from others, concealed, secret" (Göle, 1992, p. 128, Bağlı, 2011, p. 184). From all of these definitions it is viable to argue that privacy decisively emerges while contacting and communicating with the third parties and it is investigated in the transmission from private to public realms. From the moment being born, we as humans have felt no need to hide within our private realm but whenever we encounter with others, we cultivate anxiety for privacy and draw our boundaries. Claiming that privacy entails an anxiety related to encountering with others and self-presentation Irwin Altman delineates this concept as "a selective control process" (Yüksel, 2009, p. 278). Altman continues such;

In contrast to generic or common perception it can be argued that individuals not only seek privacy but at the same time they aim to establish human relations and they may willfully share personal information during any social interaction. In another saying, they not only want to exclude others and seek solidarity, but also desire to spend time with others. With this quality, privacy can be defined as a dialectic playground between the desire of being on one's own and being with others. (Gifford, 1997, pp. 173-175, Yüksel, 2009, p. 278).

The process that offered a contextual meaning to privacy that specifies the desire to be alone or be with others has emerged with modernism. The reason why privacy had not such a loaded meaning before modernizm is defined by Yüksel (2003, p. 182) such;

...in order to evoke a private realm before everything else, it is essential to foreground "individual" as a concept. In pre-modern or traditional societies in which individuals are deemed as members of the larger community or group and with not an independent entity or identity from the social composition they live in it is almost impossible to mention "individual" and "private life realm" or "private realm" of the individual.

Modernism, with the strict boundaries it drew between public realm/private realm, brought with itself our questions about what to keep in private realm or what to expose in public realm. Toprak vd. (2014, p. 140), Noting that architecture also contributed to the birth of privacy in modern sense, it can be argued that enclosed setting of inner-home area would help to be away from being monitored hence the individual could be free to perform behaviors hidden from public realm thereby giving birth to privacy.

In modern age the sharp boundary that modernism draws between public realm and private realm has been removed via new media tools. In another saying controlling this boundary has totally been initiated by the person. An individual alone can command security settings in both displaying inner-home realm to people outside and be invisible in public realm. Widespread use of new media tools and their full integration with daily life has, just as the case for our beliefs in many other cases, transformed our privacy perception too. David Lyon (2013) argues that our consent and desire to watch and being watched has emerged prior to the digital age with the spread of mass communication tools:

In this modern age, watching has been a phenomenon that influenced daily life, thus they have become a solid part of popular culture by unexpected exposure in novels, songs, movies and other communication tools and domains. That being the case, peeking, stalking, monitoring, sharing the views of stalkers and our understanding of watching have been formed partially by popular communication tools. Consumers of these tools, at the same time, have become the subjects of watching experience too. (Lyon, 2013, p. 223).

With the introduction of social media, watching and being watched has become legitimate and desirable situations. Hall Niedzviecki (2010) argued that in this modern age, a culture of peeking has become dominant. Niedzviecki also stated that; "it is as if there is a hypnotizing thought secretly whispered to our ear and same words are reiterated: you need to know and be known" (Niedzviecki, 2010, p. 11) so as to explain the instinct that makes peeking so intense. Zgymunt Bauman and David Lyon on the other hand reported that in near future, the end awaiting those minority people not having a social media account is like a 'social suicide' (Bauman & Lyon, 2013, p. 36).

In literature review on social media and privacy it is witnessed that most of the studies focused on the way social media transformed privacy conceptions of users (Dinev vd, 2009; Proudfoot vd., 2018; Beigi & Liu, 2018; Chou vd, 2019; Pingo & Narayan, 2018; Tifferet, 2018; Oz, 2014; Zengin vd, 2015). Dinev et al.'s research of which scale has been used in this study consisted of a questionnaire entailing 218 social media users and a new model was then suggested to measure how users' privacy perception has evolved. It was concluded that perceived information control and perceived trust are remarkably effective in drawing the borders of privacy in social media (Dinev vd, 2009: p. 7). In the study of Proudfoot et al. among 244 Facebook users (Proudfoot vd, 2018, p. 16), impression management (perception/impression management) referred to the balanced state between the benefits of social media and protecting privacy. In its Turkish equivalence, it is called privacy settings. According to a research the trust confided in a social network provider impacted one's worries on privacy, social benefits and perceived impression management relations.

Beigi and Liu in their research (2018) analyzed social media and privacy concepts with respect to two dimensions as the vulnerabilities and risks of social media. In the study conducted by Chou (2019) et al. on social-media usage and privacy perception the attitudes of 9-17 year old students in Taiwan were investigated. In the research of Pingo and Narayan on the other hand a structured interview was administered to 21 participants in order to examine how digital prints of social media users were tracked technically. It was then underlined that one's right for privacy is a human right that relies on the need to be on one's own. In a different study of Tifferet (2019), a research was conducted on the gender differences regarding the search for privacy in social network sites. Findings of the research demonstrated that compared to male users in social network sites, women were more disposed towards anxiety for privacy and secrecy. There was a wide-open gender difference in performances such as activating privacy settings and tagging the photos, privacy concerns and sharing of personal information (Tifferet, 2019: 1).

In Turkish literature there is a myriad of studies on the same issue (Toprak vd, 2014; Korkmaz, 2013; Aydın, 2013; Zengin vd, 2015; Oz, 2014; Akyazı, 2018; Barkuş & Koç, 2019). Zengin et al., with a questionnaire, examined privacy notion of 145 university students who also used Facebook. In their study it was concluded that participants were aware that shared information on Facebook could lead to negative results but they were not fully informed about the dimensions and origins of these potential risks. In a different study Oz analyzed concerns of 373 university students about the exhibition of private life on social media. He concluded that there was a positive relation between young people's usage of social media and other media tools and their awareness on privacy problems. (Oz, 2014: 6099) A heightened awareness level led the users to conceal their private profile settings. In a different study (Akyazı, 2019), private sharings of 40 celebrities on their Instagram accounts were analyzed and it was then concluded that although celebrities tended to share their privacy followers had no motivation to follow these private sharings. Barkuş and Koç's study (2019) has been another critical one that compiled studies on digital privacy as for Turkey.

Studies in relevant literature manifested that social network sites have triggered a vital change on the notion of privacy. Those who meet on social sites for the first time have no concerns on privacy but the moment they become aware of the risks in these sites the characteristics of their sharing may alter. In the same vein the aim of this study is to investigate if university students are aware of the risks in social sites and the way these risks can impact their preferences in sharing.

METHODOLOGY

Method and Sampling

Population of the study consists of Faculty of Communication students in Ege University. The motives for choosing university students as the population are their attraction with new technologies and being the majority portion of social media users. 164 students answered the questionnaire, 10 questionnaires were deemed invalid thus the survey was administered to 154 students. Convenience sampling method was selected. In the study Dinev et al's original work (Dinev, Heng Xu Pennslylvania, H. Jeff Smith'in "Information Privacy Values, Beliefs and Attitudes: An Empirical Analysis of Web 2.0 Privacy") was translated by experts and upon conducting equivalence (cultural compatibility) analysis the questionnaire form was compiled. In addition while questionnaire form was being formed Turkish literature was also examined (Zengin vd., 2015, Oz, 2014) and a 7-dimensional questionnaire form was devised. After forming of the questionnaire form a pretest for the form was administered to 50 people. Pretest data were analyzed via SPSS 15.0 program and at the end of the analysis it was detected that cultural adaptation of the scale was reliable(C. Alpha = 0,830). Data collection process was resumed in light of this finding. Dimensions on the questionnaire form are such; motives of users for using the social media, features of their sharings on social media, their knowledge awareness/anxiety for privacy regarding social sites, trust towards social media sites, perceived information control, exposure to an adverse experience and opinions on social media sites.

Prior to conducting the analysis reliability of the items was measured and Cronbach's Alpha value of the scale with 46 items was computed as 0,830. In social-science studies Cronbach Alfa value must be 0.70 and above, hence it is safe to claim that the scale is reliable (Hairc vd, 2014, p. 107; Kayış, 2010, p. 405)

In order to administer parametric tests, data set is expected to manifest a normal distribution. Hence, before the analysis, normalcy of data set was checked via measuring their Kurtosis and Skewness values. Thus, to accept the data to be in normal distribution, their kurtosis and skewness values must fit into +-- 1,5 range or +-- 2 range (Tabachnick & Fidell, 2007, D. George & Mallery, 2010, Karaatlı, 2010, p. 6). Findings of normalcy test posit that distribution of data is normal and applicable for parametric tests. In light of these data, the aim of this study is to seek answers to the questions below:

- What is the frequency and motives for using social sharing platforms among participants?
- What kind of sharings do these participants post on social networks?
- Have the participants ever experienced an adverse event on the internet/social media?
- What kind of a statistical relation exists between experiencing an adverse event and anxiety for privacy?

- Is there a gender-based difference with respect to anxiety for privacy?
- What are the views of participants on social networks and digital domains?
- Is there a gender-based difference with respect to the dimensions of questionnaire form?
- With whom do these participants share the information on their social sharing platforms?

Research Findings

As the findings of this research are examined it surfaces that as seen in the beginning in Table 1, gender-based distribution of descriptor statistical information was analyzed followed by gender-based distribution of the 7 dimensions. In order to analyze the relation between gender variable and dimensions independent sampling T test was administered. One Way Anova test was administered to measure the relation between dimensions and factors such as; actively used social media platforms, length of subscription on social media platforms, average time spent on the internet in a day and daily time-segments spent on the social media. In order to determine the difference between groups, Tukey test was administered and significance was accepted as p<0,05. In Table 1 descriptor statistical data are presented.

Table 1. Descriptor statistical data

	N	%		N	%
Sexuality			***Subscribed social sharing platforms***		
Female	68	44,2	Facebook	136	88,3
Male	86	55,8	Twitter	113	73,4
Age			Instagram	132	85,7
16-20	101	65,6	WhatsApp	142	92,2
21-25	47	30,5	Other	69	44,8
26-30	4	2,6	***Number of friends on Social media***		
30 ve üstü	2	1,3	Below 50 friends	3	1,9
Department			50-99 people	17	11,0
Journalism	70	45,45	100-199 people	26	16,9
RTS	84	54,55	200-299 people	21	13,6
Which tool used to access social media			300 and above	87	56,5
Mobile phone	144	93,5	***How long being a social media member***		
Tablet	1	0,6	Less than 1 year	0	0
Computer	9	5,8	1-3 years	22	14,3
How frequently social media is used			3-5 years	34	22,1
Whenever I have a chance	82	53,2	5 years and above	98	63,6
4 times and more in a day	48	31,2	***Whose friendship requests are accepted***		
3 times in a day	12	7,8	Only people I known	94	61,0
1 time in a day	6	3,9	Everyone that I have mutual friends	29	18,8
A few times in a week	5	3,2	My friends	19	12,3
A few times in a month	1	0,6	Anyone who sends a request	12	7,8

Table 1 reveals that the survey was administered among 68 female and 86 male students. It was conducted among Journalism and Radio-Television-Cinema Department students in Ege University, Faculty of Communication. A vast majority of participants accessed social media via their mobile phone. The highest percentage to the question of how frequent social media is used belonged to 53,2% with the choice whenever I have a chance. Social sharing platforms with the highest membership ratio were respectively WhatsApp, Facebook and Instagram. Among the participants of survey 56% reported to have 300 and higher quantities of friends. 63% of survey respondents reported to use social media for more than 5 years. 61% of survey participants answered to have accepted the requests of only those they knew beforehand whereas only 7% reported to accept the requests of everyone's friendship request.

Relation between Descriptor Statistical Data and Dimensions of the Questionnaire

After descriptor statistical data the relation between gender variable and seven dimensions of the questionnaire was investigated. T-test was administered to measure the relation between gender variable and dimensions among male and female groups ($p>0,05$). In Table 2 there is an analysis to see if a difference existed among male and female participants with respect to the dimensions of scale.

Table 2 shows that between male and female participants not a significant difference existed with respect to dimensions of; 'motives for using social media', 'exposure to an adverse experience' and 'opinions on social media'. Among these differences the most obvious one is in the dimension of "exposure to an adverse experience' of the statements "the things that I wanted to keep private were shared' (p: 0,039) and 'my accounts were hacked' (0,048) regarding the difference with respect to gender ($p<0,05$). Average score of the statement 'The things that I wanted to keep private were shared' was 1,81% among females and 2,24% among males. Average score of the statement 'My accounts were hacked' was 1,85% among female participants and 2,27% among males. Other average scores are almost identical and it was identified that compared to females males had larger quantities of adverse experiences.

In line with the objectives of this study the relation between other items of the descriptor statistical data and dimensions of the scale was analyzed. Results of One Way Anova analysis on the relation between subscribed social media platform and dimensions of the scale are as below:

Table 2. An analysis on the difference with respect to the Dimensions of Scale among Male and Female Participants Results of T –test

	Female Participants			Male Participants			
	N	Ort.	S.S.	N	Ort.	S.S.	p
Motives for using social media	68	3,23	0,75	86	2,98	0,76	0,041
Features of Sharings	68	3,11	0,87	86	3,11	0,84	0,992
Information Awareness/Anxiety for privacy	68	3,72	0,87	86	3,56	0,97	0,291
Perceived trust	68	2,28	0,84	86	2,57	0,98	0,060
Perceived information control	68	3,45	0,900	86	3,27	0,100	0,235
Adverse Experience	68	2,02	1,02	86	2,35	0,904	0,035
Opinions on SM	68	2,89	0,78	86	3,17	0,900	0,041

Table 3 indicates that in the very first three dimensions in Twitter a significant difference was measured. In terms of motives for using social media and features of sharings Twitter is deemed to be a strongly politicizing domain, hence obtained result is attributable to this finding. In Table 4 detailed information on anxiety for privacy dimension is given. In WhatsApp the only significant difference was measured among the groups in the dimension of perceived information control dimension. WhatsApp users, compared to nonusers, reported to have a greater control in managing their social sharing platform. These users believe that by employing privacy settings and private domain certain domains are under their authority.

An analysis of Table 4 reveals that Twitter users constitute the group believing that privacy can be violated despite their disapproval.

Among the descriptor statistical data an analysis of the relation between the variable of frequency in social media usage and questionnaire's dimensions displays that there is a significant difference among groups only with respect to the dimension of motives for using social media (p: 0,000). It then became obvious that people using social media at every opportunity or 4 times or above and 3 times in a day

Table 3. One way Anova table on the relation between subscribed social media platform and dimensions of the scale

Dimensions of the Scale	Facebook p value	Twitter p value	Instagram p value	WhatsApp p value
Motives for using Social media	,065	,000	,000	,647
Features of Sharings	,094	,006	,011	,247
Anxiety for privacy/ Awareness	,454	,008	,439	,963
Perceived trust towards Social media	,317	,359	,535	,602
Perceived information control	,207	,397	,088	,017
Adverse Experience	,853	,694	,315	,975
Opinions on SM	,719	,502	,563	,845

Table 4. Relation between awareness/anxiety for privacy and used social sharing platform via one way Anova results

	Twitter N:113 Mean	Facebook N: 135 Mean	İnstagram N: 132 Mean	Whatsapp N: 141 Mean
Social media sharings may threaten my personal security.	3,83	3,81	3,85	3,80
Social media applications may access my personal information.	3,70	3,58	3,53	3,54
If I sign privacy setting of one of my social media sharing as *available for everyone*, I am aware that everyone on the Internet can see my sharing.	4,01	3,89	3,86	3,82
Although I may activate privacy settings on my account, others can still access my social media sharings.	3,49	3,33	3,31	3,32
Social media sharings may cause trouble for us.	4,04	3,98	3,98	3,96
My personal information may be used in inappropriate ways/ contexts.	3,46	3,36	3,43	3,39

achieved higher mean scores in their motives for using social media or in other words a wider diversity could be measured in their motives for using social media.

The relation between the number of friends in their most commonly used social media account and scale's dimensions reveals that a significant difference existed among groups with respect to dimensions of motives for sharing on social media (p: 0,004) and features of sharings (0,031). Anova table manifests that the higher the number of friends the higher is the diversity in motives for social-media usage (commerce etc.) and likewise the higher the number of friends the higher is the number of sharings related to private realm (religious faith, contact information, political view).

An analysis of the relation between total hours spent one day on the internet and scale manifested that a significant difference existed among groups only with respect to anxiety for privacy dimension. (p: 0,042) Accordingly the shorter time spent on the internet the lesser trust was confided in social sharing platforms and those who spent shorter time had greater levels of anxiety for privacy. It was also witnessed that those having spent longer time had smaller levels of anxiety for privacy compared to those spending 30 minutes and 30 minutes to 60 minutes. (Mean values: 4,16 for those users shorter than 30 minutes. For others it is around 3).

Regarding the question of the length of subscription to social media platforms a significant difference was measured in the dimension of perceived information control (p: 0,025). As their mean scores are analyzed it was evidenced that those who subscribed longer than 5 years had a stronger belief that they were the ones controlling this domain. Mean scores revealed that those who have been members longer than 5 years felt more competent in controlling their social media. It was also witnessed that those who have been members for 1-3 years had weaker belief in exerting their control.

As for the question about whose friendship request on social media was accepted, not any significant difference was measured between male and female genders. Mean score of female participants was 1,54%, male participants' mean score was; 1,91%. Sig: 0,057. 69,1% of women reported to accept the invitation of only people they know while 54,7% of men agreed with this statement. 4,4% of women reported to accept all friendship requests while 10,5% of men agreed with this statement.

Relation between Dimensions of the Questionnaire and Gender

In the data analysis of this research upon examining the relation between descriptor statistical data and dimensions of the questionnaire, comprehensive One Way Anova tables were drawn to expose the relation between dimensions of the questionnaire and gender.

Table 5 reveals that with respect to statements; 'I use social media to be updated about my friends and my social group' and 'I use social media to share my photos/videos' there was not a significant difference between male and female participants. In terms of both statements alike it was seen that compared to males, females used social media for a higher diversity of purposes. One Way Anova table about the relation between features of sharings dimension and gender is as exhibited below.

In this dimension the only significant difference was measured in 'I feel comfortable in sharing my contact information on social media' statement. Male participants were witnessed to be less anxious towards sharings posts compared to females. One Way Anova table on the relation between anxiety for privacy/awareness dimension and gender is as displayed below:

Table 5. Relation between dimension of motives for using social media (1st dimension) and gender one way Anova

Statements in the Dimension of Motives for using Social media	Female Mean	Male Mean	p value
I use social media to update my personal profile.	3,78	3,43	,078
I use social media to be updated about my friends and my social group.	4,07	3,71	,037
I use social media to post my views on current politics and agenda.	3,43	3,15	,248
I use social media to share my photos/videos.	4,24	3,60	,001
I use social media to tag the places I visited.	3,07	2,71	,106
I use social media to notify my status/mood.	2,40	2,49	,679
I use social media to shop/trade.	3,10	2,71	,081
I use social media to share all the happenings in my life.	1,81	2,10	,131

Table 6. Relation between features of sharings dimension and gender one way Anova

	Female Mean	Male Mean	p
I feel comfortable in sharing my religious belief on social media.	3,78	3,55	,289
I feel comfortable in sharing my political view on social media.	3,51	3,22	,220
I feel comfortable in sharing my contact information on social media.	1,91	2,86	,000
I feel comfortable in sharing my personal interests on social media.	3,94	3,67	,190
I feel comfortable in sharing my inner-home views on social media.	2,41	2,58	,425
I feel comfortable in sharing my family images on social media.	3,19	2,83	,091

Table 7. Relation between Anxiety for privacy/Awareness dimension and gender one way Anova

	Female Mean	Male Mean	p
Social media sharings may threaten my personal security.	3,97	3,71	,140
Social media applications may access my personal information.	3,46	3,60	,462
If I sign privacy setting of one of my social media sharing as *available for everyone*, I am aware that everyone on the Internet can see my sharing.	3,94	3,69	,228
Although I may activate privacy settings on my account, others can still access my social media sharings.	3,31	3,31	,981
Social media sharings may cause trouble for us.	4,15	3,84	,078
My personal information may be used in inappropriate ways/contexts.	3,53	3,27	,226

As for the relation between anxiety for privacy and gender not any significant difference was measured in any of the statements. Yet compared to male respondents the answers provided indicated that as for mean scores females were more aware of the fact that on social media privacy could be violated. Detailed table on the relation between perceived trust towards social media and gender is as given:

Table 8 posited that a significant difference existed with respect to 'I feel okay if social media sites obtain my personal information' statement. Compared to women men felt less uncomfortable. Detailed table on the relation between the dimension of perceived information control on social media and gender is as below:

Table 9 delineates that in the statement; 'Some people on my social media friend-list are either restricted or blocked' there is a significant difference. Women were reported to use restriction and blocking feature more frequently than men. Relation between the dimension of exposure to an adverse experience and gender is in Table 10.

Table 10 reveals that a significant difference existed in the statements of 'My account was hacked by others' and 'The things I wanted to keep private about my personal life were shared'. Although men had more adverse experiences on this issue compared to women they still had less anxiety for privacy.

Table 8. Relation between perceived trust dimension and Gender one way Anova

	Female Mean	Male Mean	p
I feel okay if social media sites obtain my personal information.	1,74	2,41	,001
I believe social media sharings are harmless.	2,66	2,65	,957
I believe privacy settings on social media are capable of protecting my privacy.	2,47	2,66	,374

Table 9. Relation between the dimension of perceived information control on social media and gender one way Anova

	Female Mean	Male Mean	p
I believe I take control about accessibility to my personal information on social media.	2,82	2,91	,706
I can restrict accessibility of others to my personal information via privacy settings.	3,33	3,24	,671
I know anytime I want I can hide some of the information on my social media accounts.	3,87	3,63	,184
Some people on my social media friend-list are either restricted or blocked.	3,81	3,33	,021

Table 10. Relation between the dimension of exposure to an adverse experience and gender one way Anova

	Female Mean	Male Mean	p
I had negative experiences because of my social media interactions.	2,12	2,28	,424
The things I shared on social media were used without my permission.	2,06	2,40	,099
The things I wanted to keep private about my personal life were shared.	1,81	2,24	,039
A fake account was opened by using the information I shared.	2,09	2,28	,339
My account was hacked by others.	1,85	2,27	,048
There are some things I wish I never shared.	2,10	2,51	,073
Some people harassed me by exploiting what they found out about me on social media.	2,06	2,49	,062

The table that shows the relation of male and female participants with respect to their opinions on social media is displayed below.

Table 11 displays that a significant difference exists between male and female participants for the statement of 'None of the things shared on social media make me feel uncomfortable'. In accordance with the aims of this study Pearson Correlation Analysis of Anxiety for Privacy, Exposure to an Adverse Experience, Perceived Trust and Perceived Information Control dimensions was conducted as displayed in Table 12.

As seen in Table 12 too correlation analysis among dimensions revealed that between perceived information control and perceived trust there is a positive-way significant relation. In other terms it is evident that when social media users are knowledgeable about the security of these websites and know that they can eliminate security vacuums by exerting control there is a heightened level of perceived trust towards these sites. It was also seen that a negative-way significant relation dominated between anxiety for privacy dimension and perceived information control. In the same vein when they believe that information control is not under their authority users' anxiety for privacy climbs. It also became clear that a negative-way significant relation existed between anxiety for privacy and perceived trust. That being said as anxiety for privacy increased, a decrease was measured in perceived trust dimension.

Finally participants were asked about with whom they shared personal information on social media and the results are as shown in Table 13.

Table 13 exhibits that personal information is mostly shared with one's friends. Another striking point is that in addition to contact information they saw no harm in sharing private information with friends

Table 11. Relation between the Dimension of Opinions on Social media and Gender one way Anova

	Female Mean	Male Mean	p
None of the things shared on social media make me feel uncomfortable.	3,12	3,26	,555
I consider some of the things shared on social media as an exposure of private life.	3,07	3,58	,245
There must be an auditing system in all social media sites for sharing.	3,59	3,48	,606
I trust in social media sites not to expose any personal information without my consent.	2,62	3,02	,058
I feel totally okay with all social media sharings.	2,07	2,56	,022

Table 12. Pearson correlation analysis of anxiety for privacy, exposure to an adverse experience, perceived trust and perceived information control dimensions

	Anxiety for privacy/ Awareness	Adverse Experience	Perceived trust towards Social media	Perceived information control
Anxiety for privacy/ Awareness	1	0,078	-0,231**	-0,026
Adverse Experience	0,078	1	0,068	-0,042
Perceived trust towards Social media	-0,231**	0,068	1	0,442**
Perceived information control	-0,026	-0,042	0,442**	1

**. Correlation is significant at the 0.01 level (2-tailed).

Table 13. With whom personal information on social media were shared

	Everyone		My friends		Friends of my friends		Only me		Only people I choose (private settings)	
	N	%	N	%	N	%	N	%	N	%
Contact information	10	**6,5**	82	**53,2**	4	**2,6**	43	**27,9**	15	**9,7**
Photographs /videos	31	**20,1**	108	**70,1**	6	**3,9**	0	**0**	9	**5,8**
Location	15	**9,7**	100	**64,9**	10	**6,5**	20	**13,0**	9	**5,8**
Political view	29	**18,8**	65	**42,2**	7	**4,5**	41	**26,6**	12	**7,8**
Religious belief	34	**22,1**	69	**44,8**	3	**1,9**	38	**24,7**	10	**6,5**
Favorite opinion column/news link	35	**22,7**	93	**60,4**	7	**4,5**	14	**9,1**	5	**3,2**
Relationship status	30	**19,5**	76	**49,4**	5	**3,2**	35	**22,7**	8	**5,2**

such as political view, location or religious belief. The second choice was saved for only me option. The information most commonly shared with everyone is one's favorite opinion column and news link.

CONCLUSION

Social media that has changed our perceptions regarding daily life and all our life events also transformed our privacy boundaries. Established by modernizm the sharp distinctions that defined inner-home domain as private and outside-home domain as public realm have been eliminated. Our public/private realms that appear transitive and inextricably intertwined like a water painting have turned into applications of which privacy settings can be regulated on social sharing platforms upon our will by a simple action of clicking. According to latest data there are about 52 million social media users currently in Turkey (https://dijilopedi.com/2019-turkiye-internet-kullanim-ve-sosyal-medya-istatistikleri/). Such ubiquitous nature of social sharing networks both in the world and in Turkey necessitated the need to be aware of the risks brought by these networks. In this study it was aimed to analyze the digital privacy awareness of university students using social sharing networks and the way this awareness influenced their social-media usage.

Research population consisted of 68 female and 86 male participants. By administering a 7 dimensional questionnaire the relation between students' anxiety for privacy, privacy awareness and social-media usage was analyzed. All of the participants are subscribed to a social sharing platform and more than 56% have friends over 300. WhatsApp is the social sharing platform with the highest percentage of membership.

Between male and female participants a significant difference was measured between their motives for using social media, exposure to an adverse experience and their opinions on social media dimension. Unlike women, men experienced a greater number of adverse experience on social media. Compared to women men reported a greater number of unwanted sharings on their private life and hacking of their accounts. Regardless of this negative outcome it was detected that compared to women, men were inclined to be more open on social media and shared more information on their private life.

Female participants were detected to have a more variety and diversity in their motives for using social media. Although participants used social media to find their friends at most, women reported to use these networks for a variety of purposes. Among all of the subscribed social media platforms it was identified that Twitter had the top ratio in anxiety for privacy dimension. Since Twitter is a more politicized domain this anxiety can be attributed to its inherent nature.

A direct proportion was measured between frequency of social media usage and motives for using social media. Among people using social media at every opportunity or 4 times in a day there was an obvious diversity in their responses about motives for using social media.

It was also determined that as the number of friends increased the variety in their motives for social-media usage (commerce etc.) climbed and in parallel with the rise in number of friends a jump in private-domain sharings was observed in features of sharings dimension. As the time spent on the internet decreased there was less trust towards social sharing platforms and those who spent less time had a higher level of anxiety for privacy. Those who spent longer time was seen to have lower anxiety than the ones spending 30 minutes and 30 minutes-1 hour. Those who have been a member for more than 5 years were reported to have felt more competent in controlling social media. In sum they scored higher values of perceived control.

Among male and female participants there was a higher frequency of responses to the choices 'only those people I know' and 'everyone that I have mutual friends' in answering the question about whose friendship requests were accepted. It was seen that female participants mostly used social media to share their photographs/videos and to find information about their friends. Among male participants the highest ratio was accumulated in the choice of finding information about their friends and social circle.

As for features of sharings dimension, the most obvious difference between male and female participants is sharing of contact information. Compared to men women were more hesitant to share their contact information.

In the last stage correlation analysis among dimensions unveiled the positive-way significant relation between perceived information control and perceived trust. In other words participants who believed that information control on the privacy of social sharing networks is under their authority possessed a higher level of perceived trust towards these networks. Between anxiety for privacy dimension and perceived information control, a negative-way significant relation was detected. Anxiety for privacy is higher among participants fearing that information could be manipulated by these sites. They also chose to share fewer details. The higher perceived information control the lower anxiety for privacy was. In the same vein a negative-way significant relation was measured between anxiety for privacy and perceived trust dimension. Corresponding to higher trust towards social networks, a lower level of anxiety for privacy takes the front stage. As privacy awareness/anxiety climbs trust towards social networks and posted sharings plummet.

Findings of this study demonstrated that compared to male students, female students are more aware of the risks in social sharing networks and hold a strong level of anxiety for privacy. Hence they choose to share less of their private sphere whereas male students feel more open to private sharings. In general it is safe to claim that students are well aware of the fact that information shared on social networks could be used anytime by the third parties.

REFERENCES

Akyazı, A. (2019). Mahremiyetin Dönüşümü: Ünlülerin Instagram Paylaşımları Üzerine Bir Araştırma. Gaziantep University Journal of Social Sciences, 18(1).

Aydın, B. (2013). Sosyal Medya Mecralarında Mahremiyet Anlayışının Dönüşümü. İstanbul Arel Üniversitesi İletişim Çalışmaları Dergisi, 3(5), 131-146.

Bağlı, M. (2011). *Modern Bilinç ve Mahremiyet*. İstanbul, Turkey: Yarın Yayınevi.

Barkuş, F., & Koç, M. (2019). Dijital mahremiyet kavramı ve ilgili çalışmalar üzerine bir derleme. Bilim, Eğitim [BEST Dergi]. *Sanat ve Teknoloji Dergisi*, *3*(1), 35–44.

Bauman, Z., & Lyon, D. (2013). *Akışkan gözetim. Çev., Elçin Yılmaz*. İstanbul, Turkey: Ayrıntı Yayınları.

Beigi, G. & Liu, H. (2018). Privacy in social media: Identification, mitigation and applications. *arXiv preprint arXiv:1808.02191*.

Chou, H. L., Liu, Y. L., & Chou, C. (2019). Privacy behavior profiles of underage Facebook users. *Computers & Education*, *128*, 473–485. doi:10.1016/j.compedu.2018.08.019

Dinev, T., Xu, H., & Smith, H. J. (2009, January). Information privacy values, beliefs and attitudes: An empirical analysis of web 2.0 privacy. In *2009 42nd Hawaii International Conference on System Sciences* (pp. 1-10). Piscataway, NJ: IEEE.

George, D. & Mallery, M. (2010). SPSS for Windows Step by Step: A Simple Guide and Reference, 17.0 update (10a ed.) Boston, MA: Pearson.

Global Digital Report. (2018). Erişim Adresi https://digitalreport.wearesocial.com/

Göle, N. (1992). Modern mahrem: medeniyet ve örtünme. Ankara, Turkey: Metis yayınları.

Hair, J. F., Jr., Tomas, G., Hult, M., Ringle, C. M., & Sarstedt, M. (2014). A Primer on Partial Least Squares Structural Equation Modeling (PLS-SEM), Thousand Oaks, CA: Sage.

Karaatlı, M. (2010). Verilerin Düzenlenmesi ve Gösterimi içinde SPSS Uygulamalı Çok Değişkenli İstatistik Teknikleri, 5. Baskı, Ed. Şeref Kalaycı, Asil Yayın Dağıtım, Ankara, ss. 3-47.

Kayış, A. (2010). Güvenilirlik Analizi içinde SPSS Uygulamalı Çok Değişkenli İstatistik Teknikleri, 5. Baskı, Ed. Şeref Kalaycı, Asil Yayın Dağıtım, Ankara, ss. 404-419.

Korkmaz, İ. (2013). Facebook ve mahremiyet: Görmek ve gözetle (n) mek. Yalova Üniversitesi Sosyal Bilimler Dergisi, 3(5).

Lyon, D. (2013). *Gözetim çalışmaları*. İstanbul, Turkey: Kalkedon.

Niedzviecki, H. (2010). *Dikizleme günlüğü. Gökçe Gündüç (Trans.)*. İstanbul, Turkey: Ayrıntı.

Oz, M. (2014). Sosyal Medya Kullanımı ve Mahremiyet Algısı: Facebook kullanıcılarının mahremiyet endişeleri ve farkındalıkları. *Journal of Yasar University*, *35*(9), 6099–6260.

Pingo, Z. & Narayan, B. (2018, September). Privacy Literacy and the Everyday Use of Social Technologies. In *European Conference on Information Literacy* (pp. 33-49). Cham, Switzerland: Springer.

Proudfoot, J. G., Wilson, D., Valacich, J. S., & Byrd, M. D. (2018). Saving face on Facebook: Privacy concerns, social benefits, and impression management. *Behaviour & Information Technology*, *37*(1), 16–37. doi:10.1080/0144929X.2017.1389988

Şener, G. (2013). *Sosyal Ağlarda Mahremiyet ve Yeni Mahremiyet Stratejileri, Yeni Medya Çalışmaları I. Ulusal Kongre Kitabı* (pp. 397–401). Eskişehir: Haz. Burak Özçetin, Gamze Göker, Günseli Bayraktutan, İdil Sayımer, Tuğrul Çomu.

Tabachnick, B. G., Fidell, L. S., & Ullman, J. B. (2007). *Using multivariate statistics* (Vol. 5). Boston, MA: Pearson.

Tifferet, S. (2018). Gender differences in privacy tendencies on social network sites: A meta-analysis. *Computers in Human Behavior*.

Toprak, A. vd (2014). Toplumsal paylaşım ağı Facebook: "görülüyorum öyleyse varım!". Kalkedon Yayınları.

Türkiye İnternet Kullanım ve Sosyal Medya İstatistikleri. (2019). Erişim Adresi https://dijilopedi.com/2019-turkiye-internet-kullanim-ve-sosyal-medya-istatistikleri/

Yüksel, M. (2003). Mahremiyet Hakkı ve Sosyo-tarihsel Gelişimi. *Ankara Üniversitesi SBF Dergisi*, *58*(1), 181–213.

Yüksel, M. (2009). Mahremiyet Hakkına ve Bireysel Özgürlüklere Felsefi Yaklaşımlar. *Ankara Üniversitesi SBF Dergisi*, *64*(01), 275–298. doi:10.1501/SBFder_0000002130

Zengin, A. M., Zengin, G., & Altunbaş, H. (2015). Sosyal medya ve değişen mahremiyet "facebook mahremiyeti. Gümüşhane Üniversitesi İletişim Fakültesi Elektronik Dergisi, 3(2).

KEY TERMS AND DEFINITIONS

Instagram: Instagram is a photo and video sharing social networking service.

New Medıa: New media are forms of media that are native to computers, computational and relying on computers for redistribution.

Prıvacy: Privacy is a concept that varies from person to person and in which individuals are not to specify its borders. Rather it is determined by socio-cultural sphere as well.

Social Medıa: Social media is a virtual communication environment where people come together with different purposes such as information, entertainment and socialization, communicate independently of time and space, create and display their own profile and access other users' profiles.

Surveillance: Surveillance is the monitoring of behavior, activities, or other changing information for the purpose of influencing, managing, directing, or protecting people.

Twitter: Twitter is an online news and social networking site where people communicate in short messages called tweets. Tweeting is posting short messages for anyone who follows you on Twitter with the hope that your messages are useful and interesting to someone in your audience.

Whatsapp: WhatsApp is a freeware cross – platform messaging service. It allows the sending of text messages and voice calls, as well as video calls, images and other media, documents, and user location.

Chapter 10
Virtual Resistance of "Çiftlik (Farm) Bank Scapegoats" and Discursive Atonement of "Being Scammed"

Gurur Sönmez
Istanbul Medipol University, Turkey

Savaş Keskin
Bayburt University, Turkey

ABSTRACT

Çiftlik Bank (Farm Bank) is an investment system based on fraud that may be described to be a Ponzi scheme as a commercial term. It reached thousands of investors in Turkey via the service it provided over the internet and attracted attention by leaving a high profit margin for its investors for some time and using some abstract concepts that are held sacred by the majority. Çiftlik Bank created an earning-oriented exclusivity for its investors, but also created suitable scapegoats for the community, along with fraud. This chapter focuses on the rhetorical conflict between the scapegoat virtual group organized with the name "Çiftlik Bank Victims" on Facebook, and the society, as well as the activities of regaining the contingent/select identity.

DOI: 10.4018/978-1-7998-1041-4.ch010

INTRODUCTION

In the post-truth age where lies and lying are not found much strange, the great shame of scamming may also be covered up or tolerated by the society. A lie, which constitutes a cover against the sharpness of truth in many situations, determines the destination of the relationship the person or the society establishes with their self-existence. While the founding lies told by societies in the form of organized rituals (Demirkent, 2017) are a part of the show that needs to be kept up with, individual or inner circle lies have a riskier aspect. This is because auxiliary lies need to show a synergy with main lies. Otherwise, auxiliary lies have the potential to disrupt not the truth, but the lies told by the collective. Moreover, the position of the existence of the lie in the system also determines the dosage of the reflexes that are developed against the lie and those who are scammed.

An 'orchestrated lie', which is bound to constantly reestablish societies under the roof of a truth, is not always connected to the truth. Although thinking of the lie by itself involves difficulties, it is needed to focus on the subjective aspect of the system of judgments that dominate truthfulness and the collaboration it forms with promises that embrace desires (Aytaç & Demirkent, 2017). It may be argued that big lies usually are not the successor of the previous truth, yet they may also be independent of the following stage of truth. The age of post-truth where especially judgments rise above objective truths and become more effective (Yanık, 2017) has social dimensions where rational action is pushed aside, justified universalities. For this reason, the final power of rhetoric is controlled for *Pathos* which shapes judgments rather than *Logos* which aims at rational utility. In the judgment society, as soon as a lie that contradicts large-scale productions of truth is noticed, the judgment of the liar and the victimization of the scammed start.

Çiftlik Bank, which is an investment system based on fraud that may be described to be a Ponzi scheme as a commercial term, reached thousands of investors in Turkey via the service it provided over the internet and attracted attention by leaving a high profit margin for its investors for some time and using some abstract concepts that are held sacred by the majority. However, in the end, the system collapsed, and the owner of the company fled abroad by saying he went bankrupt, eventually being included in the wanted list of Interpol. At the first stage of the system, Çiftlik Bank managers resorted to orchestrated rhetorical maneuvers and developed discourses that would shake spirituality and the spirit of nationalism that are much outside an economic system. Çiftlik Bank discourses, which were constructed between the myth of being national (Alpman, 2016, p. 2) and the *Pathos* of the lust felt towards getting rich, easily grabbed the attention of people of the capitalist regime who competed for hitting the jackpot in an instant. When the urge to be close to masses and crowds was added to the process, the size of the masses that were scammed by extravagant shows and lies of profits increased like a snowball. When the impression of elitism that was gained with the rhetoric of the cheater turned into the sin of being cheated, the story of the counter-rhetoric of being scammed started. This situation is an indication that the investors formed arguments over two different phenomena in the times when it had not yet been understood that Çiftlik Bank was a fraud system and in later times. The investing mass, who adopted the Çiftlik Bank rhetoric where the existing orchestrated lies were produced before the time of getting scammed, developed a new counter-rhetoric that aimed to be saved from the victimhood attached to it after getting scammed which was not very moderate. The people who got organized through the channels of new media that used to be investors and now were victims created atonement with the arguments they produced and went into activities of rubbing off the traces of the purification ritual directed towards them and wriggling themselves out of the disgraceful situation they found themselves in. These joint attempts towards developing identity again occasionally uncovered the potential of the victim on the stage of othering to produce hostile attitudes.

This study focuses on the rhetorical conflict between the scapegoat virtual group organized with the name "Çiftlik Bank Victims" on Facebook and the society, as well as the activities of regaining the contingent/select identity. The assumption that what happens in the group shows a greater deviation than what happened outside the group and differentiates was established by experience. The sets of discourses that were created by examining the posts of the victimized users were analyzed, and the qualities of the arguments they produced in their verbal fight against the people who did not invest / tried to victimize them in the group they were crowded in were investigated. An analysis was carried out in parallel to the concepts of rhetorism and scapegoat that were used to create a new "counter-other" to get out of the position of the "other". The purpose of this study is to discuss how the predetermined relationship forms of being a scapegoat are transformed with social media in societies where traditional heritages and cultural accumulations are replicated in several fields. This is because social media provides a suitable setting for being able to conflict within a means and conditions that are close to the fronted parties. This setting is an ambiguous sphere where all types of speech are encountered, there are representations from all groups, and community powers are used more effectively in the relationship with the other. Discourses and other culture examples that flow inside each other as fluids reduces the absolute dominance of a judgment over another and distributes it to autonomous spheres. Çiftlik Bank Victims is an establishment that harbors the resistance motivations of individuals who produce self-utility from fragmented socializations and gather around certain 'sacred' objectives against being victimized.

COMMUNALITY OF SIN AND SACRED MISSION OF THE VICTIM

The chain of lies seen in the case of Çiftlik Bank has a victimizing force that facilitates the establishment of the dominant lie. This is because one of the ways of expressing the reality of the dominant lie may be to select synthetic liars from within the society and judge them with a 'harmful' type of deception. The Çiftlik Bank mechanism that transforms the discourses of localism and being national by populist politics for universalizing its unique lie was noticed for its profit model that disturbed the existing savings regime and liberal policies. This profit model that threatened the class-related mechanisms of the system was marked with inspection instruments and the collective attitude of the media by the time it was exposed to the attention of masses. After this point, the profiters and elites of the Çiftlik Bank lie turned into the victims and scapegoats of the purification ritual performed by social judgment.

Being scapegoats, which plays an appropriate compensation system against the crises and controversies created by social lying, takes on the responsibility of a cleansing process where sins about which people used to be silent start to be spoken. Members of the society who stay silent against many deviances and see self-deception as a skill start speaking and lynching when they encounter a person or group in which sin emerges. The lynching culture is related to not innocence but heavy sin. For example, the content of an event whose cross-section is presented in the New Testament provides an explanation of the execution relationship established by the society with sinners. As a requirement of the punishment of stoning to death seen in societies that have adopted Judaism, Mary Magdalene who was accused of prostitution is brought to Jesus. Clergymen said that they wanted to stone this woman based on what is ordered by celestial laws and asked Jesus what he thought about this issue. Jesus provokes the distorted mechanism of the society by saying 'let him who is without sin cast the first stone' (New Testament, John 8), and the crowd is dispersed beginning with the elderly. Scamming activities that also cover the Çiftlik Bank issue result in such a type of judgment and execution. Sinners who do not display radicalism while the society is being scammed may actually desire judging their own sins in the place of sinners that emerge in the time of a crisis.

In fact, usage of accusation and victimization in the creation of social elitism is not unique to this period, but it is a historical and archetypical fact. Likewise, according to Campbell (2013: 15), in the case of the 'apple' which is a symbolic description of the first sin, while Adam accused Eve, Eve accused the Serpent. This accusation has constructed the truth that has been dealt by the entire history of humanity since the beginning where the founding relationships of identity has been shaped. As accusation requires the construction of the guilt and the guilty, the dream of a society that is purified of guilt does not seem much reasonable. The social motivation behind such an attention received by the guilty that are represented in reality shows that are broadcast at noon is similar to the motivation of being purified as a community and attaching the guilt to others. Similarly, being a scapegoat, which is examined in terms of its roots in the Torah (Old Testament), shows itself in the 'Day of Atonement' tradition of Israelites (Leviticus, 16, 20-22). This important day which is realized once in every year and gathers all sinners around the same objective requires expression of sins out loud by placing the sinner's hand on the head of an adult goat and releasing the goat in the desert so that it would take away the sins with it. As the goat is considered to be a sinner as soon as it takes on the sins, it is ostracized from the society, and the society which turns away from its own sins is purified. The thing that distinguishes a Scapegoat from a regular sacrifice or votive goat is that it is allowed to live. As the existence of the Scapegoat is the preliminary condition of the existence of the society, it is recalled with a marginality that is stuck between being decapitated and allowed to live. The irony of being from a society but kept outside comes from the issue that the fearsomeness and threat of an 'other' that is cursed, shamed and demonized come from within the heart of the society. Therefore, Scapegoats take on the qualities of a society that the society does not want to see in itself, and in actuality, carry the self of the elitism that is called 'us' (Kearney, 2012, p. 98).

As in the case of Çiftlik Bank, scapegoats either emerge with a crisis or cover up a crisis. Campbell (2013, p. 40), who emphasized that scapegoats still have their societal function in modern times, talked about two types of scapegoats as unknowing and knowing. The first type involves victimization of individuals who are selected to be the source, responsible or reason for a crisis in the time of a crisis. The society, which looks for someone guilty in these extraordinary situations, calms down as the effects of the crisis fade away, and the job of the victim also ends. The Çiftlik Bank 'victims' appear to have been defined after a crisis that served the sensationalist practices of conventional media. This is because there is a question of a chronology of a tangible situation and the reactions that followed. However, it is also possible to observe a strong proximity of the Çiftlik Bank case with the second type of being a scapegoat. This is because, the second type involves identity groups that are knowingly created who are convicted by deliberate accusations for certain problems to be covered up and the society to purify itself. These types of scapegoats are the potential target of accusations, that is, they carry the possibility of having a direct or indirect connection to the guilt. However, when the actual source of guilt is different, scapegoats that are deliberately selected are left in the center of the bad state of affairs. For example, while public order crimes or unemployment problems have carried tensions that are rooted very deep for a long time, directing the accusative collective movements to a single target by the Syrian immigration (Göker & Keskin, 2015) is not a coincidence. Likewise, the Çiftlik Bank case also has a service that creates the distinction of not having been scammed. Individuals who are not involved in the system apply the ritual of belittling the other and raising themselves through what is seen low by witnessing the 'victimhood' and 'scammed' nature of those who are involved. This issue, which provided a material for the entertainment culture of the cyberbullies and trolls on social media, attracted widespread discrediting operations. Moreover, according to Girard (2005, pp. 33-34), operations that contain such systematic attitudes are actions that are applied to load the responsibility of crises onto the victims and ostracize them from the society by accusing them of desecration. Lynching operations that are familiar in the social media culture have also leaked into the Çiftlik Bank Victims page. The resistance of this community is determined by the success of those who are scammed in demanding the sympathy of soothing each other and being more righteous than others, that is, having been scammed. However, as victimization does not leave a door open for coming to terms with the victim, the severity of the tension mostly increases with the hostile attitudes of the victim.

INSTRUMENTS OF THE ÇIFTLIK BANK RHETORIC AND EXAMPLES

In Çiftlik Bank, which was inspired by the social media game FarmVille, the players were able to purchase various farm animals with different values in gold. It was needed to make feed and storage expenses for the animals which had lifespans of 365 days, and to meet these costs and earn money in exchange for the virtual animals that were grown, digital payment systems were used. It was promised that the animals that are purchased and production that is made in the game would be turned into reality at different locations in Turkey in the form of farms and facilities that produce meat and milk. The products that were claimed to have been obtained from such facilities were opened for sales in the branches that were opened.

A highly important rule of perception management and manipulation is that the content is based on "reality". With the facilities it opened up in 4-5 locations, the firm started to create the perception that this way not just a virtual game, but it was based on real investments, too. Nevertheless, these facilities have been emphasized frequently in the support videos of the victims. On the other hand, the investors deposited their money to the game they played over the internet, and they did not have any legal rights on these facilities that were established.

The key concepts on which the Çiftlik Bank propaganda was based on may be summarized as: "Local, National and Religious." Not only in declarations of support but also in the speeches of the managers of the firm, all questions and accusations about the firm were explained as the plot of foreign powers "that do not want Turkey to grow." This argument coded Çiftlik Bank as a "national/religious struggle" and called its investors to fight against these "plots". After this stage, in addition to a "fine trade" that gives 2.5 to 1, Çiftlik Bank started to be perceived as a "transcendent/divine" jihad.

At the opening ceremony of a physical farm, the presenter introduced the founder of this system, the 25-year-old Mehmet Aydın, as the following by referring to his age:

My young man, when you receive the sign from your ancestors

You will walk, and the nation will walk behind you,

I brought you greetings from Ulubatlı Hasan,

You are in the hands, in the tongue, you are in the heart and mind,

You are at the age when Fatih (Mehmed II) conquered Istanbul

As the young appearance of Mehmet Aydın, who was at a very young age and seen by the public as a "child, chubby chap," could lead to a perception of inexperience, and for the investments to not be interrupted, a rhetoric was created based on the age and success of a historical figure who changed history by his achievements at a young age, and the investors' feelings of trust were manipulated. This argument may be interpreted as a variation or a simple imitation of a rhetoric whose influence in the political sphere has been proven.

This argument has been recognized as a discourse since the lessons of Aristotle and writing of Cicero. It has guided teachers and authors in their creation of different arguments for centuries. One of the known aspects of arguments is the argument of rhetoric, and it has been designed to determine the truth or falsity of people's attitudes and the accuracy of judgments about a behavior. Negotiation arguments have been designed to determine the attractiveness of conducting or not conducting a certain act. Finally, explanatory arguments have been designed to show that a person deserved honoring and praise. All forms of arguments have common objectives such as getting others to accept a belief, changing attitudes and calling into action in the form of discourse (McCroskey, 2015: 9).

In some cases, all discourses (verbal and written) are arguments. When we speak or write (in diaries and newspapers), it is aimed to get attention to what is expressed. While attention is also paid to discourses that are found to be worth paying attention to by listeners or readers, it is ensured that our audience believes that what is expressed is provable, it is supported by data, there is a good reason for expressing it, they are watching reliable people for finding data, and these data are presented in a just way (Fahnestock & Secor, 2004, p. 7). For example, whether it contains familiarized details of life or announces a fearsome emergency situation, a friendly letter written to those who are close to us asks them to believe in the truth of what is expressed by the continuing mental health and passion of the author. While readers or listeners are asked to pay attention, any author or speaker promises not only that the discourse will be reliable but that it will provide benefits. In summary, they present an argument.

Discourses on being local and national serve the function of filling a substantial part of the political communication sphere of Turkey in recent years. Nevertheless, considering that the existing situation is articulated by the discourse of 'Survival', it may be claimed that it also performs the discursive derivatives of being local/national. This is because the discursive layer of being local that is a highly rich and fertile field is a state of happening whose contents are filled up and not left empty in several fields from economics to politics, culture to science. The appearances related to this happening turn into various persuasion myths and rhetorical instruments between the constructive power of discourse and the emotional condensations of the 'judgments society' (Baker, 2017) against truths. Aytaç (2017) argued that the discourse of being local is one of the building blocks of the process of reconstruction and metamorphosis. Such that, the discourse of being national may be considered as a prosthetic of being local that is adapted into the context of a nation-state. At this point, it is needed to distinguish being local and locality from each other with a definite line and exclude them from each other. This is because, while locality refers to the universal uniqueness of a locality against others, when being local is in question, a need emerges to direct the powers of a community towards the presence of a stranger or an image of an enemy. In the existing conjuncture where such a discourse displays established behaviors, opportunists who aim to spread the persuasive motifs of being local into marketing and use them alongside scamming elements emerge. The Çiftlik Bank case is a highly discursive issue where the actors expressed their scamming activities in the form of expressions of being local and a theological rhetoric.

The discourses of Çiftlik Bank managers and their officials concerned with press communication may be exemplified as the following.

We did a lot of great things in a short time (...) The power you need is closer to you than your carotid artery. (Çiftlik Bank Advertisement)

If your concern is the homeland, you need to stand up straight like the letter 'alif'. ...but do not forget that they will put hurdles in front of you, they will tackle you down. (Çiftlik Bank Advertisement)

Nowadays, there are some plots planned against our country. We are receiving some threats coming from abroad. Especially from London... We are establishing the largest dairy and animal production facility of Europe. As this disturbs some people, they issue such news stories. (Line TV interview with the founder of Çiftlik Bank, Mehmet Aydın)

Jerusalem is our red line. The target is Turkey. Çiftlik Bank is destroying the plot prepared by the 5 that claims they are greater than the world." (Mehmet Çevik's speech – from the TV show Diriliş Ertuğrul at the opening ceremony of the Deli Demiri - Çiftlik Bank facilities)

[As it was] in the 15 July process... I will ask you to call the name of Allah (Tekbir). Tekbiiir. (Allahuekber!) Tekbiiir! (Allahuekber!) Tekbiiir! (Allahuekber!) They are very afraid of our tekbirs. (slogans called for by the religious figure that led the prayer at the opening ceremony of a Çiftlik Bank facility after the ribbon cutting process)

Dear Mehmet Aydın, I would like to thank you for your invitations for the groundbreaking ceremony for the cattle dairy and production farm. I would like to state that I am sorry to be not able to attend due to my busy schedule. (From the telegraph sent by the period's Minister of Food, Agriculture and Livestock, Ahmet Eşref Fakıbaba for the invitation to the groundbreaking ceremony for Çiftlik Bank's Konya facilities)

In your presence, I would like to thank Mr. [Mehmet] Aydın very much as he transformed national values into national production. We need more of people like him.(From the speech of Nurettin Akbuğa, who was the mayor of Tuzlukçu, Konya in the period)

May Allah protect you and our brother Mehmet Aydın, who is the founder of such a significant initiative, from accidents, harm and evil eyes. May Allah protect all of us from the jealousy of the envious. (From the speech of Abdurrahman Fidancı, the Konya Provincial Director of Press and Information, the Office of the Prime Minister at the groundbreaking ceremony)

The first example is a move towards covering the society by integrating one of the most striking discursive patterns in Mustafa Kemal Atatürk's Address to the Youth and the religious discourse of "God being closer to us than our carotid artery." In addition to this, the statements such as "foreign axes, those who are jealous of us" that are marked as the others are expressions towards the so-called enemy of Çiftlik Bank for the person who used to be an investor but now became a victim on social media. By filling the contents of a game that is played for earning money with dominant discourses, the way of perceiving the truth is broken, and it is partly achieved that an ordinary trade evolves into a holy struggle. The most important emphases in the discourse patterns contain concepts that determine the course of the conversations of the current agenda on which the current politics spends efforts. The mediatic individuals of the age where the truth is operational in the cycle of a judgment may aim at the prestige of taking a side without much questioning on whether or not the existing discourses are related to macro issues. With these calls, the thing that is promised is not only a financial gain, but it refers to getting a position in a holy defense of the nation and heroism. For this, associations are made with politics and the heroism stories of the media. The public resistance that has been symbolized against the 15 July coup attempt is harmed in the attempts of marketing and becomes a part of the show that is dedicated to mobilizing the emotions of the target audience. Rhetorical examples that are similar to political discourses are not just associated with one political establishment. In addition to the potential of association with all political establishments, bureaucracy and gatekeepers, they have the ability to coopt religious and national values in their favor. Moreover, this rhetoric was spread from the Çiftlik Bank layer to the investors layer, and it had its place among the strongest defenses of truth by its supporters for a period of time.

METHODOLOGICAL DESIGN

The findings of this study which was modelled based on the paradigm of qualitative research were established with the discourse analysis approach. Being focused on the rhetoric of 'being scammed' requires reading of the description and sense-making relationship between the discourse and the truth. Therefore, the ways of establishment of the discourse of Çiftlik Bank Victims coursing through the collectivism of social media were investigated, and all symbolic meaning generators were interpreted as different layers of the discourse.

The discourse analysis approach that was used in the study is a form of the critical discourse analysis by van Dijk that was adapted to Facebook. This adapted model that was developed by Aygül (2013) to analyze the hate speech on Facebook was strengthened by the studies of Bayraktutan et al. (2013), and it provided effective results. According to the model that analyzes the interface allocated to the community on Facebook as a text, the thematic framework in the macro structure of discourse consists of the name of the group, description and profile picture, while its schematic framework consists of the language of narration, background information and context information (Aygül, 2013, p. 116). In this study, where micro-analysis was excluded to keep the scope of the study restricted, the macro elements of the discourse on the Çiftlik Bank Victims page were taken into analysis. This is because the objective was to understand not the details of linguistic elements but the higher layers where macro meanings were established and determine the functioning of the counter-rhetoric.

The purpose of this study is to analyze the discursive integrity that feeds the identity of 'scammed victim' produced in the Çiftlik Bank Victims group and determine the reflections of the counterarguments as a tactic of resistance against being scapegoats on the level of interaction. Another objective was to define the positioning of this interaction between the opposite ends of reconciliation and conflict and describe the ways of establishment of the counter-rhetoric. In the study, answers were sought for the following questions;

- How is the counter-rhetoric as a resistance and atonement ritual against being scapegoats constructed on the Çiftlik Bank Victims page?
- How do Çiftlik Bank Victims direct the hesitant attitude of discourse from their side and which arguments do they use to direct their atonement discourses on social media?
- In which themes does the discursive conflict between those that are scammed and those that are not gets intensified and magnified?

The "Çiftlik Bank Mağdurları" (Çiftlik Bank Victims) group which constituted the field of the study (facebook.com/groups/1582158098546652/) was selected from among several other similar groups as it had the highest level of interactions with its approximately 27,000 members. Other groups were also examined at the exploration stage of the study, and purposive sampling was carried out by considering criteria such as number of members, interaction intensity and coverage of radical discourse performances. Nevertheless, the sample is an establishment that provides a basis for the discourses of not only victims but also the victimizing mechanism and provides arguments suitable for the submission relationship between the parties and the concept of being a scapegoat.

The data were collected by document analysis and coded by the analysis categories of the discourse analysis approach. Post and comments under posts were recorded with the method of taking screenshots by computers and smartphones. The data that were collected were broadened by interpretation based on the theme whose analysis steps are presented above, and the rhetorical dimension of the meanings established by the discourse was analyzed.

THE MACRO RHETORIC OF SCAPEGOATS: THE ARGUMENT 'WE HAVE BEEN SCAMMED!' AND ORDINARY VICTIMHOOD

Usage of the political rhetoric in which the discourses of being local and national penetrate as an argument for a marketing lie provides its investors as long as the lie is protected, while an ordinary discourse develops at a time of crisis where the truth is reversed: 'We have been scammed!' This argument which

Figure 1. Profile photo of the çiftlik bank victims group
Translation: The text on the image means that 'Farm Bank Victims Rebellion'

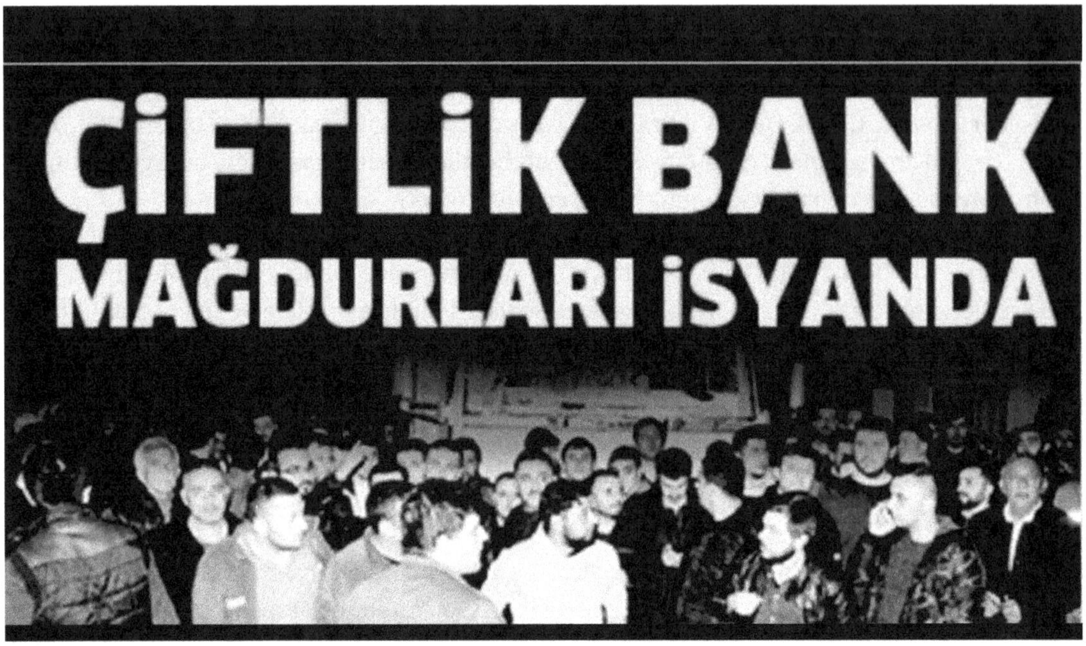

seems simple is the most linear tactic of alleviating the pressure against the unbearable weight of being a scapegoat and self-atonement. This is because this tactic is a tolerance threshold that is accepted by the audience that acts alongside the elitism founded by the lie of scamming and may be considered to be collective instead of joining the guilt. Being victimized instead of being guilty also changes the endurance of the mass and the structure of the sin. The discourse of being scammed is a denial mechanism that produces the meaning of not being in cooperation with those who scam, and the most preserved discourse created by this mechanism is 'victimhood'. The naming 'Çiftlik Bank Victims' in the macro discourse of the group which aims to brush of guilt and the pressures brought by being scapegoats in the name of being victims is a pioneering and comprehensive linguistic symbol that is transmitted to the other side. This symbol shows itself in another macro discourse of the group, namely its profile photo.

Two main meanings carried by the discourse constructed by the visual language should be discussed. First of these reflects the attitude of the victims towards the oppressive and undermining discourse towards themselves. The victims are rebelling, and they will not stay passive. This stance carries the possibility of evolution of the counter-rhetoric in the group into an attach by exceeding the limits of only victimhood and acceptance, as well as the possibility of tension-filled confrontations. As in the case of danger signs in the form of 'Caution!', the approach of 'reliable for friends and fearsome for enemies' is coded in the visual. The second main meaning is that the masses that are placed in the visual are open to be interpreted as a show of strength and solidarity. This is because the designation of victimhood is also included in the name of the group in plural form, and the emphasis of a crowd is emphasized. In addition to turning towards a goal such as creating community networks and broadening solidarity sets by gathering with others that are like them (Özmen & Keskin, 2018), identity groups also have an objective to show strength to the general public with a message of power and greatness. The multi-purpose nature of being crowded is knowingly or unknowingly carried into the macro discourse of the group. However, it may be stated that the primary reason for the group's existence serves the solidarity of those who are inside. Likewise, the type of the group on the profile page is designated as '*support*', and participation from outside was restricted by selection of a '*closed group*'.

There is also a similar solidarity vision in the last pillar that announces the macro discourse of the Çiftlik Bank Victims group. In the introductory text where the group provides information about itself, several code details that show solidarity may be found.

[This group] was created with the purpose of minimizing the losses of our siblings who have money here and help them find support and solidarity when the investigation process ends negatively for Çiftlik Bank.

Interestingly, the introductory statement starts with the consideration that the investigation on Çiftlik Bank may end positively. While people used to be proud of being members of it, Çiftlik Bank is not a stranger and an enemy element. Investors are as close as being 'siblings', and the indicator of siblinghood is having been scammed. In the Siblinghood of the Scammed, a mutual discourse that will balance losses and achieve support and solidarity is constructed. The mutuality of interests in this group is, in fact, derived from a mutuality of losses. This is because it is seen that the group develops a discourse that is completely against losing without regards to winning or creating additional value.

THE MICRO RHETORIC OF SCAPEGOATS: ATONEMENT DISCOURSES

The possibility of success is very low for open denial and rebellion, and inclusion of diversion techniques is inevitable. The main issue is to attach the guilt to another and create a scapegoat. Another issue is to find a crisis which superior dimensions and interests. Considering this last factor, Raven and Rubin (1976, p. 124) reminded that the effect of a mutual threat is as strong as to discredit very powerful negative feelings and prejudices, and after this point, three stages take place. First of all, a situation that creates tension and presents constant and normal problem-solving reactions emerges. Then, normal reaction patterns cannot fight against the situation, and the tension increases. Finally, this increased tension necessities getting above the normal reaction formula and leads to situations of re-defining the problem, discovering new coping methods or directing priorities towards mutual objectives. These situations result in the solution or avoidance of the problem. If the problem is not solved, the level of disappointment will push the individual to a level of disarrangement. This disarrangement will lead to trying relaxation methods that could previously never be accepted such as diverting the guilt and responsibility to an innocent person.

[D]o you know what actually offends me? The people of my country kick people when they are down. It is true. It is my mistake, maybe from my ignorance or my helplessness. The people of my country apparently needed a buffoon. Additionally, those who have not sunk so far do not understand. It is easy to speak from afar, but when you put yourself in that person's shoes, how could you know what they are dealing with inside? NO ONE KNOWS WHAT WILL HAPPEN TO ANYONE AND WHEN. HAVE FUN, MY TURKEY #ÇiftlikBank #MehmetAydın (A Group Member)

This micro discourse claims that the real guilt is in the pressure created on the user by internal and external threats in an anonymous victim's status of having been scammed by this system. Negative issues such as 'helplessness, debt and enforcement' are in the person's own responsibility, and their presentation as a reason for being scammed is questioned. Regardless the type of the rhetoric, all of its stages somehow include the strategy to target emotions. The agitation codes of the scapegoat loaded with emotionality are expressed silently and with lowercase letters in the process of confessing the sin, while they are written in 'uppercase' when a threat is in question, and the identity on the other side that is in power is threatened with the potential of being a scapegoat. This situation has similarities to the search for a contingent position by a disabled individual with the argument that they are 'everyone's disabled candidate'. While the person is in a fight against the abnormality of being disabled and defending being normal, calling for an empathy that takes being abnormal as a reference is an indicator of an axial shift. Nevertheless, Çiftlik Bank Victims, in the atonement ritual they aim to achieve in their discourses, try to establish a fine balance between shoeing off being scapegoats and normalizing it. They do not want the concept of being a scapegoat to be forgotten, because they have a debt to collect from Çiftlik Bank. They also want this concept to be normalized, because they do not want to live indebted to the society. This normalization develops in parallel to the accusation efforts of a pathos that includes the unpleasant aspects of empathy. This process may also be applied in cases where, instead of an individual, a group is in focus. The fear of being punished is among the main fears of humanity. Without the fear of being punished, most people would show fewer social behaviors than normal. This is because their superegos do not completely replace authorized real persons (Alexander 1948:144).

[C]ome on, is there any morality left? We are bending over backwards; they collect taxes from our underwear for a fracture. Of course, we will take risk, jump for money. As if the prosperity level is very high in this country, but we are not informed about it... No one opens up a shop or makes investment for fun or loses money for fun. It is needed to fail several times, get scammed, tackled for success. It is very natural, men who get 30 thousand lira a month by sitting in a chair are arguing. (A Group Member)

As in the case of any other discourse, the main factor in the emergence of such a defense mechanism is the imbalance between the difficulty and significance of the problem and the resources that are immediately available for coping with the problem. This is why an imbalanced element is shown as a target as a reference to this imbalance (Caplan 1964:39). Showing another problem to justify one problem is a new reflective attitude against the accusing attitude. The defense is set by directing the phenomenon of guilt and the accusation of sin that accompanies it towards an inequality and guilt. This approach is a tactic that is established upon covering up guilt by rubbing the judgers' noses in their own sings. Making accusations to counter accusations may be shown as one of the most noticeable characteristics of counter-rhetoric.

[...] I examined the company. I am just trying to warn you it is no different than other companies. Some of you will get your money. You will even profit. Most of you will bear losses. (A Critical Member of the Group)

[A]lright, fine, but are all these people stupid, and are you the only smart one? Why are you doing this? What is your purpose? People who participated in this are 18 or older, to what and whom are you serving? Say it. You think you are so smart by completing some school like METU [Middle East Technical University], right? We sat in the front row in the school of life. Everything about you is great in theory, but you are nowhere in practice (A Group Member).

[I] am happy for you. We opened this page with the hope that we could save a few people. You are free to behave as you like. We are just trying to warn people. (A Critical Member of the Group).

Aronson (1980) believed that prejudices produce creation of modern scapegoat and proposed the theory on 'scapegoats for prejudices' that transformed this belief into a theoretical assumption. His thesis is mainly based on replacement of people who are in an inexpressible position in relation to the reasons of their disappointment and reflecting this disappointment in groups that are relatively weaker where dissatisfaction emerges before disappointment. This thesis was interested in the reasons for processes of naming scapegoats as much as the form of selecting victims. As seen in the discourse, the trauma created by disappointment calls for the aggressiveness of being accused, and the deed of humiliation changes sides. When the counter-rhetoric adopts responding with the language of the powerful, it is important to determine who is stronger rather than the ways of solution. This is because discourses that progress towards requited accusations against the radicality of the discourse may result in the announcement of the truth of the more dominant side. The culture of the street against intellectual experience is presented as a revolutionary criticism against the undermining point of view of the accuser. However, in an indefinite relationship where the intellectual level of the accuser cannot be determined, being a graduate of 'METU' has a meaning that is associated with guilt. This is because being from METU is being isolated from life, while victims have the right to stagger by facing the realities of life. This argument shows that the populist discourses of polarization have also surrounded the intellectual being and reduced it to certain templates.

PARODY OF OTHERING, EXAGGERATION OF SIN AND POLITICIZATION OF GUILT

In the victims group, it may be seen that the investors who say they stand by the investments they made by establishing a rhetoric, that is, the victims that own their sin and take their chances for conflict, keep standing in parallel to the persuasion discourses of the 'scamming' founders. The dominant arguments provided by the investors against those who warned them when they were at the peak of elitism and power before news articles about the bankruptcy of Çiftlik bank are being exaggerated and have become a material for humor for some users. Some users who could not be determined to be investors or not and rather resemble trolls take part in irony by exaggerating their templated and slogan discourses and produce humor by caricaturizing the discourses of the investors that have written posts as those given above. Such that, this humor is nothing more than a discrediting process where othering is strongly felt most of the time.

[I]s it possible, is it possible? This dear Mehmet brother of ours would not con anyone. This is the plot of foreign powers so that we do not develop. Our local and national brothers withdraw money from their banks and deposit it here so that our state develops and gains some money for itself. ...but you are discrediting such a man, because you are an enemy of the state. It is obvious that a person who can deposit so much money cannot be stupid or gullible. Everyone thought of their state and gave their money, but you American spies, Russian spies, Israel spies, English spies, German spies, Chinese spies, Zimbabwe spies, Somalia spies, Papua New Guinea spies and Cape Verde spies are trying to oppose it so that our state would not develop. Damn these spies! Long live Mehmet Chief! It is also a sin in terms of religion to not deposit money into Çiftlik Bank. (A Troll)

Statements such as "Israel spies, American spies, English spies, it is a sin to not deposit money, enemy of the state" are usually the main parts of the arguments created by some investors of Çiftlik Bank in the early period of the system against individuals who told them that the investment they were making was a scam. This accusing rhetoric that used to be the discourse of elitism and the preliminary condition of persuasion has become a material for parody by being exaggerated and reinforced by other discourses in several Facebook comments.

[Y]ou dirty traitors are the plots of foreign powers. You are trying to pass Çanakkale which you could not in the past by now taking part in such plots. When you understood you could not beat us in war, you are trying to undermine us by tackling holy and spiritual institutions such as Çiftlik Bank, but no chance. We are no longer Çiftlik Bank members who were the slaves of the mandate-loving mentality of the past. Our leader is the chief Mehmet Aydın, and we are the h... in his a.... We were indebted to the IMF before Çiftlik Bank. We used to be held hostage at hospitals, and Mehmet Aydın saved us while we were melting in queues for propane tanks and bread. Do not believe claims like he got the money and fled the country. Mehmet Chief is working perfectly. Do not believe these claims even if they are published in the Official Gazette, they are all western spies. (A Troll)

While the group includes the shares of victims about their victimhood, such sarcastic comments are also frequently shared. The thing that does not change is that this rhetoric always emerges alongside the material existence of the other and accusation. In fact, it may be argued that political parody is reshaped as a type of discourse via the political discourses coopted by Çiftlik Bank. This is because an important part of the comments or posts shared specifically about the topic in question refer to the political structures that are satirized without targeting. A crisis of marketing is adapted to the ideological criticisms on social media where politics is a strong instrument.

[I] warned someone to not deposit their money. They told me I am a plot of foreign powers and I did not want our state and nation to rise. (A Troll)

In an environment where criticisms are shared aggressively from opposite sides, the synergy of conflict created by politicization brings about parodies that get behind a reasonable appearance. As in this example, it may be aimed to hide parody by applying a theme of rationality. The political dimensions of the Çiftlik Bank persuasion discourse are also satirized by small-scale parody discourses. Such that, for this satire, events that cannot be understood to have actually occurred or not may be presented as a truth.

[L]ook, brother, you say at most 10 percent. A grocer works with 30 percent profit, a seller in a market works with 50 percent profit. You are talking about new members, but there are no new member entries now. There is no reference system. We are still getting our money. Whose man are you as long as we are not victimized? What is it to you, what is your purpose? No one can say they are a victim. For which Zionist are you working? You and those like you are attacking because you are afraid these investments will develop Turkey, but Turkey will grow in spite of you. (A Group Member).

[B]rother, a grocer sells a product they buy with a 30% profit. The initial cost for a grocer is hundreds of thousands of lira. No grocer can get this money in 4 months. Even if they could, they cannot do it indefinitely. I want Turkey to grow as much as you do. Please stop treating people who do not share the same view as you as traitors. (A Critical Member of the Group)

It is seen that, rather than creating policies towards the solution of political, social, economic and cultural problems and conflicts that emerge in the public sphere, the new media environment is unjustly labeled as the 'scapegoat of all sins of the society', and policies that will improve the will and participation of the citizens are obstructed this way. This is because these fields where scapegoats are clustered are also marked as the fields where guilt is produced. Hence, it is likely for social media to produce a judgment that is associating with platforming sin. The tradition of cursing spaces by association with guilt and sin may also lead to perception of social media itself as a scapegoat.

CONCLUSION

Abusing religious and national values for the sake of marketing lies is not an issue that is unfamiliar in Turkey's tradition. Such that, a person named Fadıl Akgündüz scammed masses several times, and he contributed to the literature with the concept of 'Fadılzede' [Fadıl's Victim]. Nevertheless, the characteristic of Çiftlik Bank Victims that separates them from others that carry the common characteristics of a tradition is that this issue involves indicators of digitalism in several aspects of the process, and it is unique to this period. Even defense mechanisms were organized on social media. In this sense, it is not a coincidence that victim groups were established on Facebook and other social networks by the time Çiftlik Bank was announced as a scam. While the names of such groups directly included the term 'Çiftlik Bank', debates have continued in investor groups with names that contain some terms related to Çiftlik Bank. This is because functions of social media that are now known by masses and transformed by masses into judgments have made it easier to perform atonement and repair activities. Groups that have discovered new ways of atonement from sin and victim selection have settled in these fields where they may be atoned again or stigmatized more.

Normally, scapegoats are silent and not expected to defend themselves. The texts that describe Jesus whose sanctity is constructed as a scapegoat use the statement '*and as a sheep before its shearers is silent, so he did not open his mouth* (Old Testament/Isaiah, 53). Indeed, there is a popular view that scapegoats silently accept the repercussions that correspond to their guilt. This is because silence is a tactic of invisibility and hiding that is developed for being protected from pressures against the scapegoat (Keskin,

2017: 194). Scapegoats that wait to be forgotten by disappearing hold themselves back to avoid being subjected to ostracization and discrimination, and they accept subordination in exchange for the sins of the society. The thing that makes the topic extraordinary in this study is the effort of the scapegoats who were expected to stay silent to transform social media into a rhetorical atonement to speak and be saved from being scapegoats. The masses that are cheated with this rhetoric want to be atoned with their own rhetoric and be saved from the pressure of ostracization and symbolic lynching by turning back to 'normal'. While doing this, they may even show a counterattack as if they are trying to exceed their opponents in violence (Goffman, 2014: 43). The counter-rhetoric and counter-otherness that is aimed to be established by the victims of Çiftlik Bank are talkative act against the assumption that 'the subordinated cannot speak' (Spivak, 1988), a resistance. This is because the Çiftlik Bank 'scapegoats' start an opposing fight by judging the society that accuses them, and they establish the power of solidarity by gathering together. This virtual agency that forms the direct relation of being the majority and being able to speak up should be considered not as a payoff against the guilt itself but as an accusation towards others. This is because masses and synthetic scapegoats do not have an initiative about the root of the guilt, and community feelings dictate directing the guilt towards others. The guilt is accepted, the one who is guilty is sought, and the guilty accuses others by denying their guilt. This way, a cosmology of discourses where back-and-forth accusations are debated is created. Storytellers who have a dominant rhetoric and are able to effectively use the cultural instruments of social media are advantaged in getting others to accept their own truth and accusation. In the opposite case, speaking up may turn into a dangerous experience and tension. While blame is usually directed towards someone who is outside and weaker, it may also be distributed among some who are inside. By culling those who are weak, the sin is sent away. This is why there is a question of the Çiftlik Bank victim solidarity movement to have a group synergy in the form of trying to suppress the anger of the powerful by choosing victims from within.

In the group dynamics, it was aimed to reduce the weight of the sin with the title of victim, but a stance that was in parallel with traditional discourses was displayed. Acceptance of victimhood is a solution that is related to tolerating the procedures that are found fit for scapegoats and reducing the burden of the crisis. This is because victimhood is the person's self-passivation against the feeling or association of guilt and their abandonment of responsibility. As the epithet of victim receives less reaction than the epithet of scammer, it is an option that is easy to adopt. The lack of background and context knowledge in the discourses absorbed the effects of the catalysts in the background of the issue and produced material for current conflicts. Discourses that have been isolated from the past or flexed based on today's conditions were separated from historicity by neglecting the processes that prepared them or ignoring the layers that are associated with. This treatment that is considered to be the general character of the post-truth age summarizes the relationship between discourses and the universe of truth that serves the moment. Discourse themes whose historical background and current context are not talked about allow today's denial of what was expressed yesterday. De-etymologized discourses support the accusative atonement initiated by parties with their fluid contents that are shaped based on the conditions of the present. This tendency which makes one question the possibility of a collective memory is a social gene that has been inherited from the producers of the traditional narrative. This is because reflection of the hereditary nature of conventional media onto the practitioners of social media is an expected outcome.

In this sense, it is possible to claim that what is discussed on social media is no different than traditional media in terms of its qualities. This is because the traditional codes that still have weight in today's post-modern world are also evidence that 'new' media is not expanding with innovations as it is considered to be. The cultural codes that carry the hereditary characteristics of the society are the explainers of the repeated production of synonymous discourses by individuals.

REFERENCES

Alexander, F. (1948). Development of the ego psychology. In S. Lorand (Ed.), *Psycho Analysis Today*. London, UK: George Allen & Unwin.

Alpman, P. S. (2016). Necip milletin millet-i hakime hassasiyeti ve yerli-yurtlu gayr-ı milliler. *Toplum ve Kuram*, *11*(Spring), 13–38.

Aronson, E. (1980). *The social animal* (3rd ed.). San Francisco, CA: Freeman & Co.

Aygül, E. (2013). Yeni medyada nefret söyleminin üretimi: bir toplumsal paylaşım ağı olarak Facebook örneği. (*Unpublicated Master Thesis)*, Gazi University, The Institute of Social Sciences, Ankara/Turkey.

Aytaç, A. M. (2017). Bir Türk'ü nereden tanırsınız. *Gazete Duvar*. Retrieved from https://www.gazeteduvar.com.tr/yazarlar/2017/11/13/bir-turku-nereden-tanirsiniz/

Aytaç, A. M. & Demirkent, D. (2017). Ahmet Murat Aytaç: "hakikatin kendinden menkul bir enerjisi yoktur; hakikati savunmak gerekir". *Ayrıntı Dergi Yalan Özel Sayısı*. Retrieved from http://ayrintidergi.com.tr/ahmet-murat-aytac-hakikatin-kendinden-menkul-bir-enerjisi-yoktur-hakikati-savunmak-gerekir/

Baker, U. (2017). *Beyin ekran*. İstanbul, Turkey: Birikim.

Bayraktutan, G., Binark, M., Çomu, T., İslamoğlu, G., Doğu, B., & Aydemir, A.T. (2013). Web 1.0'dan Web 2.0'a Barış ve Demokrasi Partisi. İletişim ve Diplomasi Dergisi, 1, 31-54.

Binark, M., & Gencel Bek, M. (2007). *Eleştirel medya okuryazarlığı*. İstanbul, Turkey: Kalkedon.

Campbell, C. (2013). *Günah keçisi, başkalarını suçlamanın tarihi. Gizem Kastamonulu (Trans.)*. İstanbul, Turkey: Ayrıntı.

Caplan, G. (1964). *Principles of preventive psychiatry*. New York, NY: Basic Books.

Demirkent, D. (2017). Siyasal toplumun kurucu yalanlarıyla ne yapacağız? *Ayrıntı Dergi Yalan Özel Sayısı*. Retrieved from http://ayrintidergi.com.tr/siyasal-toplumun-kurucu-yalanlariyla-ne-yapacagiz/ adresinden ulaşıldı.

Fahnestock, J., & Secor, M. (2004). *A rhetoric of argument: a text and reader*. New York, NY: McGraw-Hill.

Girard, R. (2005). *Günah keçisi. Işık Ergüden (Trans.)*. İstanbul, Turkey: Kanat Kitap.

Goffman, E. (2014). *Damga: örselenmiş kimliğin idare edilişi üzerine notlar. Ş. Geniş at all (Trans.).* Ankara, Turkey: Heretik.

Göker, G. & Keskin, S. (2015). haber medyası ve mülteciler: suriyeli mültecilerin türk yazılı basınındaki temsili. İletişim Kuram ve Araştırma Dergisi, 46, 229-256.

https://www.facebook.com/groups/1582158098546652/

Kearney, R. (2012). *Yabancılar, tanrılar ve canavarlar. Barış Özkul (Trans.).* İstanbul, Turkey: Metis.

Keskin, S. (2017). Dini ötekilik ve iletişimsel pratikler: din değiştirenler üzerine bir araştırma. (*Unpublicated Master Thesis)*, Fırat University, Intitute of Social Sciences, The Department of Communication Sciences, Elazığ/Turkey.

McCroskey, J. C. (2015). *An introduction to rhetorical communication.* New York, NY: Routledge.

New Testament/John. Retrieved from https://incil.info/kitap/Yuhanna/8

Old Testament/Isaiah. Retrieved from https://incil.info/kitap/Yesaya/53

Old Testament/Leviticus. Retrieved from https://incil.info/kitap/Levililer/16

Özmen, S., & Keskin, S. (2018). Sosyal medyada öz-temsil ve ötekiliğin 'öteki' boyutu: 'karikateist' toplumsalı üzerine inceleme. *IntJCSS*, *4*(2), 533–558.

Spivak, G. C. (1988). Can the subaltern speak? In C. Nelson & L. Grossberg (Eds.), *Marxism and The Interpretation of Culture* (pp. 271–313). Basingstoke, UK: Macmillan Education. doi:10.1007/978-1-349-19059-1_20

Yanık, A. (2017). Popülizm (V): Post-Truth. *Birikim Dergi*. Retrieved from http://www.birikimdergisi.com/haftalik/8463/populizm-v-post-truth#.XJFJdCgzZPY

ADDITIONAL READING

Brannon, L., & Brock, T. (1994). The subliminal persuasion controversy: reality, enduring fable, and Polonius's weasel. In Psychological Insights and Perspectives. S. Shavitt ve T. C. Brock. (Eds.). Persuasion: Allyn & Bacon.

Girard, R. (1989). *The Scapegoat. Yvonne Freccero (Trans.).* Baltimore: Johns Hopkins University.

Messaris, P. (1994). *Visual literacy*. Boulder, CO: Westview Press.

Mirzoeff, N. (2013). *The visual culture reader* (3rd ed.). New York: Routledge.

KEY TERMS AND DEFINITIONS

Çiftlik (Farm) Bank: It is basically a fraud system offering investment opportunities (!) to users through a virtual game.

Post-Truth: It is a social production in which convictions and beliefs are more effective than objective reality.

Scapegoats: It is the category of identity that creates social cleanses and is inculpated the delinquency which is producing in the bosom of society.

Social Media: It is the new media derivative that provides interaction, empathy and participation production opportunities.

Victimizing: It is the process of building a subaltern identity category, which is chosen by the society and deliberately accused against the sins of society.

Virtual Atonement: repair activities in groups of people who experience identity and recognition problems in traditional spaces.

Virtual Community: It is social formations where users on digital platforms come together.

Virtual Identity: Digital based identities acquired by users in virtual communities.

Chapter 11
Identity Design and Identities Exhibited in Social Networks:
A Review Based on Instagram Influencers

Mehmet Ferhat Sönmez
Fırat University, Turkey

ABSTRACT

Identity emerges as a flexible, multidimensional, variable, and slippery concept that cannot be defined through the processes of discussion and understanding. The new construction area of this concept, which is regarded as a process constructed on the social plane, is the social networking platforms. This is because these platforms are the most common communication environments where people and their lifestyles are presented to the outside world, in addition to the cheap and rapid satisfaction of their needs for information and entertainment. Face-to-face communication and language practices are not sufficient enough in the identity presentation anymore. Individuals choose to design and update their identities through social networks and to perform an image-based identity manifestation. This chapter examines how identity was established and manifested through social networks, and analyzes the identities the popular people in these networks designed and exhibited.

DOI: 10.4018/978-1-7998-1041-4.ch011

INTRODUCTION

As of 2019, 4.3 billion people are Internet users, while 3.4 billion people (about 45% of the world's population) are social media users *(Global Overview Report* https://p.widencdn.net/kqy7ii/Digital2019-Report-en). Even this statistic alone will suffice to determine the position and importance of social media today. These platforms, which have been integrated into everyday life with the information age, have been easily accepted by masses. Even though they do not have a long history, they have been adopted in a short period of time. Every area of social life, from politics to social movements, from entertainment to education, has been influenced by the Internet and social networks.

Social Media and Social Networks

Social networking sites are applications that provide information and interaction to users through network technologies (Boyd and Ellison, 2007). In the Merriam-Webster dictionary, social media is defined as "forms of electronic communication (such as websites for social networking and microblogging) through which users create online communities to share information, ideas, personal messages, and other content (such as videos)." According to Mayfield, it is a new type of online media where a high level of sharing occurs and has the following properties (Mayfield, 2008: 5):

- **Participants:** Support and encourage individuals to contribute to the content and provide feedback.
- **Openness:** Social media platforms are open to feedback, they actively allow information sharing, and users can make comments there.
- **Conversation:** They allow bi-directional communication.
- **Community:** They pave the way for the formation of communities in a very short period of time.
- **Connectedness:** They allow links to be created to other pages and media related to topics that interest users.

Unlike traditional media, social media has its own characteristics. These can be summarized as the *determination of the content by the user, lack of time and space limitations, being in an interactive structure and the fact that users are independent of any publisher* (Erkul, 2009: 3). Social networks are the big living spaces within the small "worlds" that people create. People interacting in the Internet environment create a small world of their own. The small-world phenomenon was first discovered by sociologist Stanley Milgram in a mail experiment in 1967. Milgram has proposed a theory claiming that everyone in the world is no more than 6 people away from each other, and he has tested it. According to the classification with six degrees that emerged in the experiment, a person can reach someone he does not know through a maximum of 5 people. Being inspired by this experiment, the first social networking site on the Internet has been called "SixDegrees" (Patch, 2004, p. 4 as cited in Onat ve Alikılıç, 2008, pp. 1116–1117).

Popularity, Popular Culture and Social Media

The etymology of the term "popular" can be traced back to the terms "*populace, population, public,* and *publication*" (Batmaz, 2006, p. 19). The term, which meant "public belonging to the people" in the late medieval period, is used to mean "*loved or chosen by many people*" in the linguistic sense nowadays (Erdoğan & Alemdar, 2005, p. 9). Williams also described the concept as "highly admired by many people and something consciously done to gain appreciation" (Storey, 2009, p. 6). Popular culture is the whole of beliefs, practices and norms that are embraced and shared by a broad circle of people — that is, by almost everyone, if not everyone (Schudson, 1999, p. 169). The middle class, which has changed and thrived through urbanization, has become the strongest bearer of popular culture (Storey, 2009, p. 13). The popular culture, which Rowe (1996, p. 20) considers a leisure practice, prioritizes the elements of entertainment and curiosity. Today, most modern media have also become an entertainment tool, and public events constitute a very insignificant part of media content. Therefore, media has become the primary means in manifesting popular cultural events (Curran, 1997, p. 146).

McQuail has stated that popular cultural content is reflected by the media and that this content is again formatted through the media (McQuail, 1994, p. 40). Popular culture, which is easily accessible to everyone, is unavoidably caught by the radar of the media because of this characteristic of it (Çağan, 2003, p. 77). Social media environments are easily accessible, just like the popular culture, in addition to having an interactive structure. Therefore, a strong flow of information occurs through these environments. Many virtual cultural environments ranging from entertainment to consumption preferences emerge, and popular cultural elements are displayed on this ground created by social networks (Karaduman, 2017, p. 12). Social media, which gathers millions of people in a room, has become a social, cultural and industrial form, not just a technological tool. Social media environments have now evolved into an administrative communication tool that shows, presents and evaluates "what is popular, what is popularized and what is intended to be made popular" (Erdoğan, 2004, p. 15).

Identity, Social Identity and Social Media

Hall defines identity as designs of belonging — continuously established, undertaken and owned by the discourses, conditions and experiences — which are used to address the need for identification and are also formed by different cultural elements (Binark, 2001, p. 75). While the identity phenomenon in the pre-modern period was a concept that was unquestionable and unnegotiable, this situation has changed with modernism, and the concept of identity has become something that is personal, mobile, and open to change and innovation (Karaduman, 2010, p. 2890). According to Kellner, the features of the modern identity are as follows (Kellner, 2001, pp. 195–196):

- The identity has been drawn into the social context and become linked to the "other." And thus, its boundaries have been expanded.
- It has evolved into a selectable, producible and reproducible form.
- In this selection and production process, social norms, social roles and expectations have settled in the position of a reference center.
- Identities now realize that they are in a position to change at any time.

With the "other" becoming prevalent in the identification of identities in the modern period, certain theories concerning the social aspect of identity have been proposed. Names such as Tajfel and Turner (1976, 1978, 1988) have proposed the social identity theory. Gofmann emphasized the identities displayed through daily life. Brewer has proposed that two basic human motives — the need to be unique and the necessity of belonging — are decisive in the formation of social identity, and said that individuals' inclusion and differentiation needs are thus met (Padilla & Perez, 2003, p. 43). Depending on the place of identity on the social plane, individuals distinguish themselves from other groups to raise their positive social judgments and their collective self-esteem. Individuals and groups who do not have satisfactory social identities will try to re-establish/acquire positive identification through "*mobility, assimilation, creativity* or *competition*" (Jenkins, 2016, pp. 125–126). Gofmann has conceptualized this situation as elements such as the highlighting of social category in interpersonal interactions, as well as the image/face that they offer based on their social status. According to Gofmann, the role of a person in the relationship with others will be perceived as an image in the mind of the opposite party (Bilgin, 2007, p. 13). Gofmann concentrates on the concept of "self" and explains how the person reviews and presents himself by updating himself in the event that he encounters the other. He explains to which means the person refers to by using the concept of "self-presentation". The strategy the person develops when making this presentation is bi-directional. On the one hand, he shapes the impressions he desires to give; on the other, he chooses to conceal the impressions that he dislikes and avoids. This situation is described as impression management. This bi-directionality in the presentation of the self is hidden in Goffman's definition of the dual self. According to him, the consensual self — the self that is formed as a result of the practices of both the performer and the observer during an interaction — and the player self implicitly coexist in each individual. This is because the moment an individual gets in touch with another, he actually goes on a stage in front of the other and interacts with the other by transferring some things in this scene, and by covering some things over some other things (Bayad, 2016, p. 83).

With postmodernism, the concept of identity would evolve once again. In the construction of postmodern identity, where unruliness and indeterminateness get to be dominant, slipperiness will prevail and the antecedent paradigm of the modern era, "the other," will be replaced by concepts such as originality, uncertainty, diversity, complexity and relativism (Karaduman, 2010, p. 2894). Bauman, questioning the causes of the need to obtain identity, claims it to be the reason to get rid of the annoying discomfort of the uncertainty of "neither this nor that." This is because the modern social statutes where belongingness is resolved are now inadequate in resolving this need. However, in today's mobility, it also points to the difficulty of identifying someone with something. The identity, described as "a clock to be removed when needed" by Weber, is constructed on the move, planned to be short term and is short-lived, again according to him (Bauman, 2017, p. 38–43).

Social media networks provide the person with a continuous identity development process through the opportunity for global communication and continuity that they offer (Bakıroğlu, 2013, p. 1049). According to Schroeder (1994, pp. 524–525), virtuality circulates the desire of people who wish to re-express themselves/express themselves by renewing.

With online social networks which have evolved with the advancement of communication technologies, the issue of time and space has disappeared, and the room for maneuver has expanded. In these networks, which are open 7/24, people open their everyday lives to the outside world — in every situation they find opportunity — and reflect/build their identities. Individuality that Niedzviecki (2011, p. 18) describes as new conformism sits on a central plane through social networking environments. Again, according to Niedzviecki, the awareness of being individual will be noticeable by being followed by others and by receiving comments (Niedzviecki, 2010, p. 37). Social networks are platforms where an individual escapes from his own reality, finds the freedom to act as he wants to, and experiences his virtual "self" as he desires. He overcomes the negative aspects of his self and the shortcomings he has felt in socialization by using the virtual self he creates and gets away from his true self by being captivated by this virtual reality. Thus, feelings of satisfaction and pleasure are experienced intensely (Akmeşe & Deniz, 2017, p. 28). Goffman's dramaturgical approach was also employed for virtual identities in social media environments. Virtual environments correspond to Goffman's metaphor of the scenes where the self is expressed by wearing a mask (Bayad, 2016, p. 90).

Online social networks allow people to design and consume multiple identities instantaneously. Identities produced on digital platforms are the identities that are intended to be presented to others and have a socially-desirable characteristic. These identities, which cannot fully be exploited in real life, still do not mean that they are not real, and can have a reality effect on the perception of both the identity creator and the person seeing this creation. Identity, which is already a complex concept in the incarnated world, becomes even more complex in the online field (Akgül & Pazarbaşı, 2018, pp. 17–19).

THE STUDY

Purpose and Method

The main objective in this study was to examine the identity exhibitions of the people who were popular in social media environments and analyze how they manifested their identities. In the study, two people who were the most popular and had the greatest number of followers in the Instagram environment were selected for analysis. The reason why Instagram was preferred was primarily for the users of this platform to attach importance to visuality. In addition to this, Instagram is among the popular social networking sites. It was established in 2010 and was soon accepted. The first reason why the application is so popular is that it makes photos look more beautiful through its 11 photo filters. Moreover, users have the opportunity to instantly, practically and quickly share on other social networks the photos they share on Instagram (https://www.brandingturkiye.com/instagram-tarihi-instagram-nedir-nasil-kullanilir-ne-ise-yarar/).

According to TRACKALYTICS data in April 2019, the people who ranked first and second can be seen in Table 1.

Table 1: The accounts of influencers on instagram with the greatest number of followers

Instagram Account of the Influencer	Number of Followers
Cristiano Ronaldo	160,096,599
Ariana Grande	150,042,569

Source: https://www.trackalytics.com/the-most-followed-instagram-profiles/page/1/

Figure 1. His social identity (His modern identity)

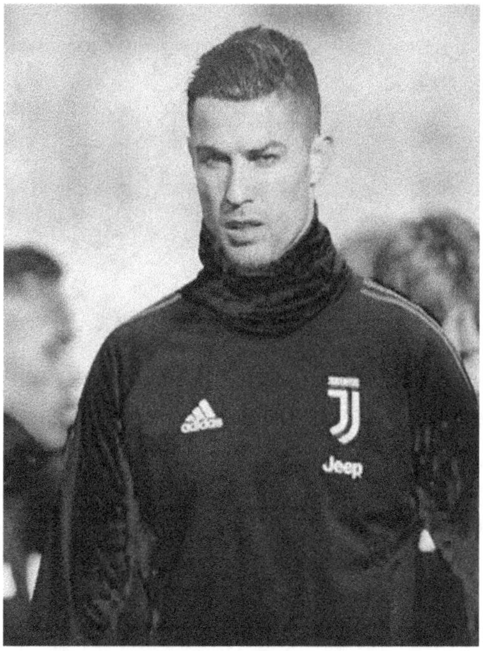

This study was centered on Goffman's impression management theory, and netnographic analysis was chosen as the method. Communication ethnography, whose area of study is the daily life of individuals (Kartarı, 2017, p. 216), aims to observe cultural value patterns by receiving supporting from anthropology and to interpret the specific codes of a culture within that culture (Özüdoğru, 2014, p. 266). The adaptation of ethnographic research techniques to online environments is called netnography. Netnographic analysis has recently been used frequently and has become a popular method especially in studies related to social networks (Mansell et al., 2015, p. 292). This method, developed by Kozinets in the 1990s, is a qualitative and interpretative research methodology (Jupp, 2006, p. 193), ensuring both that research environments are examined in more natural forms, as well as reaching richer content (Kozinets, 2002, p. 62; Langer & Beckman, 2005, p. 200).

Findings and Analysis

When the Instagram account of Cristiano Ronaldo, who ranked first in terms of the number of followers, was examined, it was possible to say that he usually shared posts that put forward his "sportsman" identity. The state of "accessible status" (Giddens, 2012, p. 181), which the individual gains by his own efforts, is evident in the posts of C. Ronaldo. C. Ronaldo reflects his sportsmanship as his prominent social identity. He has also strengthened his identity with competitive, ambitious and dominant roles (Fougère & Moulettes, 2007, p. 17), which coincides with Hofstede's masculinity dimension of the cultural structure. This pronounced image is related to the social context one belongs to. C. Ronaldo perceives to

Figure 2. His self-oriented identity (His post-modern identity) Sportsman Strong sportsman/invincibility "myth"

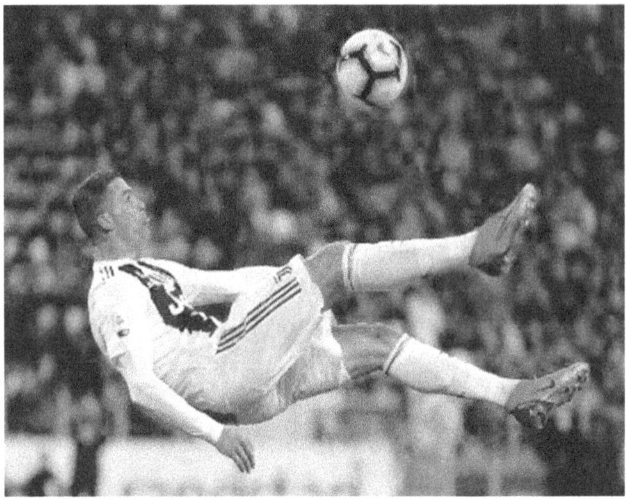

be/feels to be belonging to the football player class and perceived so by those who watch the show. On the other hand, a new type of identity is emerging, with the submission of stationary social structures to today's post modernity. This identity typology, defined by Funk (2007, pp. 7, 12) as a "post-modern self-oriented" personality, has become more prominent at the point of shaping the thinking and action forms of people under the influence of their living conditions and living spaces (İlhan, 2013, p. 240). Again, according to Funk, this personality type is based on a powerful "self" (Funk, 2007, pp. 55–56). The next phase of this construction is the desire and necessity to create a unique and sparkling myth of "self". The individual who cannot fulfill this need and cannot resolve the necessity will not be able to wriggle oneself out of being modern and reach post modernity. The only functional weapon of the person who wants to sculpt his own myth is his visibility. This is because "what is visible is good, and what is good is visible" and in today's world, which Debord described as the society of the spectacle, of course the laws of the show will be decisive (Debord, 1996, p. 16). A person who wants to reach satisfaction by placing his self in the center will present demonstrations supported by his biological and physical characteristics.

Identities exhibited on social media platforms may not always overlap with reality. Posts on personal accounts prioritize the satisfaction of psychological needs such as appreciation, recognition and being noticed. In other words, there may be a mismatch between the image that the person has and the image he wants to reach. However, the identity that a person reveals to his social environment can also be undistorted, reflecting every moment of his life as it is (Sabuncuoğlu, 2015, p. 373).

Another social identity that Cristiano Ronaldo exhibited on stage is having a high-income level, meaning his being "rich." As a status symbol, wealth stands out in the posts of the Instagram influencer. Wealth as a social identity is shared voluntarily and overlaps with reality — considering that he is at the top of the list of sportsmen who earn the most money.

Figure 3. His social identity (his modern identity) rich

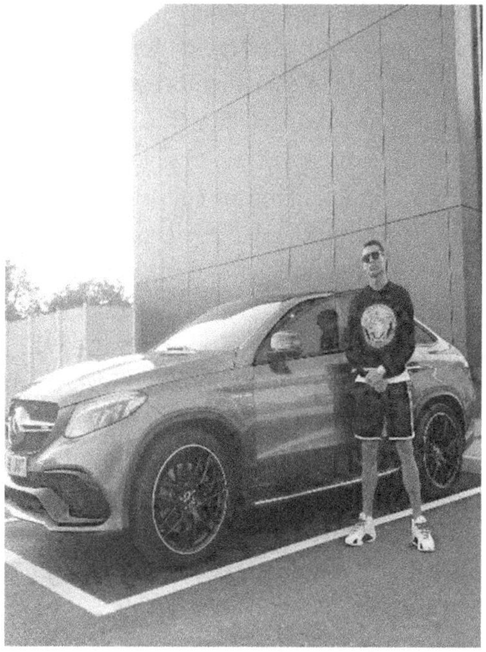

Figure 4. His self-oriented identity (post-modern identity) ultra-rich/unreachable myth

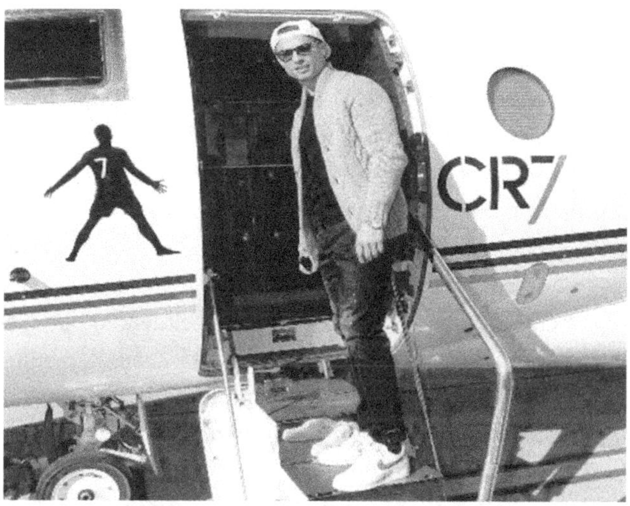

It was to strengthen the self-esteem with these posts where tangible assets were exhibited, and it was aimed to be noticed. Another reason why individuals seek a social identity is to raise self-esteem and get self-respect. According to Scitovsky, it is sometimes not sufficient for an individual to desire to become a member of a particular group. What become a priority in a consumption-centered world is what people have, not who they are (Odabaşı, 1999, p. 95). Tangible assets are assessed according to a social structure that prioritizes power and prestige rather than the necessities that meet vital needs (Sabuncuoğlu, 2015, p. 371). Baudrillard has also stated that the primary purpose of today's people is to pursue the existence through vanity and wealth (Baudrillard, 1997, pp. 193–194). When we take into account that consumption is a way of life, it is possible to say that consumers' search for "drawing attention" is a fundamental need for consumption. This pursuit will be resolved by the exposition of brands and products on virtually any platform (Clark et al., 2007, p. 46, Gökaliler et al., 2011, p. 38). Virtual platforms are the largest of these exhibition halls, and these environments offer an unlimited space for people to show and prove their social identities to the environment (Sözen, 1991, p. 94), liberalizing them at the point of creating exciting identities (Denizci, 2009, p. 59).

Another person who has the highest number of followers on Instagram is the singer and actress Ariana Grande. The first notable feature of the shares of Ariana Grande is that they are the photographs that reflect beauty, aesthetic and physical attraction. These posts triggered by the motivation for appreciation originate from women's feeling that they are under surveillance and that this process becomes a situation that encircles them (Oğuz, 2010: 184). As a matter of fact, according to Berger, a woman's perception of her own existence is complemented by someone else's sense of appreciation of her (Berger, 1990, pp. 46–47). With these shares of her where visual appeal is idealized, A. Grande aims to put her own self into a cognitive comparison process in order to be placed in the category of beautiful/attractive women in the minds of others. In this context, Bocock also highlights that the practice of creating identity through an image and the body is not a simple reflex of consumption, but a process in which desires are embedded (Bocock, 1997, p. 107). Berger conceptualizes this situation through "being watched" (Berger, 1990, p. 447):

Men are however they behave, and women are however they appear. Men watch women. Women watch their being watched. This does not only determine the relationship between men and women, but also the relationship between women and themselves. The observer inside a woman is a male, but the observed is a female. So, the woman turns herself into an object — especially a visual object —, something to be spectated.

When the significance of being watched is combined with the ideal of being beautiful and impeccable, the woman who creates her own mirror and her own myth will internalize being a goddess and put it on stage. If we cite Goffman's (2018, p. 65) quote from Simone de Beauvoir: "*... even the least sophisticated of women does not present herself to observers anymore after she gets dressed. She is a tool that implies a character such as a painting, a sculpture, or an actor on a stage, that is, a character that she represents but not there. What satisfies her is to identify with something that is unreal, unchanging, and perfect such as a novel hero, a portrait or a bust; she strives to identify with this figure, and thereby, to see herself to have stabilized and be legitimized in her own glory.*"

In virtual platforms, the fact that motivations such as desire, being watched and being noticed have become attractive is because these platforms have an untouchable and fictional nature. It is almost impossible to get disappointed in this fictionality. The demonstration scene that is suitable for the irresistible appeal of the satisfaction of desires, the fulfillment of being noticed, and the watching (being watched) is these network environments (Robins, 2013, pp. 40–41).

Figure 5. Her social identity (her modern identity) being an animal lover

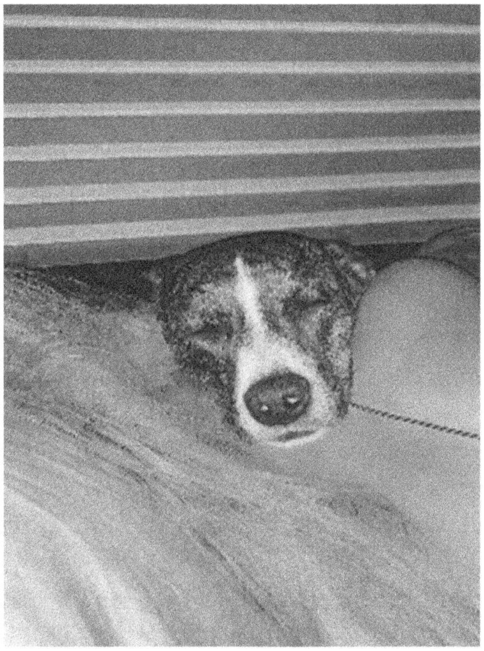

Figure 6. Her self-oriented identity) her post modern identity) beauty attraction/goddess myth

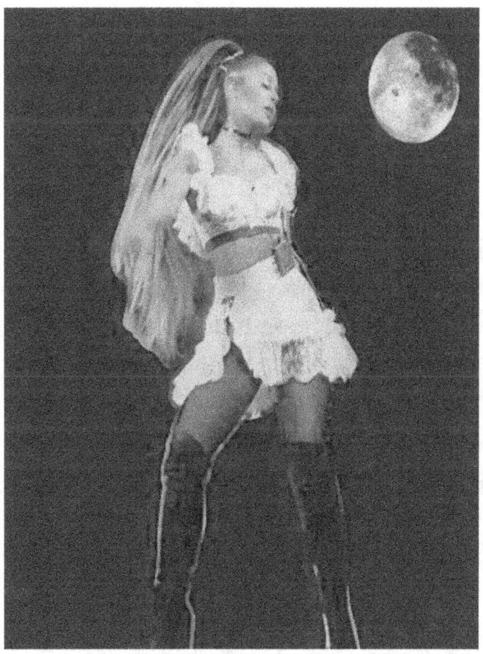

Table 2. Male influencer and female influencer features

MALE INFLUENCER	FEMALE INFLUENCER
Masculine characteristic	Feminine characteristic
Being unique	Being attractive
Desire to be noticed	Desire to draw attention
Defiance of the body	Attractiveness of the body
The myth of invincibility and God	The myth of beauty and Goddess

On Ariana Grande's Instagram posts, we also see pictures taken with animals. Based on this fact, it is possible to say that she has adopted to be an "animal lover" as a social identity and that she is sensitive to animal rights. Grande, who has also preferred veganism, said in an interview related to this topic, "*I am a person who believes in herbal nutrition that prolongs life and will make you a happier person.*" The "exaggeration effect," which is one of the main proposals of social identity theory directly coincides with this situation. It is an undeniable fact that vegan nutrition is essential for health, but the debate continues in the field of medicine as to whether it is the only criterion. Grande has glorified the perception practices of her group when making comparisons between the social group where she belonged/ she had the sense of belonging and other groups, as well as favoring her ingroup. Additionally, Grande's preferring and staging of the animal-lover identity is also related to the elements of compassion, mercy and grace, which are regarded as indicators of the feminine culture.

DISCUSSION AND CONCLUSION

When we carry out a general assessment of the posts of the two people that are popular on Instagram, it is possible to say the following:

- The "sportsman" and "rich" identities that the male influencer chose and exhibited as an identity overlap with the elements of the masculine culture. He tried to prove that he was stronger in nature based on his sportsman identity, and demonstrated his personality that was defiant and aspired to succeed. His family man identity represents masculinity that protects and envelopes. It was seen that he had a desire to be noticed, based on his posts that revealed he had a high-income level.
- The "animal-lower" identity, which was preferred by the female influencer, contains elements such as mercy, soft compassion and mercy, which are the characteristics of the feminine culture.
- In the Instagram posts of both influencers, there was virtually no room for the universal spaces that became symbols and brands. In our opinion, the reason for this situation is due to the photography-based nature of Instagram. As mentioned in Lacan's mirror stage metaphor, the pleasure of seeing and being seen is primarily experienced by the individual himself. (Gündüz et al., 2018, p. 1873). Therefore, the ultimate urge on the basis of the posts is to lure the eyes and attract attention by being attracted to peeking/being peeked.

- Apart from the identities that we analyzed in the Instagram profiles of the both people, we also found posts where they presented their different identities, but these occupied a very small space in the showcase. Goffman suggests that this is due to the assumption that the personality that is reflected covers everything about the person reflecting it because the routine that is staged is perceived to be the only routine or the most important routine (Goffman, 2018, p. 56). According to VanDick, with the emergence of online media, direct experience has been replaced by mediated perception. Thus, an iconic and symbol-based perception form has emerged, and experiencing has been replaced by witnessing (VanDick, 2006, p. 212). When we look from this angle, the posts, which are indicative of the performance identities — that the people who were the subjects of this study prioritized and ranked first — will also be perceived as a holistic reality in the minds of those who watch the scene.
- The posts of the both people were based on self-centered personal content. The subject, rising with post modernism, has come to be both a watchable and observable entity synchronously with the emergence of social networks, and a structure that everyone is watching everyone else has risen (Uluç & Yarcı, 2017, p. 91). What is really important is to demonstrate performances to be applauded while being aware that you are being watched, as well as designing what role to play by watching.
- When we remember that the elements that trigger identity is the need to be unique and to belong to, it is possible to say that the male identity reflects the uniqueness side and the female reflects the belongingness side.

The reference point for designing both personal and social identities involves psychological motivations such as appreciation, approval and strengthening of self. In essence, there is no major difference between online environment and offline in the construction of identity. The binary structure of identity, based on the self and the internal-external dynamics, is in place but has just transformed into a new form. Users regenerate their offline selves on online platforms, but do not include all of their offline identities in this construction process. Only the appropriate portions of the offline identity, which has many different sorts of content, are manifested. Although it is a matter of debate to reach an induction such as that personal or social identities are to be established with gender codes and that male users will feed on masculine and female users will feed on feminine codes, what is not to be discussed is the fact that "performances" will continue to be exhibited (Morva, 2014, p. 238). Social networks are like black holes that instantly pull every object in their orbit towards them and swallow it with the power of gravity they create. Just as the theory suggests that if an object swallowed by a black hole will not be lost but is assumed to change its dimension, social media environments also have people experience a similar metaphorical situation. The perception/interpretation skills and self-esteem of the person who goes under the influence of social networks will also be transformed — just as black holes change the structure of substances through their force of gravity that is almost infinite. From the moment any person enters social networks, he becomes subject to a mental and cognitive transformation whether he shares any posts or not. The first phenomenon to be influenced by this transformation is the self. This is because the self, that is, the selfhood, is the first station of new quests. The mind that internalizes virtuality will also virtualize the self, and the order of social networks will begin to function.

REFERENCES

Akgül, S. K. ve Pazarbaşı, B. (2018). Küresel Ağlar Odağında Kültür, Kimlik ve Mekan Tartışmaları, Hiperyayın, İstanbul.

Akmeşe, Z. ve Deniz, K. (2017). Stalk, Benliğin İzini Sürmek. *Yeni Düşünceler*, *8*, 23–32.

Bakıroğlu, C. T. (2013). *Sosyalleşme ve Kimlik İnşası Ekseninde Sosyal Paylaşım Ağları, Akademik Bilişim, XV*. Akademik Bilişim Konferansı Bildirileri.

Batmaz, V., (1981). Popüler Kültür Üzerine Değişik Kuramsal Yaklaşımlar, İletişim 1981/1, Ankara, Turkey: AGTGA Gazetecilik ve Halkla İlişkiler Yüksek Okulu Yayını No:2, 163-192.

Bauman, Z. (2017). *Kimlik*. Ankara, Turkey: Heretik.

Bayad, A., (2016). Erving Goffman'ın Benlik Kavramı ve İnsan Doğası Yaklaşımı, Psikoloji Çalışmaları / Studies in Psychology, 36-1 (81-93).

Berger, J. (1990). *Görme Biçimleri, Çev.; Salman, Y*. İstanbul, Turkey: Metis.

Bilgin, N. (2007). *Aşina Kitaplar*. İzmir, Turkey.

Binark, M. (2001). Kadının Sesi Radyo Programı ve Kimliği Konumlandırma Stratejisi. In *Toplumbilim, Sayı:14*. Ankara, Turkey: Bağlam Yayınları.

Bocock, R. (1997). *Tüketim, Çev., Kutluk, İ*. Ankara, Turkey: Dost Yay.

Boyd, D. M., & Ellison, N. B. (2007, Oct. 1). Social Network Sites: Definition, History, and Scholarship. *Journal of Computer-Mediated Communication*, *13*(1), 210–230. doi:10.1111/j.1083-6101.2007.00393.x

Çağan, K. (2003). *Popüler Kültür ve Sanat*. Ankara, Turkey: Altınküre.

Curran, J. (1997). Medya ve Demokrasi, 139-197. In Medya, Kültür, Siyaset, (Eds: Süleyman İrvan), Ark Yayınları, Ankara, Turkey.

Demirtaş, H. A. (2003). Sosyal Kimlik Kuramı, Temel Kavram ve Varsayımlar. İletişim Araştırmaları, 1(1), 123-144.

Erdoğan, İ. (2004). Popüler Kültürün Ne Olduğu Üzerine, Bilim ve Aklın Aydınlığında Eğitim Dergisi: Popüler Kültür ve Gençlik. *Sayı*, *57*, 7–19.

Erdoğan, İ. ve Alemdar K. (2005). Popüler Kültür ve İletişim, Erk, Ankara.

Erkul, R. E. (2009). Sosyal medya araçlarının (web 2.0) kamu hizmetleri ve uygulamalarında kullanılabilirliği. Türkiye Bilişim Derneği-Bilişim Dergisi (116).

Fougère, M., & Moulettes, A. (2007). The Construction of the Modern West and the Backward Rest: Studying the Discourse of Hofstede's Culture's Consequences. *Journal of Multicultural Discourses*, *2*(1), 1–19.

Giddens, A. (2012). *Sosyoloji*. İstanbul, Turkey: Kırmızı Yay.

Goffman, E. (2018). *Gündelik Yaşamda Benliğin Sunumu*. İstanbul, Turkey: Metis.

Gündüz, A., Ertong Attar, G., & Altun, A. (2018). Üniversite Öğrencilerinin Instagram'da Benlik Sunumları. *DTCF Dergisi*, *58*(2), 1862–1895. doi:10.33171/dtcfjournal.2018.58.2.32

Jenkins, R. (2016). *Bir Kavramın Anatomisi Sosyal Kimlik*. İstanbul, Turkey: Everest Yay.

Jupp, V. (n.d.). The Sage Dictionary of Social Research Methods. London, UK: Sage.

Karaduman, N. (2017). Popüler Kültürün Oluşmasında ve Aktarılmasında Sosyal Medyanın Rolü, Erciyes Üniversitesi Sosyal Bilimler Enstitüsü Dergisi XLIII, 2017/2, 7-27.

Karaduman, S. (2010). Modernizmden Postmodernizme Kimliğin Yapısal Dönüşümü. *Journal of Yasar University*, *17*(5), 2886–2899.

Kartarı, A. (2017). Nitel Düşünce ve Etnografi: Etnografik Yönteme Düşünsel Bir Yaklaşım. Moment Dergi, 4(1), 207-220. Retrieved from http://dergipark.gov.tr/moment/issue/36383/411586

Kellner, D. (2001). Popüler Kültür ve Postmodern Kimliklerin İnşası, Çev: Gülcan Seçkin, Doğu Batı, Sayı: 15.

Kozinets, R. V. (2002). The Field Behind the Screen: Using Netnography for Marketing Research in Online Communities. *JMR, Journal of Marketing Research*, *39*(February), 61–72. doi:10.1509/jmkr.39.1.61.18935

Langer, R., & Beckman, S. C. (2005). Sensitive Research Topics: Netnography Revisited. *Qualitative Market Research*, *8*(2), 189–203. doi:10.1108/13522750510592454

Mansell, R. vd., (2015), The International Encyclopedia of Digital Communication and Society. UK: Wiley-Blackwell.

Mayfield, A. (2008). What is Social Media. iCrossing. e-book. Retrieved from http://www.icrossing.com/sites/default/files/what-is-socialmedia-uk.pdf

McQuail, D. (1994). *Kitle İletişim Kuramı: Giriş. (Çev. Ahmet Haluk Yüksel)*. Eskişehir: Kibele Sanat Merkezi Yayınları.

Mlicki, P. P., & Naomi, E. (1996). Being Different or Being Better? National Stereotypes and Identifications of Polish and Dutch Students. *European Journal of Social Psychology*.

Morva, O. (2014). In S. Çakır (Ed.), *Goffman'ın Dramaturjik Yaklaşımı ve Dijital Ortamda Kimlik Tasarımı: Sosyal Paylaşım Ağı Facebook Üzerine Bir İnceleme, Medya ve Tasarım* (pp. 231–255). İstanbul, Turkey: Urzeni.

Niedzviecki, H. (2010). *Dikizleme Günlüğü, Çev.: Gündüz, G.* İstanbul, Turkey: Ayrıntı.

Niedzviecki, H. (2011). *Ben özelim; Bireylik Nasıl Yeni Konformizm Haline Geldi, Çev.; Erduman, S.* İstanbul, Turkey: Ayrıntı.

Oğuz, G. Y. (2010). Güzellik Kadınlar İçin Nasıl Vaade Dönüşür: Kadın Dergilerindeki Kozmetik Reklamları Üzerine Bir İnceleme, Selçuk iletişim, C.: 6, S.: 3, (184-195).

Onat, F. ve Alikılıç, Ö. A. (2008). Sosyal Ağ Sitelerinin Reklam ve Halkla İlişkiler Ortamları. *Journal of Yaşar University*, *3*(9), 1111–1143.

Özüdoğru, Ş. (2014). Nitel Araştırmanın İletişim Araştırmalarında Rol ve Önemi Üzerine Bir Deneme, Global Media Journal, S. 4(8), 260-275.

Padilla, A. M., & Perez, W. (2003, February). Acculturation, Social Identity, and Social Cognition: A New Perspective. *Hispanic Journal of Behavioral Sciences*, *25*(1), 35–55. doi:10.1177/0739986303251694

Robins, K. (2013). *İmaj Görmenin Kültür ve Politikası, (N. Türkoğlu, Çev)*. İstanbul, Turkey: Ayrıntı Yayınları.

Rowe, D. (1996). *Popüler Kültürler. (Çev. Mehmet Küçük)*. İstanbul, Turkey: Ayrıntı Yayınları.

Schroeder, R. (1994). Cyberculture, cyborg post-modernism and the sociology of virtual reality Technologies. *Future*, *26*(5), 519–528. doi:10.1016/0016-3287(94)90133-3

Schudson, M. (1999). *Popüler Kültürün Yeni Gerçekliği: Akademik Bilinçlilik ve Duyarlılık, Popüler Kültür ve İktidar, (Derleyen: Nazife Güngör)*. Ankara, Turkey: Vadi Yayınları.

Storey, J. (2009). *Cultural Theory and Popular Culture* (5th ed.). Harlow, UK: Pearson.

Tajfel, H. (1982). Social psychology of intergroup relations. *Annual Review of Psychology*, *33*(1), 1–39. doi:10.1146/annurev.ps.33.020182.000245

Uluç, G., ve Yarcı, A. (2017), Sosyal Medya Kültürü, Dumlupınar Ünv., Sosyal Bilimler Dergisi, S.: 52, 88-102.

Van Dick, J. (2006). *The Network Society Social Aspects of New Media*. London, UK: Sage.

ADDITIONAL READING

Ellemers, N., Spears, R., & Doosje, B. (2002). Self and Social Identity. *Annual Review of Psychology*, *53*(1), 161–186. doi:10.1146/annurev.psych.53.100901.135228 PMID:11752483

Featherstone, M. (2007). *Consumer Culture and Postmodernism*. Sage Publications.

Hall, S. (2000). Who Needs 'Identity. In A. Reader, P. du Gay, J. Evans, & P. Redman (Eds.), *Identity* (pp. 15–30). Sage Publications.

Turkle, S. (1995). *Life on the Screen: Identity in the Age of the Internet*. New York: Simon & Schuster.

KEY TERMS AND DEFINITIONS

Identity: It is a collection of signs, qualities and features that show what a person is as a social being.

Instagram: Instagram, a photo sharing application created by the endeavors of two entrepreneurs at the end of 2010, mainly enables the sharing of photos in mobile devices on social networks.

Masculinity and Femininity: Express roles imposed on genders.

Popular Culture: It can be defined as a type of culture that is based on the pleasures of ordinary people, not a trained elite.

Popularity: It is the state or condition of being liked, admired, or supported by many people.

Postmodern Identities: Postmodernism always envisions how to live if it is felt rather than pre-determined strict rules. Postmodernism considers diversity and differences in identity and diversified identities oppose the monopoly of meta-narrative and teachings.

Self-Presentation: Presenting himself/herself in a way that leaves the desired image in accordance with socially and culturally accepted norms of action and behavior.

Social Media: is an online network that publishes and publishes its own content. Social media is actively used by many people and institutions. In this way, quick access is easier, users can view the contents, articles, news, thoughts, daily events, photos by social media through social media.

Chapter 12

Real-Time Marketing as a New Marketing Approach in the Digital Age:

A Study on the Brands' Social Media Sharing in Turkey

Selçuk Bazarcı
Ege University, Turkey

ABSTRACT

Nowadays, in order for brands to respond to consumer expectations, digital media efforts need to be involved in the brand communication process. Brands have a unique way to remind their names in a consumer's mind with real-time marketing. In addition, real-time marketing offers a way to make it easier for marketers to reach their target audiences at a low cost when increasing the speed and functionality of information. In this chapter, real-time marketing posts that have high user interaction on Twitter are handled in the context of their process, content features, and message appeal. Examined were 185 tweets. According to the data obtained, brands are trying to create positive brand image for consumers. Besides, it has been determined that both informational and emotional appeals are used intensively in order to create brand awareness.

DOI: 10.4018/978-1-7998-1041-4.ch012

INTRODUCTION

Developments and practical applications in digital media have an encouraging feature for brands to find alternative ways to reach consumers. Due to the changes in consumer expectations, marketing communication efforts must be re-established through digital platforms. Because today's consumers have tended towards brands that hold interesting features in which make a difference rather than searching for new and quality ones. At this point, it is crucial that the brand diversifies its marketing communication efforts through alternative applications and identifies an integrated brand strategy to create a positive perception.

One of the most important elements of consumer interest in the brand is to capture the moment. In this context, real-time marketing is one of the applications that brands perform in connection with current practices on digital media platforms. Real-time marketing which allows a brand to establish an effective connection with consumers is an alternative digital marketing communication effort to meet consumer expectations. It is also a Web 2.0-based social media application forms online moves against competing brands.

Real-time marketing possibly addresses a process in which consumer interaction that involves engagement and communication oriented structuring in the definition. Additionally, the brand organizes activities to take active dynamism and action on time.

Real-time marketing, a strategy which brand-generated content simultaneously transmits to the user, creates an organic link between the brand and the consumer. Besides it ensures consumers to engage in an interactive process.

MEDIA AND MARKETING COMMUNİCATION FROM TRADITIONAL TO DIGITAL

One of the most distinctive principles for the survival and socialization of humanity is the sharing of messages and ideas with the dynamics that make up society. In this context, the communication which is a product of the way of human being's existence and which has evolved according to the developments in the form of survival is a human-specific fact (Oskay, 2001, p. 1).

The development of technology and the adaptation of humanity to these changes have changed the dimension of communication. Due to this rapid change, it is possible to say that a significant part of communication practices has shifted to digital environments. With the help of web-based applications, interactive connections has changed the structure of communication. In particular, users' content and the ability to share these contents, causes significant changes in the functioning of communication. The different practices that took place on an individual basis together with the transformation have led to significant changes in the social sense can be stated.

The rapid progress of technology makes it easier to include in social practices. The evolution of media into a digital dimension supports this social change. When defining the concept of digital media, it is important to focus on the transformations and the effects of societies on the transition from traditional media to digital. Every society has influenced by this process of transformation which has penetrated the communication practices. McLuhan (1964) argues that the speed, scale or type of change in any media or technology has a profound effect on human relations. In this respect, there is a social change in the transition from traditional media to digital. Castells (2005, p. 441) states that this new communication system, which has global reach, brings all communication tools together and has the potential of mutual interaction, causes serious differences on the culture and this will continue rapidly. Dijck (2013, pp. 17-18) says that the media has historically changed with the masses that use it. According to Dijck, in the last two centuries media technologies have become mature as a part of daily social practices. It is an example of this through mobile devices that interact with people on a virtual plane, transforming digital communication into community routines or cultural practices.

Media evolving from traditional to digital has a feature that facilitates the realization of interactive, participatory and speed-oriented activities. As with many consumer-oriented structuring, a digital transformation is necessary for marketing. Instead, marketing approaches around the features of new media and digital marketing should be radically changed and reapplied. In this respect, it is necessary to demonstrate the dynamics of how consumers are affected by digital channels when basic marketing principles such as positioning and segmentation of marketing continue (Wertime & Fenwick, 2008, pp. 29-30).

REAL-TIME MARKETING

Nowadays, as the technology and internet connected applications change the structure of communication, brands are turning to alternative marketing efforts. Web 2.0 generally considered as the basis of the modern communication system particularly, is important in every aspect of communication. Regardless of time and space, Web 2.0 creates a virtual network environment where users can create their content and share quickly. In this context, the production of information becomes accessible with the new network environment can be assumed. Furthermore, the dynamics enabling the spread of new knowledge transformed into an efficient structure.

For a long time, different methods for marketing communication studies have been discussed by experts who are interested in creating a strong communication between brands and consumers. Especially communication technologies that provide convenience in the collection and processing of information reveal new trends in marketing communication. With the systemic use of these new trends by brands, it has become possible to build an effective connection with consumers globally (Macy & Thompson, 2011, pp. 57-58).

Real-time marketing is the creation and execution of an instantaneous marketing message in response to and in conjunction with an occurrence during a live event (Clow & Baack, 2016, p. 287). Since real-time marketing as an instant marketing communication work that is produced by applied content in social networks over an existing event or situation over time, there are important strategies that web-based marketing efforts have in positively influencing brands to gain a place in consumer mind and purchasing decisions. Real-time marketing produced relatively the topics and situations raised by social media users. At this point, it is not possible to think of real-time marketing apart from content marketing (Odden, 2012, pp. 15-16). It is critical to present the content at the right time in the consumer purchase cycle. Before a real-time study is carried out, a specific strategic infrastructure must be provided to control the process within the control. Especially, brands follow real-time trends to strengthen their real-time marketing efforts, create dynamic and customized content, increase audience engagement, and ensure full understanding of digital analytics.

Real-time marketing issues have different characteristics. Kerns (2014, p. 32) called these two different contents 'known' and 'unknown' (Figure 1):

Figure 1. The real-time marketing matrix, Kerns (2014, p. 32)

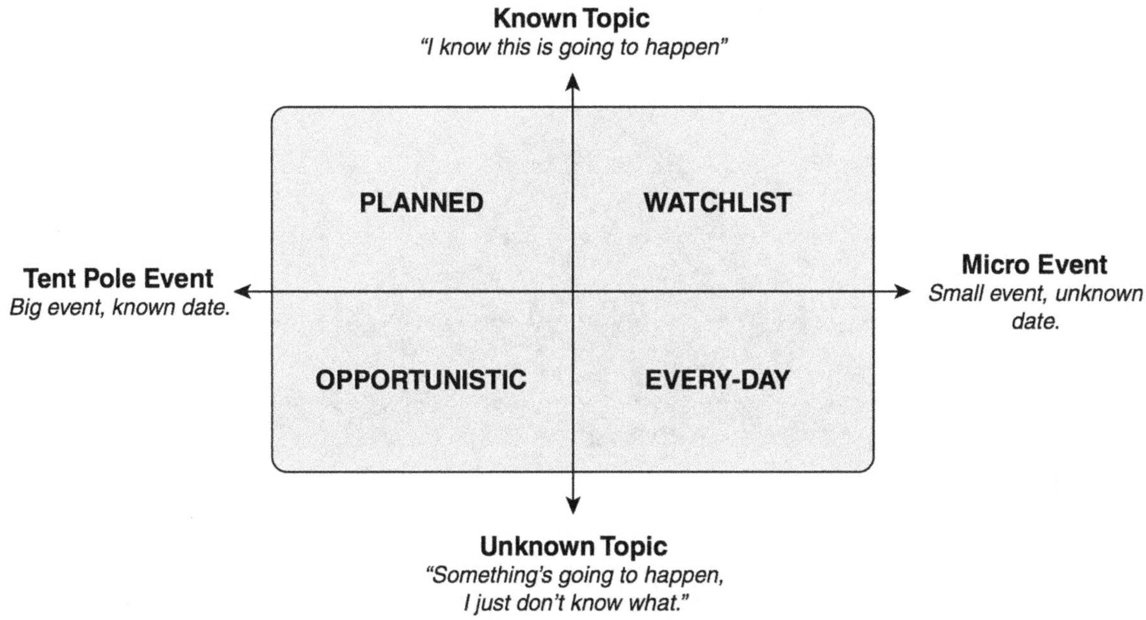

Known real-time marketing content is meant to be posted in real-time as a large event is taking place, but it can be planned ahead of time due to the predictability around certain parts of the event. The planned and watchlist categories in the top of the matrix are two zones that allow content creation and approval before an event occurs. Unknown real-time marketing content is unpredictable, therefore must be created in a reactive manner after a problem has become a trend. Hence, a topic in social media shall be organized reactively by its creator and presented on digital media platforms until it turns a new trend. Creating real-time marketing content requires a quick reactive process. Therefore, the current short time is for brands creating opportunities to share (Kerns, 2014, p. 32).

Today, brands must understand consumers' expectations. Sain-Dieguez (2015) emphasizes the importance of consumer-wise for the brand to compete with its competitors. According to Sain-Dieguez, the real opportunity of real-time marketing, which provides efficient marketing communication in consumer transformation, is that it attracts the attention of things that are of interest to the consumer rather than the trend. On the other hand, giving target audiences what they need in a manner that is relevant, timely, and useful, is a successful long-term strategy that will keep them coming back again and again. The key is to focus on the three R's of RTM and find conversations where relevancy for business, customer, and time intersect (Sain-Dieguez, 2015).

One important factor in a successful sharing is that the content is business-related. Users should be able to perceive the brand in content. The main goal in real-time marketing is to attract attention and direct the consumer to the brand. For this reason, information about the business must be found in the shares. Another component is related to the user. It is important to organize content for the target audi-

Figure 2. The three r's of real-time marketing (Sain-Dieguez, 2015)

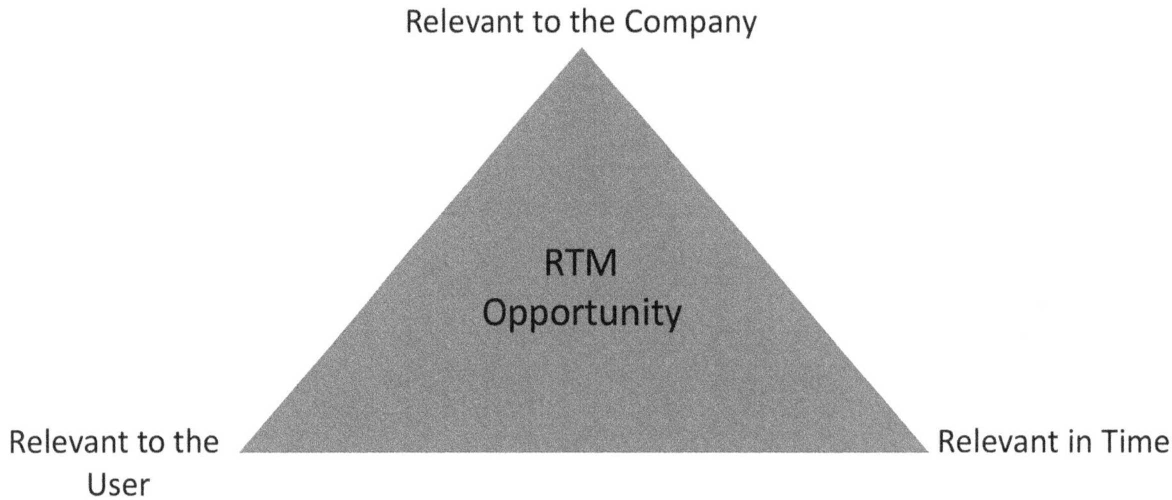

ences. Because if users do not like sharing, they cannot turn to the brand. The last component is related to time. Time is limited to create a content and deliver it to target audiences. Due to the fact that it is a strategy based on instant reactions, the right content must be presented to the users at the right time.

Before a real-time content has been created, a specific strategic work must be carried out in order to maintain the process within the control. The occurrence of any unexpected events may cause the brand to encounter a crisis For this reason, it is important to determine all processes from consumer analysis to the structure of current developments. Sanches and Restrepo (2015, pp. 174-178) emphasize four key steps for brands to strengthen their real-time marketing efforts:

1. **Tracking Trends in Real Time:** Trends are generated each day in social networks and the possibilities for brands to participate and being part of the conversation are innumerable, however it should be evaluated in a professional way which ones are related to their audience and specific niche.
2. **Dynamic and Customized Content:** Contents should be fast, creative and attractive enough to go viral through the web a practice known as word of mouth marketing. This content is not usually planned in advance but generated in real time with an organized marketing plan.
3. **Engagement with the Audience:** Today, brands are concerned about creating an emotional link to consumers in their digital marketing efforts. In this relationship, which is defined as engagement, consumers who interact with the brand voluntarily share their brand content as well as serve as a shield against negative comments coming to the brand. This defense mechanism means more than just 'likes' or 'retweets'. The brands now need to reveal their views, perceptions, needs or preferences more clearly.

4. **Fully Understanding of Digital Analytics:** It is important that the analysis of social networks and digital media is carried out effectively. The superb amount of information from the Internet and diversity of social networks and virtual communities makes it impossible to analyze without a specialized software to monitor trends, markets and audiences. In this respect, it is very important to use analytical data to carry out real-time studies successfully.

Consumers expect to find opportunities to interact with the brand on social media platforms. For this reason, it is essential that the brand controls the changing consumer reactions and the perceptions about the brand. Through real-time marketing, it is possible for the brand to keep both the consumer-related feedback cycle updated and to create a positive brand perception in the consumer with a real-time response system.

METHOD

In this research, we obtained the descriptive research method. It is possible to state descriptive research as a qualification study to examine, define and highlight details on a subject (Ethridge, 2004, p. 24). The goal of descriptive research is to describe a phenomenon and its characteristics (Nassaji, 2015, p. 129).

The Problems of the Research

- What are the content features of real-time marketing shares with high user interaction?
- What are the general characteristics of the intended strategies in real-time marketing?
- How does the message appeals of the posts show with their formal and contextual dimensions?

The Purpose of the Research

The purpose of this study is to demonstrate the formal and contextual features of real-time marketing posts which high user interaction.

DATA ANALYSIS

Research data are assessed by the content analysis method. The main purpose of the content analysis is to create a framework to reveal the relevant concepts and relations to explain the data obtained (Yıldırım & Şimşek, 2013, p. 259). One of the reasons for choosing content analysis in the research is that content analysis is appropriate for the nature of research in defining the structural features of Twitter shares. Moreover, the content analysis provides a correct framework for the researcher to summarize and standardize the data obtained sensibly.

SAMPLE

According to the study, brands' real-time marketing shares on Twitter constitute the research universe. The sample of the study is the most valuable brands in the report of Brand Finance, an independent brand valuation organization (www.brandfinance.com). The report obtains one hundred most valuable brands in 2016 of Turkey. Accordance with high user interaction 185 tweets of the brands examined.

Table 1. Brands sharing frequency

	Brand Name	Followers	Frequency	Percent	Known RTM	Unknown RTM
1	Turkish Airlines	1,16 Mn	46	24,9	35	9
2	Turkcell	624 B	28	15,1	21	7
3	Teknosa	531 B	3	1,6	3	-
4	Ford Otosan	335 B	3	1,6	3	-
5	Türk Telekom	333 B	8	4,3	8	-
6	Migros	280 B	3	1,6	3	-
7	Garanti Bankası	263 B	9	4,9	9	-
8	Ziraat Bankası	252 B	12	6,5	9	1
9	Pegasus Airlines	236 B	2	1,1	2	-
10	Yapı Kredi	188 B	11	5,9	11	-
11	İş Bankası	168 B	10	5,4	9	1
12	Ülker	154 B	15	8,1	12	3
13	CarrefourSa	136 B	7	3,8	6	1
14	LCWaikiki	120 B	1	0,5	1	-
15	Koton	105 B	3	1,6	3	-
16	Halkbank	92 B	1	0,5	1	-
17	Qnb Finansbank	85 B	5	2,7	3	2
18	Vestel	84 B	2	1,1	1	1
19	AlbarakaTürk	55 B	3	1,6	2	1
20	İnci Akü	48 B	1	0,5	1	-
21	Boyner	34 B	2	1,1	2	-
22	Opet	21 B	1	0,5	1	-
23	Sinpaş Gyo	18 B	1	0,5	1	-
24	Pınar	13 B	4	2,2	4	-
25	Petlas	6 B	3	1,6	2	1
	Total		**185**	**100**	**156**	**29**

CREATION OF CATEGORIES AND COLLECTION OF DATA

The tables containing the convenient code and theme list in line with the purpose determined within the scope of the research were created after the important studies accepted in the literature were evaluated. The categories are shown in Tab.2, have been adapted from Elden and Özden's (2015) study. The categories are shown in Tab.3, are adapted from Akyol's (2011, p. 211) doctoral thesis. The categories are shown in Tab.5, are adapted from Yeniçıktı's (2006, pp. 105-106) article. The categories are shown in Tab.6 are from Ashley and Tuten's (2015, p. 21) study. Finally, the categories identified in Tab.4 and Tab.7, in which the characteristic of post and advertising appeals have been developed using Elden and Bakır's (2010, p. 87-91) study.

RELIABILITY BETWEEN CODERS

Important studies in the literature were examined in the preparation of the tables which form the code and theme list in the research. After the tables created, all are checked by two experts with necessary arrangements. Two independent encoders studied for reliability. Accordance with the analysis made to measure the consistency between the encoders, the data obtained in this study was found to have a consistency of *0.874*. According to the interpretation of the Kappa analysis, the relevant study has excellent consistency (Viera & Garrett, 2005, pp. 362-363).

RESULTS AND DISCUSSION

One of the most important elements in capturing high user interaction is the creation of content that attracts users. It is also important to deliver the message quickly without disturbing the consumer. Although Twitter is predominantly a text-based platform, almost all of the high user interaction shares have visual content. Visual content is used in approximately 98% of shares (Table 2). At this point, it is possible to say that the visual content used in real-time marketing studies is an important factor in obtaining interaction.

When the table containing the structure of the post is considered, it is seen that informative contents are preferred frequently (42,2%). In particular, it can be said that informative narrative forms are frequently included because of the nature of the special days and announcements. By the data obtained from the table, brands prefer an impressive and sensual expression style after informative content in real-time content sharing (29,2%). Real-time content needs constructing to create high user interaction with unique shares.

Table 2. General content of the post

General Content of the Post	Frequency	Percent
Text	5	2,7
Visual	5	2,7
Text and visual combination	175	94,6
Total	**185**	**100,0**

Table 3. Structure of the post

Structure of the Post	Frequency	Percent
Informative	78	42,2
Impressive, sensual	54	29,2
Action oriented	20	10,8
Question sentence	11	5,9
Peremptory	4	2,2
Conditional sentence	1	0,5
Storyteller	16	8,6
Other	1	0,5
Total	**185**	**100,0**

Table 4. The characteristic of the post

The Characteristic of the Post	Frequency	Percent
Wishing success (to someone)	15	8,1
Motivating	18	9,7
Thanking	1	0,5
Wishing	5	2,7
Challenging	6	3,2
Identifying with a hero	7	3,8
Integrating with a brand	47	25,4
Celebration, remembering	79	42,7
Other	7	3,8
Total	**185**	**100,0**

Real-time marketing content includes both current unpredictable events and the works that are already known and arranged within a plan. Recently, consumers see brands as an organism. For this reason, consumers' expectations from brands are not only the introduction of their products and services but also they have a dynamic understanding that can react to current events and situations. In this context, it is seen that special days, which are important for the target audience, provide convenience to brands in providing high user interaction (42,7%). In addition, generating content for helping consumers to commune with the brands is essential. When the obtained data analyzed, this feature used in 25.4% of the shares is seen.

Brands need to create a structure that can reach their target audience quickly. Besides, it is important to create content to increase brand image and brand awareness. Real-time marketing is one of the digital marketing communication pieces, preferred by brands as a complementary strategy. According to the table, it is determined that brands generate content for special days 40% of the shares to catch high user interaction (Tab.5). In addition, 14.1% of the shares are made for the purpose of sponsorship. 13.5% of the shares are informative and announcement-oriented. 11.9% of the posts are intended to target audience interaction, while 8.6% event and competition, 7% advertising.

Table 5. Intended strategies

Intended Strategies	Frekans	Yüzde
Information, announcement	25	13,5
Product introduction	2	1,1
Promotion	1	0,5
Advertising	13	7
Social responsibility	2	1,1
Target audience interaction	22	11,9
Special day	74	40
Sponsorship	26	14,1
Sales-promotion	1	0,5
Campaign (Corporate)	2	1,1
Event, Competition	16	8,6
Other	1	0,5
Total	**185**	**100,0**

Table 6. Message strategies

Message Strategies	Frequency	Percent
Integrated content	-	-
Interactivity	32	17,3
Functional appeal	3	1,6
Emotional appeal	78	42,2
Experiental appeal	9	4,9
Unique selling proposition (USP)	3	1,6
Comparative	2	1,1
Resonance	-	-
User Image	1	0,5
Social cause	11	5,9
Exclusivity	7	3,8
Animation	-	-
Spokescharacter/spokesperson	5	2,7
Other	34	18,4
Total	**185**	**100,0**

In order to meet the content of the brand with consumers, it is necessary to construct the structure of the messages by reflecting a certain integrity. It is a platform where social media content is consumed quickly. Therefore it is important to ensure that the brand is active in the consumer's mind. One of the primary methods to be catchy and attractive is the use of emotional elements in the posts. Message

Table 7. Advertising appeals

Advertising Appeals	Frequency	Percent
Informative	27	14,6
Collective value	31	16,8
Activity	17	9,2
Kindness	14	7,6
Humor	6	3,2
Identification	11	5,9
Unique	4	2,2
Popularity	5	2,7
Pride	10	5,4
Prestige	4	2,2
Responsibility	5	2,7
Family	10	5,4
Excitement	8	4,3
Curiosity	4	2,2
Cheer	9	4,9
Superiority	4	2,2
Other	20	10,8
Total	**185**	**100,0**

strategies in the posts support that. Message strategies with emotional appeal constitute 42,2% of the total share. According to the data obtained, interaction is another element that brands prefer as a message strategy (17,3%). It is important to create interaction in the sense of branding because the involvement of consumers in the sharing process by establishing an organic link makes it possible for other consumers to participate. Another remarkable category is the other (18,4%). This category includes informative content that celebrations, commemorations, and sponsorships.

There are many types of advertising appeals that affect the content of the post. Some of them are preferred by brands more by considering the data obtained. At the beginning of the most preferred message appeals is collective content. Collective messages are used by brands 16,8% of the shares. One way to mobilize consumers with real-time marketing content is to use a collective message tone for sharing. Collective messages that affect people deeply provide some convenience to brands in order to ensure high user interaction. In addition to the collective message appeals, which is part of the emotional narrative style, the informational message appeals are also preferred by the brands. With the informative appeals that constitute 14.6% of the shares, brands forward a message to consumers with announcement content. In addition to these, activity (9,2%), kindness (7,6%), identification (5,9%) and pride (5,4%) message appeals are preferred by brands.

CONCLUSION

One of the most important goals of social media sharing is to create content in a way that attracts users. Real-time marketing activities are based on attracting consumer interest by creating quality and original content. In doing so, it is intended that individuals interact with the brand and contribute to the dissemination of shared content. The reason why the shares are created over the visual content is to increase the interaction power of the sharing. When checking at the shares in this research, almost all of the created content preferred for visual studies is seen. At this point, it is possible to indicate that the visual content for real-time posts is an important element in achieving high user interaction.

Another issue that draws attention in the study is that almost half of the shares are edited with an informative content feature. Particularly, the creation of a significant part of the shares with high user interaction for special days and announcements is the reason for the frequent use of information contents in the sharing. According to the findings, a significant part of the posts is shaped by an impressive and sensual structure. It is important to use creative and interesting features frequently in order for real-time marketing to be successful. In this context, it is highly possible to use dynamics with emotional appeal in sharing to create brand awareness on the consumer.

Consumers' expectation from brands is to demonstrate the necessary reactions to current developments rather than promoting their products and services. In particular, it is expected that the brand will make a meaningful sharing about the agenda issues that consumers are interested in and on special days concerning social memory. When the findings obtained from the research are examined, it is seen that the brands prepare their contents near to half of the shares for special days. It is determined that brands frequently use emotional-based appeals in the construction of these contents. In terms of the strategies aimed at this point, it can be said that the content which is prepared for special days and composed of emotional appeals is important in terms of real-time marketing studies.

ACKNOWLEDGEMENT

This study is adapted from author's master thesis.

I would like to thank Assoc. Prof. Dr. Uğur Bakır, my thesis advisor, who contributed greatly to the emergence of this study and shared his scientific knowledge

REFERENCES

Akyol, A. Ç. (2011). *Reklam Mesajlarında Bilgilendirici İçerik: Dergi Reklamları Üzerine Bir İçerik Analizi*, (Unpublished doctoral thesis), Selçuk University/Social Sciences Institute, Konya, Turkey.

Ashley, C., & Tuten, T. (2015). Creative strategies in social media marketing: An exploratory study of branded social content and consumer engagement. *Psychology and Marketing*, *32*(1), 15–27.

Castells, M. (2004). *The network society: a cross-cultural perspective*. Cheltenham, UK: Edward Elgar Publishing. doi:10.4337/9781845421663

Clow, K. E., & Baack, D. (2017). *Integrated advertising, promotion and marketing communication. (8th edit.)*. Harlow, UK: Pearson Education Limited.

Dijck, J. (2013). *The culture of connectivity: A critical history of social media*. New York, NY: Oxford University Press. doi:10.1093/acprof:oso/9780199970773.001.0001

Elden, M., & Bakır, U. (2010). *Reklam Çekicilikleri: Cinsellik, Mizah, Korku*. İstanbul, Turkey: İletişim Publications.

Elden, M., & Özdem, Ö. O. (2015). *Reklamda Görsel Tasarım - Yaratıcılık ve Sanat*. İstanbul, Turkey: Say Publications.

Eltantawy, N., & Wiest, J. B. (2011). Social media in the Egyptian Revolution: Reconsidering resource mobilization theory. *International Journal of Communication*, (5), 1207–1224.

Ethridge, D. E. (2004). *Research methodology in applied economics*. New York, NY: John Wiley & Sons.

Jenkins, H., Purushotma, R., Weigel, M., Clinton, K., & Robinson, A. (2009). *Confronting the challenges of participatory culture: media education for the 21st century*. London, UK: MIT. doi:10.7551/mitpress/8435.001.0001

Kerns, C. (2014). *Trendology: building an advantage through data-driven real-time marketing*. New York, NY: Palgrave MacMillan. doi:10.1057/9781137479563

Macy, B., & Thompson, T. (2011). *The Power of Real-Time Social Media Marketing*. New York, NY: McGraw Hill.

McLuhan, M. (1964). *Understanding media: The extensions of man*. London, UK: Routledge.

Nassaji, H. (2015). Qualitative and descriptive research: Data type versus data analysis. *Language Teaching Research*, *19*(2), 129–132. doi:10.1177/1362168815572747

Odden, L. (2012). *How to Attract and Engage More Customers by Integrating Seo, Social Media, and Content Marketing*. Hoboken, NJ: John Wiley and Sons.

Oskay, Ü. (2001). *İletişimin ABC'si*. İstanbul, Turkey: Simav Publications.

Sain-Dieguez, V. (2015). Are you overlooking the most valuable real-time marketing strategy?. Retrieved from http://www.convinceandconvert.com/digital-marketing/are-you-overlooking-the-most-valuable-real-time-marketing-strategy/

Sanches, C. ve Restrepo, J. C. (2015). Strategic real-time marketing. Advances in the Area of Marketing and Business Communication, pp. 164-184.

Viera, A. J., & Garrett, J. M. (2005). Understanding interobserver agreement: The kappa statistic. *Family Medicine, 37*(5), 360–363. PMID:15883903

Wertime, K., & Fenwick, I. (2008). *Digimarketing: The Essential Guide to New Media and Digital Marketing. Hoboken, NJ:* John Wiley & Sons.

Yeniçıktı, N. T. (2016). Halkla ilişkiler aracı olarak Instagram: Sosyal medya kullanan 50 şirket üzerine bir araştırma. *Selçuk İletişim, 9*(2), 92–115.

Yıldırım, A. & Şimşek, H. (2013). *Sosyal Bilimlerde Nitel Araştırma Yöntemleri,* Ankara: Seçkin Publications. Retrieved from https://brandfinance.com/knowledge-centre/reports/brand-finance-turkey-100-2016/

ADDITIONAL READING

Dodson, I. (2016). *The Art of Digital Marketing*. New York: John Wiley&Sons, Inc.

Koçyiğit, M. (2015). *Sosyal Ağ Pazarlaması: Marka Bağlılığı Oluşturmada Yeni Bir Pazarlama Stratejisi*. Konya: Eğitim Publications.

Manovic, L. (2001). *The Language of New Media*. London: The Mit Press.

Öztürk, G. (2013). *Dijital Reklamcılık*. İstanbul: Beta Publications.

Palacios-Marques, D., & Saldaña, A. ve Vila, J. (2013). What are the relationships among Web 2.0, market orientation and innovativeness? *Kybernetes, Vol, 5*(42), 754–765. doi:10.1108/K-03-2013-0057

Scott, D. M. (2011). *Real-Time Marketing and PR*. New Jersey: John Wiley&Sons, Inc.

Zahay, D. (2015). Digital Marketing Management: A Handbook fort he Current (or Future) CEO, V. L. Crittenden (Ed.), Digital and Social Media Marketing and Advertising Collection. New York: Business Expert Press.

KEY TERMS AND DEFINITIONS

Advertising Appeals: Advertising appeals are communication strategies that marketing and advertising professionals use to grab attention and persuade people to buy or act. Advertising appeals aim to influence the way consumers view themselves and how buying certain products can prove to be beneficial for them.

Brand Awareness: Brand awareness is the level of consumer consciousness of a company. It measures a potential customer's ability to not only recognize a brand image but to also associate it with a certain company's product or service. Brand awareness is the probability that consumers are familiar with the life and availability of the product. It is the degree to which consumers precisely associate the brand with a specific product.

Content Marketing: Content marketing is a strategic marketing approach focused on creating and distributing valuable, relevant, and consistent content to attract and retain a clearly defined audience — and, ultimately, to drive profitable customer action.

Digital Conversion: It means adapting the digital media to the brand communication process to make it easier for brands to work and to increase their productivity.

Digital Media: Includes any format or device used to convey content using digital signals. Digital media is digitized content that can be transmitted over the internet or computer networks. This can include text, audio, video, and graphics. This means that news from a TV network, newspaper, magazine, etc. that is presented on a Web site or blog can fall into this category. Most digital media are based on translating analog data into digital data.

Marketing Communication: Can be defined as the methodologies and tactics adopted by the brands to convey the messages in a unique and creative manner to their existing and prospective customers about their offerings of products and services.

Real-Time Marketing: Real-time marketing is an important digital marketing communication strategy that enables brand and consumer to make an active meeting through social media. RTM is the ability to engage with brand's customers or fans instantly based on real-time information - like their actions and behavior, changes to brands own data or external news or events.

Social Media Marketing: Refers to the process of using social media such as Twitter, Facebook and YouTube as a means of interacting with target audiences. Social media marketing means marketing on social networks, or the promotion of goods and services through digital media. By utilizing the social aspect of the web, social media marketing is able to connect and interact on a much more personalized and dynamic level than through traditional marketing.

Chapter 13
Social Networks:
The New Medium of Advertising – Instagram Case

Zuhal Akmeşe
Dicle University, Turkey

ABSTRACT

The development of technology at an incredible speed today and the fact that the internet has become an important area of social life has led to differentiation in the structure of mass communication and content production, too. This differentiation has stimulated advertisers and companies to reach the target audience through social networks with many users and different characteristics. Companies employ different strategies to be effective in these platforms. One of these strategies is collaboration with social media phenomenon. The relationship between the social networks considered as the new medium of advertising, social media phenomenon identified as influencer in these networks, and advertising is examined within the scope of this chapter. In this context, data obtained from interviews with 50 Instagram phenomenon by using semi-structured interview technique, which is a qualitative research method, were analyzed and advertising collaborations with influencers in social networks were evaluated.

DOI: 10.4018/978-1-7998-1041-4.ch013

INTRODUCTION

It is possible to define advertising as a communication tool that informs individuals forming the target audience of a brand and product and persuades them to purchase those products and services of this brand. The main purpose of the advertising is to stimulate the target audience for a product or service via messages and to ensure that this product and service is purchased (Kocabaş & Elden, 1997, p. 9). The presence of a fierce competition environment of our today's world has created the necessity of creating more effective marketing areas for brands' products and services (Gürel & Alem, 2014, p. 5). Brands need new advertising channels and new strategies to make a difference in the message bombardment of numerous products and services and to reach the target audience. Companies that use the advertising strategies intensively to bring their brands and products together with the target audience have focused on social networks in this sense.

The fact that the concept of social media becomes a part of our daily lives by means of developing technologies, widespread use of social networks, and enabling the use of social networks by technological communication tools at any time by eliminating space and time barriers have made these platforms a functional space for advertising. Based on the idea that digital technologies are reshaping relationship and personal experiences with all dimensions in this sense and transforming current lives (McLuhan, 2005, p. 8), it is possible to say that advertising strategies are affected by this change and social networks are used as new platforms for brands and products in terms of marketing. Today, almost all brands take part in social networks through accounts of their brands and products and implement their digital advertising strategies by using different social networks in order to strengthen their brand and corporate image and to increase the reputation of their brand and corporate. In order to reach the target audience that is outside the range of their social media accounts, companies use the people identified as social media phenomenon for their companies' promotion and advertising works by cooperating with them and benefiting from the target audience of these people who have a certain number of followers in social media and are called as social media celebrity.

This study focuses on Instagram, where advertising and marketing strategies are used extensively and the collaborations of companies with social media phenomenon. In this context, the data obtained from interviews with 50 social media phenomenon were analyzed and the relationship between advertising and phenomenon in social media was evaluated.

Social Media Use and Instagram

Social networks have become an important element of everyday life as a result of widespread use of Internet. It is possible to define social networks as platforms that allow individuals to connect with other individuals and masses. The fact that social network applications eliminate space and time barriers in terms of communication, enable an interactive interaction and have easily accessible features are effective factors in expanding these networks and increasing the number of users each passing day. Social networks that appeared in social life after 2000s have been met with great interest because of their features such as sharing video and photo, making friends and expressing ideas. The fact that content in social networks is generated by users and social networks provide an opportunity to users to meet around their interests, common purposes, common good tastes or common benefits, these factors are other factors in the spread of these networks.

Figure 1. Countries' Instagram use data

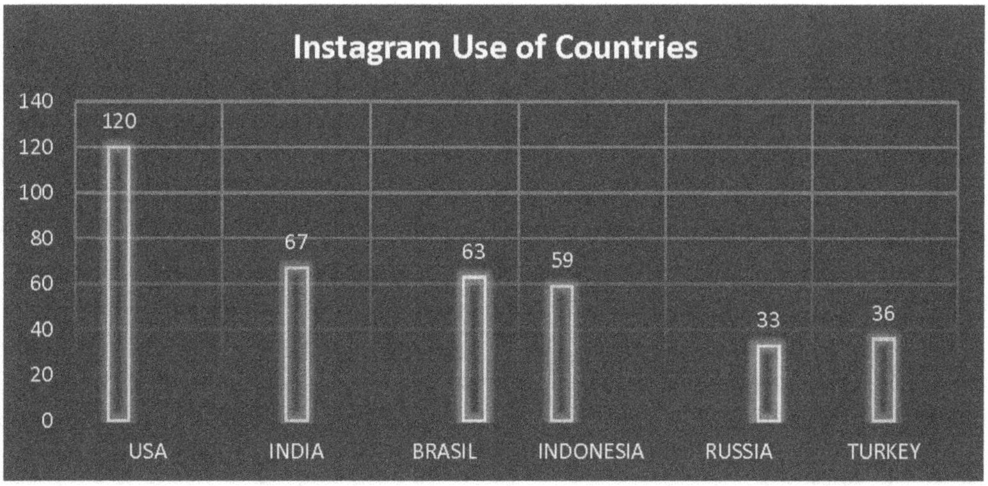

When the report for 2018 data of "We Are Social" concerning the use of internet and social media in the world analyzed; it is seen that 4.08 billion people (54% of the world's population) are internet users, 3.29 billion people (43% of the world's population) are social media users and 5.06 billion people (66% of the world population) are mobile users. When data examined, it is possible to express that the use of the Internet and social networks is very common throughout the world and that the interaction area of these platforms is very wide. Among these platforms, one of the most widely used applications is Instagram (Yegen, 2015).

Instagram was released in 2010 as a mobile photo sharing application first developed for IOS by Kevin Systrom and Mike Krieger. The name of the application originates from the combination of "instant" and "telegram". After Instagram has attracted attention all over the world in a short time, Facebook, one of the social media giants, purchased the application for a billion dollars in 2012. According to researches, it is stated that 30% of Instagram users are between 18-24 years old, the rates of female and male users are almost equal, 80% of users follow at least one company or business account and 50% of the companies followed use the "Instagram stories" feature regularly. According to July 2018 data, it is indicated that Instagram has more than one billion active users. Looking at the distribution of users around the world, while the US ranks first with about 120 million users, there are 67 million active Instagram users in India, 63 million in Brazil, 59 million in Indonesia, and 33 million in Russia(Yegen, 2015). When Instagram use in Turkey examined, it is seen that Turkey is ranked as the fifth country among the countries which use Instagram most in the world with 36 million active users.

When the data is evaluated in terms of the population of the countries and the number of Instagram users, it is possible to say that Turkey takes place near the top in Instagram use. Worldwide use gives an idea about how wide the domain of the shares made on this platform are. In this context, it is possible to say that Instagram is an effective medium for advertising and this platform interested by companies is highly functional for advertising and promotion.

Features and Use of Instagram

Developed for instant photo sharing on smart phones, this application has attracted attention in a short time and the number of users has increased by means of its features such as photo filtering, video and photo sharing. Its photo and video filtering feature enabled the application to become one of the most widely used applications on social media (gezginlerdunyasi.com). In addition to providing photo sharing opportunity and making photos more aesthetic and enjoyable with the filters it develops, the fact that it has a fast, practical and easy interface and it corresponds with social media platforms has been effective in increasing the number of users and the widespread use of the application. With its continuous updates and new added features, the application has become a promotion, marketing and e-commerce platform in addition to its main purpose of photo sharing. This made Instagram attractive for companies. When the occurrence adventure of the application is examined, released in 2010 as an application that enables photo sharing on smartphones, it is seen that new updates are added to Instagram with the feature to send direct messages to other users and share video (Huey & Yazdanifard, 2015, p. 3); besides, by the latest updates, the application has expanded its usage area with status sharing (Story), live broadcast and tagging features. By adding a tagging feature (#) with its update in 2011, it allows users to interact not only with friends on their network, but also anyone who shares using the same tag. In the same year, rewarded by TechCrunch in the category of "the best mobile application", Instagram has announced that the number of its users has reached 15 million members and 400 million photos (brandingturkiye.com). The increase of active users' number day by day, Instagram use of celebrities (artists, politicians, etc.), very large user network, and its features such as being applicable for visual and audio sharing and enabling interactive communication make the Instagram even more attractive; furthermore, in 2013, it enabled users' news feed to meet with "sponsored content" that concerns small businesses, boutiques and corporate brands based on their interests. Instagram, with its users 400 million in 2014, is increasing the number of its users by releasing new updates every day. In 2015, Instagram introduced the Layout collage, which allows you to share several photos in one frame, and users have the opportunity to experience a new and different application experience with this feature. Instagram, which included Boomerang in its features in 2015, continues to attract all attention with its features such as live broadcast, photo video sharing, story sharing and tagging (brandingturkiye.com). All these developments have taken Instagram away from its purpose of occurrence, i.e. photo sharing, and have loaded new missions to the application such as advertising, product marketing and shopping (Yegen and Yanık, 2015, p. 384). This change and conversion has also provided a basis for the emergence of Instagram phenomenon.

Celebrities of the Digital World, "Phenomenon" and "Influencers"

When looked at the origin of the word phenomenon, it is defined as "event, phenomen" (tdk.gov.tr) in Turkish dictionary. In the philosophy dictionary, the concept of phenomena is defined as a philosophy term that comes from the word *phainomenon* which means "appearance" in ancient Greek; besides, it is defined in the situated philosophy language as a philosophical term describing all things that can be thought of in terms of quality, object, relationship, location, and event that can be perceived, observed, heard, experienced or revealed by sense organs (Güçlü, Uzun, Uzun, & Hüsrev Yolsal, 2010, p. 610).

The concept of the phenomenon, transferring from English to Turkish, is a word that has transferred from Latin to English. The *phenomenon*, which means *famous* and *known*, has become a title that is used for people who are well-known and have more followers in the internet world. However, the concept of phenomenon used for social media is used for people who became famous via digital platforms and have a certain number of followers. Celebrities, defined as "*a well-known person by his/her acknowledger*" (Turner, 2004, p. 5), can be expressed as people who increase their recognition level on social media platforms and who stand out from the people they compete with in a different way. It is seen that social media celebrities, defined as phenomena, are generally defined by using 'Instagram phenomenon' and 'YouTube phenomenon' (teknoloskop.net). In this context, it is possible to define the "phenomenon" as celebrities of social media or the digital world. Phenomenon increase their popularity by means of the profiles they create through video, photo sharing, blogs or social networks (Senft, 2008, p. 25). Phenomenon present themselves by framing themselves on a particular topic of interest (fashion, motherhood, food, loving animals, humor, sightseeing etc.), or with a feature that they highlight, and in this subject or interest, they come to the fore as a "most" in a sense. They make efforts to continuously increase the number of their followers by making shares related to the relevant field, to reach out to everyone who is interested in that field and try to get more recognition. It is possible to say that phenomenon, with these images and attitudes and behaviors they create, they first transform themselves into a brand in a sense (personal branding), after that, they play opinion leader role through their popularity by influencing their target audiences (follower). Recently resulting around the world, Instagram mothers set an example for that. Instagram mothers, or with their widespread use, Instamoms, are used for people who have reached a certain number of followers in the Instagram application and who have created their identities on digital platforms through their motherhood identity. Instagram mothers who identify themselves as Instamom, first make their own branding by transforming themselves into a brand through their motherhood identity, and then, they regularly share their experiences about their children, starting from the pregnancy process to the birth process, up to the basic needs of the child, such as child care, child health and child nutrition through their personal blogs and social media profiles with their target audiences; in these shares, they act as a kind of natural opinion leader with the suggestions and information they provide to the mother and the people who are interested in motherhood.

The concept of social media phenomenon has brought out the concept of influencer. The masses of these people who are identified as social media celebrities have attracted the attention of brands in this sense, and this situation has led to the use of these phenomenon in product promotion and campaigns on social media platforms. This event was effective in the spread of influencer campaigns and in making social media celebrities, i.e. influencers, more visible. Influencer is used for a person or groups who engaged in promotional and marketing activities by sharing their experiences about any product or brand with their followers through their social media and digital platform profiles and has a certain audience and has the potential to influence and direct this audience (tipeffect.com.tr). As part of influencer campaigns, phenomenon is considered as influencer in terms of promotion and marketing of products and services related to the areas where they build their own identities and they share the marketing of the related brands on their personal pages for the fees or products given to them.

Table 1. Demographic characteristics of the phenomenon

Gender Distribution	Female	41
	Male	9
	Total	50
Age ranges	15-25	18
	26-35	29
	36-35	2
	46-55	1
Educational status	Elementary	
	High School	5
	Associate Degree	3
	Undergraduate	36
	Postgraduate	5
	Doctorate	1
Areas of shared posts	Mother-Baby and childcare	12
	Sightseeing-travel	5
	Body care-health-aesthetic-beauty	27
	Recipes	3
	Fashion-design	2
	Video sharing (humor)	1
	Other	0
	Total	50

Social Networks, Advertising Relationship and Advertising on Instagram

Intensive use of social networks and their potential to reach a very larger audience have been effective of them in transforming into an important platform for public relations and advertising areas. The fact that social networks eliminate time and space barriers enables the promotion and advertising made through digital platforms to reach the target audience instantly. In our today's world where the concept of convergence comes to the fore, it is possible to interact at any time by means of smart phones and applications used in these technological tools. Moving advertising and promotional activities to digital platforms as a result of developing technologies has led to the reorganization of the sector, to a change in advertising strategies and applications such as social networking, e-marketing and online brand communication and social media planning. As a result of seriously orientation of advertising and promotion activities to digital platforms, social media users are also subject to a serious advertising bombardment. Users exposed to advertisements based on their interest even in their personal use such as social network profiles and e-mail are affected by these ads and become consumers after a certain period of time. Nowadays, rather than going to places such as shopping malls and stores, the tendency to make purchases from digital addresses of these brands and stores is on the rise. Therefore, it is seen that consumption habits entirely change (Yegen, Yanık, 2015, pp. 367-368) digital platforms come into prominence in terms of shopping preferences. It is seen that all social networks and applications have become advertising channels in this sense and companies that produce brands and services effectively use social media for activities such as promotion, reaching the target audience and increasing brand awareness and reputation. As a highly functional platform for advertising, Instagram allows many features to be used, which are released as applications, in addition, maximizing user interaction via some other features such as tagging, providing sale from profile, sharing visual files and story offers great advantages both for advertisers and for the phenomenon who create their own personal branding.

Results of the Study and Assessment of the Results

Semi-structured interview technique was used in the study. The interview was conducted with 50 phenomenon with followers between 10.000 and 200.000 and carried out by categorizing the answers given to the questionnaire and analyzing these answers. The demographic characteristics of the phenomenon (gender, age, educational status) included in the study and the areas they shared were considered and evaluated. Data on gender distribution, age ranges and educational status of the phenomenon and the fields they share according to their interests are shown in Table 1.

When the data in the tables are examined, it is seen that, in terms of gender distribution, 41 of the Instagram phenomenon are female and 9 of them are male, in terms of age range, 18 of them are between the ages of 15-25, 29 of them between 26-35, 2 of them between 36-45 and one of them is between 46-55 years old. The data show that the phenomenon included in the study are mostly female and more than half of them are in the age range of 26-35.

When the educational status of the phenomenon is analyzed, it is seen that 5 out of 50 phenomenon participating in the study are high school graduates, 3 of them associate degree, 36 of them undergraduate, 5 of them postgraduate and 1 one them doctorate graduate. The data show that the phenomenon included in the study are mostly composed of highly educated individuals.

Within the research, when the shared topics are examined, it is seen that Instagram phenomenon mostly share posts in the areas of care, health, aesthetics and beauty (27). This category is followed by posting on mother-baby and childcare (12), sightseeing-travel (5), recipes (3), fashion-design (2) and humorous video (1). In this sense, it is possible to say that people who share posts in the topics of beauty, aesthetics and personal care attract intensive attention. The topic of motherhood can also be considered as one of the important points. Mothers, expressed with concepts such as Instamom and Socialmom and presented themselves through motherhood identities on all social media platforms, Instagram in particular, are among the most popular phenomenon in the digital world.

When the answers to the questions of the study analyzed, the question that constitutes the starting point of the study and related to the relationship between phenomenon-advertising, "Are you the one who choose the brands and products you prefer to advertise, or are you being made an offer by company?" was answered as follow: 36 participants stated that the company contacted with them and offered a collaboration; 10 of the phenomenon expressed that they both received offers from the companies and made offers to the companies for chosen brands and products; 4 of them stated that they made an offer to the company by themselves because they think these companies reflect their own style. When the answers are analyzed, it is seen that brands mostly determine the phenomena who can promote their products and services, and immediately afterwards make contact with them.

Regarding the question posed to phenomenon within the scope of the study as "Are the products and services of the companies and brands you advertise suitable for your interests, or are these products and brands operating in different fields?", the given answers are as follow: all phenomenon sharing posts in the area of mother-baby and childcare (12) stated that they received offers about product and advertisement promotion related to their interests and topic they share posts; however, 38 phenomenon in other areas stated that they received offers from both their own area and different areas. In this sense, apart from some specific issues (motherhood, baby care etc.), it is possible to say that the companies notice the area of interest of phenomenon, besides, companies give an advertisement to phenomenon considering their number of followers, their likes and their target audience. More followers, daily interaction rate and the size of the target audience are important criteria for the company.

The question that measures whether there is a pricing standard for the fees and promotional service fees for partnership between brands and phenomena was answered as follow: 6 of the phenomenon stated that "Yes, there is a standard", 14 of them stated that they benefit from brand and product services rather than a specific pricing and they received free products and services in exchange for advertising or promotion.

The question of whether the advertiser interferes in the sharing posts of the advertised product and services was answered as follow: 46 phenomenon stated that there was no interference, but only a certain number of sharing criteria during the day; 3 phenomenon stated that the contents were intervened from time to time. 1 phenomenon stated that he contacts with company representative before sharing a post and gets their opinion about choosing among the offered alternatives. When the answers are analyzed, it is possible to say that companies mostly leave the choice to the phenomenon of content used in advertising and promotion activities during the partnership process.

Regarding the question of whether phenomenon get a professional support in determining sharing hours and contents, all of the phenomenon stated that they pay attention to the sharing hours and intervals. In terms of professional support, while 28 of the phenomenon stated that they do not work with a team, they make natural sharing and they think that this is sincerer to the followers, 22 of them stated that they received professional support.

CONCLUSION AND ASSESSMENT

Current situation of today's technology and widespread use of social media accounts have brought about the change of advertising and marketing strategies of companies and brands and therefore, this has resulted in reaching consumer masses through social media platforms. In online social networks, which enable the user/consumer both to produce content and to be included in these networks with these produced content and are produced entirely with digital codes, while individuals construct an ideal identity for themselves in a virtual environment, they also have the opportunity to share their requests, likes, moods and their feelings, thoughts and reactions to political events through these networks. The individual who creates profiles through social networks such as Facebook, Twitter, Instagram, Foursquare and etc. transforms himself/herself into an object that can be watched, followed and accessed at any time. All these developments lead to changes in the social structure, too. The transfer of communication to digital platforms has also created radical changes in life and relationship styles. Virtual platforms, which take the place of face-to-face communication, have become new media where social life is performed. Individuals are reconstructing themselves in social networks, and they prefer to meet all their needs such as friendship relations, relations of private life, consumption habits, shopping, expressing their political thoughts and their music taste through these networks. In a sense, these platforms turn into new areas of existence and self-actualization. These platforms and new technologies also lead to the emergence of new professions. In recent years, some new professions such as YouTuber, social media phenomenon and blogger have emerged and it is possible to say that these professions take their place among the most interesting professions. It is possible to say that this situation is effective in the presence of examples that have come to the forefront in these platforms and earned very serious money and reputation. In short, new technologies and the spread of social networks have led to many changes. This situation brought about the necessity of companies to exist in these areas, so companies focus on social media to bring their products and services together with the target audience, increase corporate reputation and ensure their own continuity. In this sense, social media phenomenon come into play as an important tool for companies. Instagram phenomenon who address to a specific audience and reach a significant number of followers attract the companies' attention, and therefore this phenomenon is preferred for product promotion. Advertising companies use phenomenon that serve as opinion leaders to promote their own brand by contacting with them. Phenomenon, who deal with companies for product promotion and advertising, make suggestions by sharing the brands and products they have agreed on through their personal accounts in order to promote products and advertisements of these brands and to increase the product sales, and they execute advertising partnerships.

When an overall assessment is carried out, with the development of technology and becoming of social networks as one of the indispensable areas of daily life, it is seen that advertising and promotion activities and strategies have been moved into digital media, it is observed that companies, in order to exist in these new platforms, try to reach the target audience by going into partnership with the phenomenon defined as the celebrities of these platforms.

REFERENCES

Güçlü, A., Uzun, E., Uzun, S., & Hüsrev Yolsal, Ü. (2003). *Felsefe sözlüğü*. Ankara, Turkey: Bilim ve Sanat Yayınları.

Gürel, E., & Alem, J. (2014). *Ürün Yerleştirme*. Ankara, Turkey: Nobel Yayıncılık.

Huey, L. S. & Yazdanıfard, R. (2014). How Instagram can be used as a tool in social networking marketing. *ResearchGate*. Retrieved from http://www.researchgate.net

Kocabaş, F., & Elden, M. (1997). *Reklam ve Yaratıcı Strateji: Konumlandırma ve Star Stratejisinin Analizi*. İstanbul, Turkey: Yayınevi Yayıncılık.

McLuhan, M. (2005). *Yaradanımız medya: Medyanın etkileri üzerine bir keşif yolculuğu. Ü. Oskay (Çev.)*. İstanbul, Turkey: Merkez Kitapçılık.

Senft, T. M. (2008). *Camgirls, Celebrity & Community in The Age of Social Networks*. Retrieved from https://books.google.com.tr/books

Turner, G. (2004). *Understanding Celebrity*. London, UK: Sage. Retrieved from http://eclass.uoa.gr/modules/document/file.php/MEDIA118/celebrity+culture/Book_understanding+celebrity_graham+turner.pdf

Yegen, C., & Yanık, H. (2015). Yeni Medya İle Değişen Tüketim Anlayışı: Kadınların İnstagram Üzerinden Alış-Veriş Pratiği. In T. Kara & E. Ozgen (Eds.), *Ağdaki Şüphe Bir Sosyal Medya Eleştirisi*. İstanbul, Turkey: Beta Yayınları.

KEY TERMS AND DEFINITIONS

Advertising: The business of trying to persuade people to buy products or services.

Influencer: Someone who affects or changes the way that other people behave, forexample through their use of social media.

Instagram: The name of a social networking service for taking, changing, and sharingphotographs and video.

Social Networks: A website or computer program that allows people to communicate and shareinformation on the internet using a computer or mobile phone.

Chapter 14
Rise of Facebook in the USA and WeChat in China: Commodification of Users

Naziat Choudhury
Department of Mass Communication and Journalism, University of Rajshahi, Bangladesh

ABSTRACT

The owners of Facebook and WeChat repeatedly promote their media as the preferred platform for people to connect. Improving social relationships was marketed as the reason for their innovation. But users' urge to unite on these OSN services alone cannot explain the success of these media in the US and China. There is a different or rather new business approach underpinning these OSN services that contribute to their success. The author argues that there is an implication of owners' profit-based interest in ensuring the popularity of their online platforms. Audience commodity analysis as discussed by Dallas W. Smythe and Christian Fuchs is employed in the contexts of the US and China to comprehend the complex factors related to online social media owners' interest and their negotiation with the government in online media's prosperity. Through archival research including examination of newspapers, policy documents from OSN-based companies, and survey results from 2015 to mid-2018, this chapter demonstrates the political economy of Facebook and WeChat.

DOI: 10.4018/978-1-7998-1041-4.ch014

INTRODUCTION

Facebook in the US and WeChat in China are two of the most popular online social networking (OSN) services in the world. Both are trying to win the top place in the OSN market. This paper tries to understand the political economy aspect of these popular online social media. Facebook has been in the news in recent times with the way they deal with their users' data (BBC, 2018). The latest report suggests that Facebook has accumulated "… millions of user passwords in plain text files" (Cuthbertson, 2019, para. 1). This raised questions about how the other social media deal with their user base. In a world where commodification processes demand that companies look for new forms of consumer dependency, the rise of OSN platforms and the power of these media to actively engage users has become a new frontier. Whilst users have been engrossed in communicating within their online social networks, company owners have been busy designing the online platforms in a way that helped to harvest these communication contents for profit. Facebook (owned by Facebook) and WeChat (owned by Tencent) have used sophisticated algorithms to transform these online platforms into automatic personal data-collecting apparatus. Some of these OSN services have spread out from their originating country to other countries in the world and are not confined to specific geographic locations. The political economy factors that were associated with the success of these online services are analysed here in the context of the US and China. It is argued that there is an implication of owners' profit-based interest in ensuring the popularity of their online social platforms.

BACKGROUND

Audience commodity analysis as discussed by Dallas W. Smythe (1981) and Christian Fuchs (2016) is employed in the contexts of the US and China to comprehend the complex factors related to online social media owners' interest as well as their negotiation with the government in online media's prosperity. It is important to focus on the influence of government regulation and surveillance of OSN services. These companies function within laws and regulations set by the government of a country. Governments design and create a market economy which structures the way a company will run. Examples of these can be the structure of taxation and private data ownership and control in the case of OSN-based companies.

The limited inquiry on the political economy perspective of OSN services in China (in the English language) was evident while conducting this research. The central focus of the majority of research papers was censorship or use practices. The Chinese Internet scenario poses unique characteristics that separate it from others; namely, the obvious state control over the Internet activities. In the case of OSN services, Benney (2014) argues that the Internet in China was another tool for state control. He further argued that the Internet interfaces were designed in a way that the users were unconsciously led to use the technology in a certain manner conformed to the Chinese state and market. This, he showed, was congruent with Sina Weibo (Benney, 2014).

This article demonstrates how the companies of OSN services are monetising and commodifying the "attention economy" of media users (Goldhaber, 1997, para. 3; Christophers, 2010; Trottier, 2016). The greater the number of users of an online platform, the greater the potential for possible revenue.

Following the work of Dallas Smythe (1981) on the audience commodity, this paper explores the way advertising has driven the development of OSN features. In turn, the advertising market is the basis for financial success with OSN services across China and the US. Christian Fuchs (2016) demonstrates that online social platforms have two forms of economies: the advertising economy and the finance economy. In the advertising economy, owners earn revenues from advertisement sales. In the finance economy, these owners of OSN services raise stock prices by "sell[ing] shares to investors" (Fuchs, 2016, p. 35).

Audience power is used by these OSN industries to gain revenue, in which audience time is referred to as a "commodity" (Smythe, 1981, p. 234). Smythe (1981) shows that audiences play the roles of being both a worker and buyer. He explains that, in the context of commercial television industry, audience's attention is sold to the advertisers and commercials are sold back to the audiences. These audiences participate in the consumption process of commercials, but in this process of buying and selling, they do not gain financial profit (Smythe, 1981). Smythe (1981) argues that audiences work to produce commodities and all the hours spent not working and not sleeping are purchased by advertisers which he describes them as audience commodity.

As key artefacts for understanding the political economy of OSN services, the policy documents of Facebook and WeChat are brought into the article for analysis purposes. Archival research is needed to analyse the OSNs' policy-related reports as well as newspaper articles. In this paper, archival research consisted of the examination of newspapers, policy documents from OSN-based companies and their miscellaneous reports and survey results. These sources provided information to document the ways OSN companies use users' contents, companies' relationships with advertisers, the companies' profit-making and government's relationships to these companies.

Archival research has treated the policy documents of the OSN companies as texts, the investigation of which reveals the curious mechanisms of the political economy of OSN services. For audience commodity-related factors, research materials from 2015 to mid-2018 were collected. Archival materials such as statistical data are collected from Facebook.com, WeChat.com, Tencent.com, Alexa.com, Socialbakers.com, Statista and the Pew Research Center. Additionally, Factiva was used to search for newspaper articles.

AUDIENCE COMMODIFICATION AND OSN SERVICES

Fuchs (2012b, 2016) discusses how the commodification process takes place on commercial OSN services. He points out three main areas of concern related to this. First, basing his discussion on Marx's analysis of capitalism, Fuchs (2012b) argues that commercial OSN services like Facebook only offer their users a platform to communicate by commodifying their personal data. There exists no financial exchange between the owners and users for such activity, despite the owners gaining profit through this commodification process. So, for Fuchs (2012b), Facebook is a place for "consumption" and "production" (p. 714).

The second area for concern for Fuchs (2012b) is the emphasis on the advertisers' interest in the content created by the users. Within this process, the argument of Smythe (1981) is evident where "the audience itself – its subjectivity and the results of its subjective creative activity – is sold as a commodity" (Fuchs, 2012b, p. 704). Fuchs (2012b) claims that in the case of OSN platforms, work is outsourced to users who provide services without any financial gain. This helps the companies to invest less, save labour costs and "exploit" the workers (Fuchs, 2012b, p. 711). By content, Smythe (1981) was referring to television programs and radio broadcasts. But in this research content will be understood as the materials produced by the users. Most of the content on these OSN services is produced by users.

Despite the various online platforms and whether a user is using mobile or computer technologies, users provide free labour while creating content online. Beverungen et al. (2015) argue that Facebook provides not simply a space where people can communicate "freely" but actually contributes to a new form of free labour (p. 480). The workers of Facebook offer dual services: mapping out the website to allow for the creation of more content by the users and hence the production of more data; and building algorithms for collecting those data. Facebook's popularity is based on its power to retain its users through the search option and the inclusion of other popular OSN platforms, thus ensuring "compulsory friendship" (Gregg, 2011, p. 96).

It is important to note that users' data not only concerns the number of them using OSN services; advertising companies also build profiles of users with details including hobbies and interests as well as their online use patterns (Fuchs, 2012c). The commercials are tailored accordingly to meet the demand of the target audiences. The equation is simple: more user engagement with an online platform means better returns for the owners of that network as the company will make a bigger profit. For example, the existence of Facebook might be jeopardised if its users do not share or create content or communicate information with other that the advertising companies can collect and then direct advertisements to the users (Fuchs, 2012b). In simple terms, the political economy perspective argues that the more free labour is available on Facebook, the greater its chance for profitability as well as the promise of future profitability. The political economy perspective is not only limited to this form of economic surveillance but also the political surveillance (Fuchs, 2012a; Sandoval, 2012). This leads to the third point of focus.

As Fuchs (2012a, 2012b) highlights, the government has an interest in the commodification process. Along with these advertising companies, the state also has an interest in the users' data. OSN users and the contents they create are of great interest to both governments and the owners of OSN services.

THE COMMODIFICATION OF USERS

Source of Revenue

In order to understand the commodification of users' process, first the source of revenue needs to be discussed, followed by the users' role in the process. Facebook's (2014) Annual Report began with two clear statements in the overview section:

Our mission is to give people the power to share and make the world more open and connected.

Our business focuses on creating value for people, marketers, and developers. (p. 5)

The second statement quoted above clearly defines the business aspect of Facebook. Facebook, which is a commercial company, makes their revenue and profit mainly through advertising, a fact which is not clearly declared within the two quoted statements. Although Facebook does not mention its reliance on advertising for profit in its overview statement, it is explained in the section on how Facebook creates value for marketers:

We generate the substantial majority of our revenue from selling advertising placements to marketers. Our ads let marketers reach people on Facebook based on a variety of factors including age, gender, location, and interests. Marketers purchase ads that can appear in multiple places including in News Feed on mobile devices and personal computers, and on the right-hand side of personal computers. (Facebook, 2014, p. 5)

Small companies found Facebook a cheaper and more convenient place to advertise than other media. Initially, it was mainly small and medium-sized businesses that used Facebook space to advertise their products (Deagon, 2015; Swartz, 2015). These advertisements on Facebook's mobile devices are strategically placed within the News Feed, while advertisements on the desktop version are on the right-hand side of the profile. Although Facebook do not release details about the companies that have advertised on the site, they do mention that large portion of the revenue comes from such small companies (Swartz, 2015). Understanding the lack of technical expertise of these small businesses, Facebook offers training regarding online advertisements. Also to support small businesses, Facebook has introduced tools such as Local Awareness ads that can lead these companies towards their targeted audiences. Through this feature, businesses can identify potential customers in a specific area belonging to particular age group and gender with specific interests (Swartz, 2015). This enables a much easier process of targeting online users with advertisement and companies are now more knowledgeable about their potential customers.

Additionally, to further support this, Facebook launched a mobile ad manager service (IOS apps) through which advertising companies can create, manage and monitor commercials from around the globe. Facebook frequently opens new avenues for advertising companies to create and reach suitable online users. All these business-boosting approaches relate to users being hit with more advertisements. Such initiatives have proven effective for Facebook, as the advertising revenue grew from US$ 4,279 million in 2012 to US$ 11,492 million in 2014 (Facebook, 2014).

Facebook's decision to focus on small and medium-size businesses by providing them with technical support encourages other companies to join. Hence, more and more companies are investing in advertising on Facebook. Facebook (2016d) reports that in 2016, the OSN service had three million companies advertising on their site, out of which 70 percent were from outside the US. Facebook might soon challenge Google's top position, as these are the only two leading companies for online advertisements. Google's online advertising market share fell from 32 percent in 2013 to 31 percent in 2014, while Facebook's share increased from 5.8 percent in 2013 to 7.8 percent in 2014 (Deagon, 2015). In terms of digital display advertising revenue, Facebook has overtaken Google. In the US market, this form of revenue for Google dropped from 13.7 percent in 2014 to 13.0 percent in 2015 (Sullivan, 2015). These figures reflect the strength of Facebook as a company. Increasing revenue provides a positive and strong image of the company to the advertisers and, due to growth in both revenue and the number of users on

Facebook, the company earned trust and satisfaction in the market that led to rise in share prices by 33 percent in 2015 (Chaykowski, 2015). This in turn encourages more companies to invest in advertising on the site. It is interesting to note that although the price of each ad increased 285 percent in the first quarter of 2015, the viewership of advertisements fell 62 percent (Goel, 2015). So, either the advertisement viewership or the prices of it did not have an impact on the number of advertisements the company received; rather, the number of users and their activities on the site did, as discussed in the next section on the commodification of users' content.

Tencent, the owner of popular OSN service WeChat in China, relies on Value Added Services (VAS) for revenue (Tencent, 2016b). The Chinese OSN services' advertising environment is dominated by two main Internet companies, Tencent and Sina Weibo. Among the two, Tencent has the bigger market, as Tencent's advertising business increased 110 per cent in year-on-year revenue in 2016 from 2015, in comparison to Sina Weibo, which saw a rise of 52 percent (Perez, 2016). Tencent's revenue in 2016 from VAS rose by 34 percent to RMB 24,964 million. VAS includes online games, content subscription services, QQ membership and sales of virtual items. The company's second highest revenue came from online advertisements, which increased 73 percent (RMB 4, 701 million), as seen in Table 1 (Tencent, 2016b). It is interesting to note here that, just like Facebook, WeChat began its journey as an ad-free destination. However, in August, 2015 the company began to roll out advertisements on its "Moments" section.

Based on the sources of revenue discussion of Facebook and Tencent, two points stand out: the dependence on advertisement for revenue and also the dependence on advertisement for survival in the online social media market. These two perspectives also indicate that marketers and advertisers are inclined to select those OSN platforms that harness more membership, based on Facebook and WeChat's large user base. This will be explained in the following section.

Users' Relationship to the Revenue-Making Process

Users' contribution to the revenue-making process aspect can be seen through the relationship between the number of users and the profit earned. As the number of Facebook users continue to increase, so does its revenue. It makes sense that advertising companies would not be interested in advertising on websites that have a limited number of users or that are losing their users. There is a clear business motivation behind Facebook's reluctance to publish details on deactivating accounts. Another significant matter missing in their data is the number of people who have multiple accounts and also details of false accounts. Fear of losing advertisers may drive the company to publish only general data on the total number of users. In doing this, Facebook is maximising the market economy approach.

Table 1: Revenue of tencent

Revenues (Unaudited)	1Q 2016 (RMB in millions)	4Q 2015 (RMB in millions)
VAS	31, 995	30, 441
Online advertising	4, 701	5, 733
Others	2, 330	1, 640

Source: (Tencent, 2016b)

Figure 1 shows Facebook's annual revenue and net income from 2007 to 2015. The figure clearly illustrates the increase of the company's revenue every year. In parallel to that, Figure 2 shows a constant rise in the number of Facebook users. The concept of "more users meant more advertisements" appears to be demonstrated here. Figures 1 and 2 are a clear indication of such a trend.

Figure 1 shows that Facebook as a company lost net income from US$ 1 billion in 2011 to US$ 53 million in 2012; however, the total revenue continued to increase in those years. In connection with the total revenue is the number of users, which continued to grow in 2012 (as seen in Figure 2). This is the time when Facebook gradually lost its stronghold position to other OSN services. During that period, new OSN apps began to emerge and online users were looking for online media that offered new and innovative features. Facebook realised that this new online media could potentially create an impediment

Figure 1. Revenue and net income of Facebook from 2007 to 2015

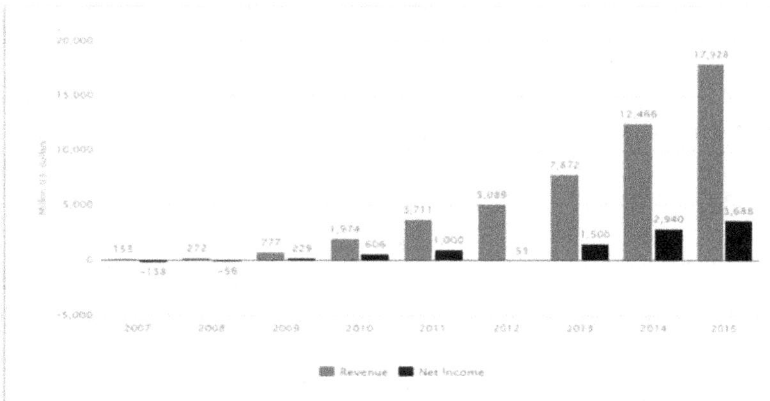

Source: Statista, 2016a

Figure 2. Number of Facebook users in millions from 2008 to 2015

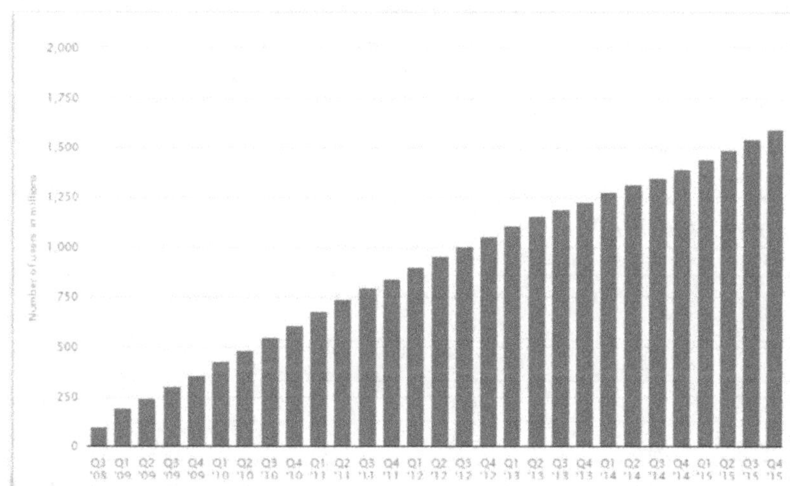

Source: Statista, 2016b

to the success of Facebook. For instance, Instagram is one such OSN service which continued to attract new users, so Facebook purchased the company for US $1 billion dollars in 2012. This merger proved to be a successful business deal, as in consecutive years Facebook had a rise in net income. Whenever Facebook faces a threat of extinction, the company quickly purchases popular mobile apps, often above their market value. The company has understood that Internet users are constantly shifting from one platform to another. From Friendster, MySpace and then Facebook, history shows that Internet users are always on the move.

WeChat's popularity also plays a role in Tencent's business revenue from advertisements. WeChat's monthly users' number is clearly correlated to revenue from advertisements, as can be seen in Table 2.

Despite being able to earn revenue successfully from advertising, Tencent's WeChat has ventured into payment services where purchases can be paid for with the app instead of using cards or cash. The mobile payment option has witnessed an incredible growth rate of 253.69 percent (2014) and 194.86 percent (2013). 90 percent of this "Code Scanning Group" is dominated by only two companies, Alipay and WeChat Payment (Chen, 2016, para. 1). In introducing this feature, WeChat appears to have taken control of the method of financial payment. In both the financial and communication sector, WeChat has retained their users. The app is designed so that users can use the one app to conduct their personal as well as financial activities. Stores also offer special discounts and promotions through the app payment option. Therefore, the retail stores as well as Tencent are being benefiting here while also providing convenience to the users. In 2015, WeChat Payment could be used at more than 300,000 retail stores, and approximately 200 million WeChat users conjoined their ATM cards with the WeChat Payment service (China Internet Watch, 2016). Part of this payment option is WeChat's successful initiative of digital red envelopes (known as *hongbao*), which could be sent through WeChat. During the Spring Festival holiday, 3.27 billion cash-filled virtual envelopes were sent in 2015 and this number jumped to 32.1 billion in 2016 (Meng, 2016b). The introduction of this feature was another attempt on the part of WeChat to retain its users and encourage more users to join.

In addition to the mobile payment option, Tencent has begun to target specific groups with WeChat. For instance, it has introduced a work-based app, known as Enterprise WeChat or Qiye Weixin, which can be used by big and small companies. This chatting option in WeChat allows users to do word-based activities including clocking in and out as well as seeking leave from the office (Meng, 2016a).

Mobile advertising is an example of how a larger user base attracts advertisers. Following the introduction of mobile apps, Facebook's number of daily active users rose to 894 million in September 2015, which was a 27 percent increase from 2014 (Facebook, 2015a). There seems to be a clear relationship between number of users and revenue collection. In the third quarter of 2015, mobile advertising revenue

Table 2. Number of monthly users and revenue collection of WeChat

Year	Number of Monthly Users (WeChat) (in millions)	Revenue from Advertisements (in millions in RMB)
2013	355	5,034
2014	500	8,308
2015	697	17,468
2016	762	4,701 (First Quarter)

Source: Statista, 2016c; Tencent, 2015, 2016a, 2016b

was around 78 per cent of total advertising revenue. This form of revenue was up from 66 percent in the third quarter of 2014 (Facebook, 2015a). This growth continued in the fourth quarter of 2015 as well. Mobile advertising revenue grew by 80 percent of the total advertisement revenue, which was also an increase from 69 percent in the fourth quarter of 2014 (Facebook, 2016a).

Tencent has also sensed the current pulse of the Internet users' reliance on mobile devices and has benefited from mobile advertising. Around 80 percent of Tencent's total advertising revenues came through mobile platforms. Tencent company's report proclaims that greater traffic on mobile-based apps and development in monetisation of advertising inventories led to this growth (Tencent, 2016b). Mobile advertising is a key area for revenue collection for these OSN services.

Facebook and Tencent embody the commercial entity in its full form and are similar to that of other conglomerates in the mass media industries around the world such as Time Warner and Disney. For instance, the Time Warner company owns businesses in different areas such as publishing (Time Inc., Little Brown and Co.), film (Warner Bros., New Line Cinema) and television production and distribution (Warner Television, WB Network) (Wasko, 2005). Disney also owns the American Broadcasting Company (ABC) television network and is a partial owner of ESPN, A & E and Lifetime channels along with its own home video, music and theme parks and resorts. Facebook and Tencent are taking the same route. Facebook owns several companies such as Instagram, WhatsApp and Onavo, Tencent holds QQ, Tencent Weibo and WeChat. Facebook and Tencent are the leading Internet-based companies along with Amazon, Alibaba and Google. Wasko (2005) argues that companies that purchase other businesses belonging to a similar group are regarded as being horizontally integrated, using Time Warner, who owns more than 140 magazines, as an example (Wasko, 2005). Thus Facebook and Tencent belong to this form of conglomerate, as their businesses concentrate mainly on information and communication technologies.

In the first phase of understanding the role that users play in the revenue process and how users are being commodified, it is important to note that Facebook and Tencent are commercial Internet-based institutions whose aim is profit-making. In this process, both users and advertisements are correlated and are crucial for the companies' survival. By capitalising on the users, the owners of these online social media, especially Facebook and Tencent, are constantly presenting new features to attract more users. These users are used as data for advertisers who can target ads to users, based on their preferences. This process clearly resonates with Smythe's (1981) concept of a "free lunch," where users are encouraged to use online social media and then these users and their contents are sold to advertisers for profit. Users are part of the profit-making process but not part of the profit-sharing.

THE COMMODIFICATION OF USERS' CONTENT

Data Collection and Management

According to data published by STRATA in 2015, advertising agencies prefer to use OSN services to advertise on and Facebook was the predominant choice (93 percent chose Facebook) (Whitman, 2015). Similarly, another survey conducted by RBC Capital Markets found that ad buyers were more interested in spending money advertising on Facebook (61 percent) in comparison to Google (53 percent) and YouTube (43 percent) (Ray, 2015). In the current Internet era, advertisements are mainly targeted. On average, Internet users use Facebook and other sites owned by the company for over 46 minutes per day. In 2015, Facebook users performed 1.5 billion searches a day and the company "indexed" around two trillion posts, according to the owner of Facebook, Mark Zuckerberg (Seetharaman, 2015). In connection with that, advertising revenue is also escalating. The average revenue per user on Facebook in 2015 was US$ 3.73, which is a 33 percent year-over-year increase (Gottfried, 2016).

In order to provide users with Internet-based advertisements, marketers or concerned companies need information about users. Marketers mainly target the young (8 to 24 years old), knowing that they hold the purchasing power of about US$ 211 billion (Montgomery, 2015). Facebook discloses its interest behind collecting data on users. One such reason is for marketing communications (Facebook, 2015b). Although what is involved in these communications is not clear, Facebook explains that personal information such as names and emails that could identify individuals is not provided to third parties (Facebook, 2015b). Without users' consent, identifiable personal information is not disclosed to advertising or Facebook's analytics partners. Instead of identifiable information, they provide them with aggregated information (Facebook, 2015b). Within that aggregated information, what type of data are collected and shared with the companies are not clear. Additionally, how in-depth the data collection processes are remains void within the company's published documents. This is an ambiguous area regarding the type of information Facebook collects.

The more information collected means that companies can better target advertisements. Facebook uses a complex tracking system to monitor and monetise users' behaviours. Facebook not only collects information but also stores it for an indefinite period. Even after user's accounts have been deleted, Facebook retains their data. Once it has information on users, the information never goes away. In other words, the data collected are recycled and reused according to the needs of the company and the advertisers. In its data policy, Facebook affirms that data would be erased if the account did not exist anymore and when they "no longer need the data to provide products and services" (Facebook, 2015b). Therefore, removing accounts from the site does not mean the complete elimination of data. The company also warns that "information that others have shared about you is not part of your account and will not be deleted when you delete your account" (Facebook, 2015b). For profit accumulation, Facebook uses invisible tracking, data monitoring and data monetising processes.

In order to efficiently collect users' information, Facebook invested in an online advertising providing company and bought the e-commerce search engine company "The Find" in 2015 to provide target-oriented advertisements to the users. This company "connects people to products" (Sloane, 2015, para. 2). "The Find" incorporated its activities within Facebook's advertising services, so that online retail stores could target shoppers more efficiently.

Chinese OSN companies are also following Facebook in the matter of purchasing new apps and adding new features to its already existing OSN service. The Chinese economy is primarily focused on exports and investments but now it has shifted to innovation and consumption (Fu, 2016). The giant success of the Internet in China has led to constant innovation, introducing new elements that would encourage more consumption and consumption-created demand for advertisements and, in turn assist, both Internet-based company owners and marketers in profit-making.

The extraordinary buying capacity and the propensity of the brand shopping of Chinese Internet users means that online media in China is a fertile land for advertisers. In 2015, there were 659 million online media users in China, which was more than in the US and Europe combined (Kemp, 2015). Along with this, the number of online shoppers in China increased from 301.89 million in 2013 to 413.25 million in 2015 (Statista, 2016c). Advertising companies would clearly like to reach this large numbers of users. Targeted advertising requires information on users which has been revealed already. Tencent's policy related to users' data collection and management is similar to Facebook, and the company has similar policies concerning collecting data for the purpose of targeted advertisements. Although WeChat has announced that users' content is not provided to third parties, the terms of use state that users' content is shared with other organisations in the name of service developments:

... in using Your Content for these purposes, we and our affiliate companies may copy, reproduce, host, store, process, adapt, modify, translate, perform, distribute and publish Your Content worldwide in all media and by all distribution methods, including those that are developed in the future;

we may share Your Content with third parties that we work with to help provide, promote, develop and improve WeChat, but these third parties (other than our affiliate companies) will not make any separate use of Your Content for their own purposes (i.e. for any purposes that are not related to WeChat). (WeChat, 2016a, para. 23)

Such claims are broad enough to encapsulate advertising companies' interest in collecting users' personal information. Although Facebook and WeChat provide targeted advertisements on users' pages, they do not clarify the type and amount of users' data collected for advertisers to provide targeted advertising. Socio-cultural or political differences have no control over how users' data are being used by these online OSN companies.

The Complex Process of Privacy Settings

The features of Facebook are designed to encourage users to provide more information about themselves. This structural design creates more involvement with the site, meaning more information for advertisers. Previously, this information about users was collected without their consent. European countries have investigated Facebook's use of personal data on the Facebook site as well as on external sites through "like" and "share" buttons. Belgium's Privacy Commission declared that, "Facebook processes the personal data of its members as well as other Internet users 'in secret,' without asking for consent or adequately explaining how the data would be used" (Fleisher and Fairless, 2015, para. 3). Due to allegations such as these, Facebook recently launched advertising preferences through which users could control the advertisements they saw. Under "Ad Choices" on Facebook, the company explained the process. Interestingly, users have to opt out from each device separately; in other words, opting out through mobile devices do not mean automatically removing advertisements from the other devices that a user uses. Introducing these options indicates that Facebook is aware of users' fear that personal data may be collected. Facebook also mentions that the users need to visit specific advertising companies to opt out i.e., Digital Advertising Alliance in the US, the Digital Advertising Alliance of Canada in Canada and the European Interactive Digital Advertising Alliance in Europe (Facebook, 2019). The association with these companies reflects Facebook's collaboration with powerful data analysis and research centres around the world to collect data on users in most effective manners.

The process for users to change their Facebook privacy settings is notoriously complex. Although the option to opt-out from advertisements and public views exists, several levels of approval automatically discourage the users from it. By default, users agree to share personal data with Facebook. When signing in to be a member of Facebook, users are inevitably roped into their tracking system. Content is posted as "public" by default. Not all Facebook users are familiar with privacy settings. In general, Internet users avoid the process of disabling tracking mechanisms due to its complexity. A survey showed that 91 percent of respondents in the US made no adjustments to their Internet or cell phone use to evade monitoring (Madden and Rainie, 2015).

Users' data are not only being shared with marketers and advertising companies but also with publishers. Facebook wants the site to become a one-stop destination where news, entertainment and communication are provided and the company is achieving this aim at the expense of users' personal data. The site allows certain news-based media houses to publish their news straight on the site and these houses "will keep all of the revenue on ads they sell directly; when the ads are sold by Facebook, revenue will be shared. Publishers will also get lots of data about how their stories are faring on Facebook" (Stelter, 2015, para. 6). Facebook has data on billions of users, which these media companies could easily access to gain insight into their audiences or readers. As Stelter (2015) mentions, no news-based companies in the USA or in other countries have such a vast array of audiences alone as Facebook does. Interestingly, the newsfeed algorithm through which Facebook controls what users see has generated controversy. News items with various perspectives usually seen on websites may not be viewed by the users. For instance, when the protests in Ferguson, Missouri in the United States were widely covered by all media and became a major domestic issue, the proof of the incident on Facebook was limited (Miller, 2015). This raises the issue of who has control over news items that are made visible to the users on Facebook.

Users in WeChat also need to undertake several levels of tasks to opt out from receiving targeted advertisements. In their privacy policy, the company declares that a privacy officer or marketing communication page should be contacted for shielding personal information from advertising companies. Guaranteed protection through such a process is not ensured, as the users may continue to receive "advertisements that are not direct marketing" (WeChat, 2016b). WeChat seems to have followed Facebook's footstep in users' information protection to pay heed to advertisers. Despite this growth, users are skeptical about storing information on the site due to fear of privacy leaks. Although 80 percent of WeChat users used their real names to register, the Internet Society of China found that around 78 percent of Chinese Internet users complained about stolen personal information including names and addresses (Ma and Cao, 2015). Due to the number of complaints such as these, Facebook and WeChat changed their privacy policies and introduced new options which involved layers of adjustments.

Surveillance by Governments

This section explores the political strategies and policies of online media, including the influence of state control on the owners of these online social media sites. The vested interests of online social media owners are linked to state interest as well. In the case of Facebook, the company's policy towards government requests for users' data states:

We may access, preserve and share your information in response to a legal request (like a search warrant, court order or subpoena) if we have a good faith belief that the law requires us to do so. This may include responding to legal requests from jurisdictions outside of the United States where we have a good faith belief that the response is required by law in that jurisdiction, affects users in that jurisdiction, and is consistent with internationally recognized standards. We may also access, preserve and share information when we have a good faith belief it is necessary to: detect, prevent and address fraud and other illegal activity; to protect ourselves, you and others, including as part of investigations; or to prevent death or imminent bodily harm. (Facebook, 2015c, para. 35)

This rather ambiguous policy has led the company to face legal battles many times within and outside of the US. A major legal action the company faced was related to users' data transfer from European Union (EU) to the US. EU was concerned about the US's data surveillance, especially as, unlike the US, they have a protection law for personal data (Fox-Brewster, 2015). EU countries have ensured their citizens have a fundamental right to know the information that companies have on them.

Through Facebook, some companies in the US were able to access European users' data through the Safe Harbour Framework (agreement between US and EU where US could transfer EU citizens' data). Due to concerns over the safety of personal information, the European Court of Justice ordered the Safe Harbour as invalid in 2015 (Fox-Brewster, 2015). Edward Snowden, the whistleblower, revealed that the US National Security Agency (NSA) was involved in online surveillance in 2013. According to him, the NSA in the US and the Government Communications Headquarter (GCHQ) in the UK collected Internet users' information by using programs such as PRISM and XKeyScore (Fuchs, 2016, p. 31). These programs allowed NSA and GCHQ to gather detailed browsing and content data on Internet users including online social media content (Fuchs, 2016). Facebook publishes a record of countries whose governments have requested users' details and the restriction to certain contents. These requests covered all Facebook services including WhatsApp. Regarding these requests, Facebook clarifies:

Every request we receive is checked for legal sufficiency. We required officials to provide a detailed description of the legal and factual basis for their request, and we push back when we find legal deficiencies or overly broad or vague demands for information. We frequently share only basic subscriber information … We have included instances in which we have removed content that governments have identified as illegal, as well as instances that may have been brought to our attention by non-government entities, such as NGOs, charities, and members of the Facebook community. (Facebook, 2016c, para. 37)

Data published by Facebook shows that the US government made more requests than any other country, which supports the view that the US government intends to have some form of control over users' data (Facebook, 2016b).

That the government of China carries out surveillance on Internet users, Internet-based companies and the content created by users is well known. In order to monitor online activities, China imposed three forms of censorship: the Great Firewall (restrictions on foreign websites), the Golden Shield (monitoring local online activities) and the keyword blocking (online content with prohibited words or phrases) (Monggilo, 2016). Moreover, each website installed up to 1,000 censors and the government has employed around 20 to 50 thousand Internet police (*wang jing*) and Internet monitors (*wang guanban*) (King et al., 2013). China has engaged hundreds of thousands of people who are known as "fifty cent party" (*wumao dang*), based on their fees per post, to write positive notes on the government (Gunitsky, 2015). In the name of security, profiles are blocked, content is deleted and access to international websites such as Facebook, Twitter and YouTube are barred by the government.

For obvious reasons, WeChat follows the Chinese government's Internet surveillance rules. As with all of the Internet companies in China, WeChat is also held liable for all the information shared through the app. This implies that WeChat has to censor the users' content that is circulated within China. Hence, the company has different terms of services for those residents (in Chinese language) and non-residents of China (in English). The Chinese privacy policy provides more detail regarding the acceptability of users' content than the English version. For instance, section 8 in the Chinese language terms of services provides a detailed guideline of the subject matters that are prohibited "in violation of state laws and regulations" (WeChat, 2016c). Hence, it can be assumed that Internet surveillance exists in similar pattern for all OSN services in China. Ng (2015) shows that political issues that are most censored in public accounts of WeChat are Bo Xilai, Hu Yaobang, Hu Jintao, the freedom of the press, demolitions and the maintenance of stability. Nevertheless, WeChat is a more favourable tool for the government of China than other micro-blogging sites and social networking sites as the design of the messenger apps matches the interests of the government. As Ng (2015) points out, other forms of online social media's capability of convenient mass reach worried the government and WeChat is far more personal. Hence, the design of WeChat is appreciated by the government of China for two reasons: it avoids clashing with the government's strict censorship policy on collective action; and it makes a profit. This was also reflected in Monggilo's (2016) statement, who writes, "the Chinese government does not want the state and its citizens into the democratic activists, but activists on liberalism or capitalism with the Internet." (p. 948). So, the Chinese government allows the expansion of its market with a strict vigilance on online activities which perform against the party's interest. Micro-blogging site Weibo was seen as a threat to the stability of the regime, as it was increasingly becoming a place of protest for many Internet users. This led to government to make requests to the site to block content. Due to such pressure, researchers such as Ng (2015) assume that WeChat's popularity might be the result of strict control on Weibo. He argues:

Whether government officials intentionally set out to attack Weibo to push users to the less viral-enabling WeChat or whether this was an unintended consequence is unclear, but intention aside, the net result was a boon to regulators and policy makers who were concerned about Weibo's role in facilitating nationwide conversations and organizing capacities. (Ng, 2015, para. 14)

Since the 1990s the number of state-owned enterprises has decreased while the number of private institutions has increased. Nearly 12 million private companies were in business by 2013 (Tse, 2015). During Mao's regime, party members were mainly farmers and labourers but now they are more likely to be businesspeople. The composition of party workers in CCP related to business has increased over the years from 13 percent in 1993 to 34 percent in 2004 (Li, 2009). The following table shows the top wealthiest businesspeople and their membership to the state party.

In the case of China, the online media owners appear to have some form of affiliation with the government party. Data in Table 3 show that Baidu and Tencent owners have membership in state party which directs towards their connections with the authorities of China. For example, by 2012, Tencent developed Party committee and the owner of Tencent became part of the 25 vice-presidents of the Internet Society of China, "an intermediary organization under CCP guidance" (Creemers, 2016, p. 4). Moreover, Ma Huateng was also one of the delegates in the Chinese People's Political Consultative Conference (CP-PCC) and the National People's Congress (NPC) in 2015 (AFP, 2015). Ma Huateng was also part of the team of businessmen who accompanied Xi Jinping on his first formal state visit to the United States.

The discussion in this section indicates that both the governments in the US and China conduct surveillance on users. The broad context of political culture in both contexts is different, as the US is democratic while China practices socialism. Despite the contrast in political culture, the government and owners share a mutual interest in users' data and their online content, which is also pointed out by Fuchs (2016).

CONCLUSION

Facebook and Tencent share certain commonalities between them in terms of audience commodity. OSN sites and messenger app companies such as these rely on advertisements for their revenue and Facebook and WeChat are the biggest revenue earners in this respect. The more users an OSN service has, the larger number of users, marketers and advertising companies can reach with minimum investments.

The next resemblance among these companies is sharing users' information and user-created content with advertising companies. These companies share such information with the advertisers in order to generate financial revenue. Companies like Facebook and Tencent need to survive in the ever-competitive market of OSN services. Without the state's direct or indirect support, these companies may not be able to continue with their businesses. Hence, these companies share users' information with the government.

The main findings of this research supports the existence of the commodification process on OSN services. Owners of these OSN companies are making profits by commodifying the users of these services. Facebook and Tencent gain profit from selling users' data to advertising and marketing companies. But,

Table 3. Businessmen's membership to state party in China

Name	Company	Wealth (in US$ trillion)
Robin Li	Baidu	14.7
Ma Huateng	Tencent Holdings Ltd.	14.4
Lei Jun	Xiaomi	13.5

Source: (Shabrina and Winarsih, 2016)

their main intention is to provide more support to advertisers than to users. To benefit the advertisers, these companies design and incorporate many features that would force users to create more content and thus allow the company to gain more information on users, while users have been provided with only one or two options to protect their information from advertisers (i.e., an opt-out option). Moreover, how the protection of personal data works is not clear. On the one hand, for monetary gains, users' data are shared with advertisers and marketers. And on the other hand, information is not delivered to the government in the name of protecting users' privacy. It is also not clear if information is provided to the government without the users' consent. Therefore, economic and government surveillance is being conducted simultaneously on these OSN services where users are being taken advantage of on both accounts. OSN sites and messenger apps in both the US and China depend on two powerful institutions for their existence: advertising companies and the government. The substantial growth of their user base reflects not only the interest of the users to build social capital on online social media but also the sites' increasing dependence on advertising that has shifted from mainstream media to online social platforms.

REFERENCES

Abutaleb, Y. & Bhattacharjee, N. (2015, Aug. 13). Facebook Struggles to Sell Advertising in India. *Reuters*. Retrieved from http://in.reuters.com/article/facebook-india-ads-idINKCN0QH0DY20150812

AFP. (2015, March 2). 5 of China's 10 wealthiest to take part in key political meetings. *Business Insider*. Retrieved from http://www.businessinsider.com/afp-chinas-wealthiest-to-take-part-in-key-political-meetings-2015-3?IR=T&

BBC. (2018, Dec. 19). Facebook's data-sharing deals exposed. Retrieved from https://www.bbc.com/news/technology-46618582

Benney, J. (2014). The Aesthetics of Chinese Microblogging: State and Market Control of Weibo. *Asiascape: Digital Asia*, *1*(3), 169–200. doi:10.1163/22142312-12340011

Beverungen, A., Böhm, S., & Land, C. (2015). Free Labour, Social Media, Management: Challenging Marxist Organization Studies. *Organization Studies*, *36*(4), 473–489. doi:10.1177/0170840614561568

Chaykowski, K. (2015, Nov. 4). Facebook Beats Third-Quarter Earnings, Revenue Estimates, Shares Rise. *Forbes*. Retrieved from https://www.forbes.com/sites/kathleenchaykowski/2015/11/04/facebook-beats-third-quarter-earnings-revenue-estimates-shares-rise/#37a0771a37a0

Chen, G. (2016). Quick Pass Churns Payment Market. Retrieved from http://en.ce.cn/Insight/201604/18/t20160418_10589502.shtml

China Internet Watch. (2016). WeChat Payment Reached over 300K Retail Stores. Retrieved from http://www.chinainternetwatch.com/17437/wechat-payment-reached-over-300k-retail-stores

Christophers, B. (2010). *Envisioning Media Power: On Capital and Geographies of Television*. Lanham, MD: Lexington Books.

Creemers, R. (2016, May 24-25). Disrupting the Chinese State: New Actors and New Factors. Paper Presented at the *Conference on Digital Disruption in Asia: Methods and Issues*, University of Leiden, The Netherlands.

Cuthbertson, A. (2019, March 22). Facebook Admits Storing Millions of User Passwords in Plain Text Files for Years. *Independent*. Retrieved from https://www.independent.co.uk/life-style/gadgets-and-tech/news/facebook-passwords-plain-text-instagram-admission-a8833941.html?utm_medium=Social&utm_source=Facebook&fbclid=IwAR35iRRkYT9ezasULRIaY0YrWdJqUicx9TEWmKvUiBD8KKwS_Ws2dgL58XQ#Echobox=1553185822

Deagon, B. (2015, March 9). Facebook Making Progress with Major Brand Advertisers. *Investor's Business Daily*. Retrieved from http://www.investors.com/facebook-targets-national-brand-advertisers/

Facebook. (2014). *Facebook Annual Report*. Retrieved from https://materials.proxyvote.com/Approved/30303M/20150413/AR_245461/#/8/

Facebook. (2015a). Facebook Reports Third Quarter 2015 Results. Retrieved from http://investor.fb.com/releasedetail.cfm?ReleaseID=940609

Facebook. (2015b). Data Policy. Retrieved from https://www.facebook.com/full_data_use_policy

Facebook. (2015c). Data Policy. Retrieved from https://www.facebook.com/privacy/explanation

Facebook. (2016a). Facebook Reports Fourth Quarter and Full Year 2015 Results. Retrieved from http://investor.fb.com/releasedetail.cfm?ReleaseID=952040

Facebook. (2016b). Government request report. Retrieved from https://govtrequests.facebook.com/#

Facebook. (2016c). Publishing Information about Government Requests to Facebook. Retrieved from https://govtrequests.facebook.com/about/

Facebook. (2016d). Three Million Business Stories. What's Yours? Retrieved from https://www.facebook.com/business/news/3-million-advertisers

Facebook. (2019). AdChoices. Retrieved from https://www.facebook.com/help/568137493302217

Fleisher, L. & Fairless, T. (2015, May 15). Belgian Watchdog Raps Facebook for Treating Personal Data 'With Contempt'. *The Wall Street Journal*. Retrieved from https://www.wsj.com/articles/belgian-watchdog-slams-facebooks-privacy-controls-1431685985

Fox-Brewster, T. (2015, Oct. 6). 'Landmark' Decision Threatens Facebook Use Of European Personal Data. *Forbes.com*. Retrieved from https://www.forbes.com/sites/thomasbrewster/2015/10/06/safe-harbour-invalid/#1d98763a1d98

Fu, J. (2016, Jan. 26). Optimism Over China's Economy Won Out at Davos. *China Daily*. Retrieved from http://europe.chinadaily.com.cn/opinion/2016-01/26/content_23244212.htm

Fuchs, C. (2012a). Critique of the Political Economy of Web 2.0 Surveillance. In C. Fuchs, K. Boersma, A. Albrechtslund, & M. Sandoval (Eds.), *Internet and Surveillance: The Challenges of Web 2.0 and Social Media* (pp. 31–70). New York, NY: Routledge.

Fuchs, C. (2012b). Dallas Smythe Today - The Audience Commodity, the Digital Labour Debate, Marxist Political Economy and Critical Theory. Prolegomena to a Digital Labour Theory of Value. *TripleC*, *10*(2), 692–740. doi:10.31269/triplec.v10i2.443

Fuchs, C. (2012c). The Political Economy of Privacy on Facebook. *Television & New Media*, *13*(2), 139–159. doi:10.1177/1527476411415699

Fuchs, C. (2016). Baidu, Weibo and Renren: The Global Political Economy of Social Media in China. *Asian Journal of Communication*, *26*(1), 14–41. doi:10.1080/01292986.2015.1041537

Goel, V. (2015, April 23). Facebook Reports Quarterly Results Dominated by Shift to Mobile and Video. *The New York Times*. Retrieved from http://www.nytimes.com/2015/04/23/technology/facebook-q1-earnings.html?_r=0

Goldhaber, M. H. (1997). The Attention Economy and the Net. *First Monday*, *2*(4). Retrieved from http://journals.uic.edu/ojs/index.php/fm/article/view/519/440

Gottfried, M. (2016, Jan. 27). Facebook: It's Hard to Argue with Results Like These; The Social Network's Results Show it Hasn't Missed a Step. *The Wall Street Journal Online*. Retrieved from https://www.wsj.com/articles/facebook-its-hard-to-argue-with-results-like-these-1453936786

Gregg, M. (2011). *Work's Intimacy*. Cambridge, UK: Polity.

Gunitsky, S. (2015). Corrupting the Cyber-Commons: Social Media as a Tool of Autocratic Stability. *Perspectives on Politics*, *13*(1), 42–54. Retrieved from http://individual.utoronto.ca/seva/corrupting_cybercommons.pdf doi:10.1017/S1537592714003120

Kemp, S. (2015). *Digital, Social and Mobile in 2015*. Retrieved from http://wearesocial.com/uk/special-reports/digital-social-mobile-worldwide-2015

King, G., Pan, J., & Roberts, M. E. (2013). How Censorship in China Allows Government Criticism but Silences Collective Expression. *The American Political Science Review*, *107*(2), 326–343. doi:10.1017/S0003055413000014

Li, C. (2009). The Chinese Communist Party: Recruiting and Controlling the New Elites. *Journal of Current Chinese Affairs*, *38*(3), 13–33. doi:10.1177/186810260903800302

Ma, S. & Cao, Y. (2015, Dec. 18). Experts Call to Keep Data Safe. *China Daily*. Retrieved from http://www.chinadaily.com.cn/business/tech/2015-12/18/content_22741317

Madden, M. & Rainie, L. (2015). *Americans' Attitudes about Privacy, Security and Surveillance*. Retrieved from http://www.pewinternet.org/files/2015/05/Privacy-and-Security-Attitudes-5.19.15_FINAL.pdf

Meng, J. (2016a, April 19). Tencent's WeChat Launches New App for Enterprises. *China Daily*. Retrieved from http://www.chinadaily.com.cn/business/tech/2016-04/19/content_24649488.htm

Meng, J. (2016b, Feb. 16). Tencent Claims Surge in Digital Red Envelopes over New Year. *China Daily*. Retrieved from http://www.chinadaily.com.cn/business/tech/2016-02/16/content_23496768.htm

Miller, C. C. (2015, May 15). How Facebook's Experiment Changes the News (and How It Doesn't). *The New York Times*. Retrieved from http://www.nytimes.com/2015/05/15/upshot/how-facebooks-experiment-changes-the-news-and-how-it-doesnt.html

Monggilo, Z. M. Z. (2016, Jan. 26-28). Internet Freedom in Asia: Case of Internet Censorship in China. Paper Presented at the *International Conference on Social Politics, Universitas Muhammadiyah Yogyakarta*, Indonesia. 10.18196/jgp.2016.0026

Montgomery, K. C. (2015). Youth and Surveillance in the Facebook Era: Policy Interventions and Social Implications. *Telecommunications Policy*, *39*(9), 771–786. doi:10.1016/j.telpol.2014.12.006

Ng, J. Q. (2015). Politics, Rumors, and Ambiguity: Tracking Censorship on WeChat's Public Accounts Platform. Retrieved from https://citizenlab.org/2015/07/tracking-censorship-on-wechat-public-accounts-platform/

Perez, B. (2016, April 4). China's Social Networks Set for Boom, As Advertising Spending Forecast to Jump 56 Per Cent to Us$5.33b This Year. *South China Morning Post*. Retrieved from http://www.scmp.com/tech/enterprises/article/1933558/chinas-social-networks-set-boom-advertising-spending-forecast-jump

Ray, T. (2015, Oct. 3). Rich Get Richer as Google and Facebook Dominate Web Ads. *Barron's Asia*. Retrieved from http://www.barrons.com/articles/rich-get-richer-as-google-and-facebook-dominate-web-ads-1443851396

Sandoval, M. (2012). A Critical Empirical Case Study of Consumer Surveillance on Web 2.0. In C. Fuchs, K. Boersma, A. Albrechtslund, & M. Sandoval (Eds.), *Internet and Surveillance: The Challenges of Web 2.0 and Social Media* (pp. 147–169). New York, NY: Routledge.

Seetharaman, D. (2015, July 30). Facebook, Google Tighten Grip on Mobile Ads; Facebook's Quarterly Revenue Jumps 39% as it Lures Big Brand Advertisers and Smartphone Users. *The Wall Street Journal Online*. Retrieved from https://www.wsj.com/articles/facebook-revenue-rises-39-1438200350

Shabrina, F., & Winarsih, A. S. (2016, Jan. 26-28). Businesspeople Co-Optation In China's Communist Party Adaption. Paper Presented at the *International Conference on Social Politics, Universitas Muhammadiyah Yogyakarta*, Indonesia.

Sloane, G. (2015, March 13). Facebook Buys a Company That Could Solve an Annoying Online Ad Problem. *Adweek*. Retrieved from http://www.adweek.com/news/technology/facebook-buys-company-could-solve-annoying-ad-problem-163475

Smythe, D. W. (1981). *Dependency Road: Communications, Capitalism, Consciousness, and Canada*. Norwood, NJ: Ablex.

Statista. (2016a). Facebook's Revenue and Net Income from 2007 to 2015 (in Million U.S. Dollars). Retrieved from http://www.statista.com/statistics/277229/facebooks-annual-revenue-and-net-income/

Statista. (2016b). Number of Monthly Active Facebook Users Worldwide as of 4th quarter 2015 (in Millions). Retrieved from http://www.statista.com/statistics/264810/number-of-monthly-active-facebook-users worldwide/

Statista. (2016c). Number of Monthly Active WeChat Users from 2nd Quarter 2010 to 4th Quarter 2015 (in Millions). Retrieved from http://www.statista.com/statistics/255778/number-of-active-wechat-messenger-accounts/

Stelter, B. (2015, May 13). Why Facebook is Starting a New Partnership with 9 News Publishers. *CNN*. Retrieved from http://money.cnn.com/2015/05/13/media/facebook-instant-articles-news-industry/

Sullivan, L. (2015, March 26). Google Takes Backseat to Facebook's Digital Display Ad Revenue. *MediaPost.com*. Retrieved from http://www.mediapost.com/publications/article/246520/google-takes-backseat-to-facebooks-digital-displa.html?utm_source=newsletter&utm_medium=email&utm_content=headline&utm_campaign=81438

Swartz, A. (2015, March 25). How Facebook is Courting Small Businesses and Their Advertising Dollars. *Silicon Valley/San Jose Business Journal Online*. Retrieved from http://www.bizjournals.com/sanjose/news/2015/03/25/how-facebook-is-courting-small-businesses-and.html

Tencent. (2015). Tencent Announces 2014 Fourth Quarter and Annual Results. Retrieved from http://www.tencent.com/en- us/content/ir/news/2015/attachments/20150318.pdf

Tencent. (2016a). Tencent Announces 2015 Fourth Quarter and Annual Results. Retrieved from http://www.tencent.com/en-us/content/ir/news/2016/attachments/20160317.pdf

Tencent. (2016b). *Tencent Announces 2016 First Quarter Results*. Retrieved from http://www.tencent.com/en-us/content/ir/news/2016/attachments/20160518.pdf

Trottier, D. (2016). *Social Media as Surveillance: Rethinking Visibility in a Convergence World*. New York, NY: Routledge. doi:10.4324/9781315609508

Tse, E. (2015). *China's Disruptors*. New York, NY: Penguin.

Wasko, J. (2005). Studying the Political Economy of Media and Information. *Comunicação e Sociedade*, *7*(0), 25–48. doi:10.17231/comsoc.7(2005).1208

WeChat. (2016a). Terms of Service. Retrieved from http://www.wechat.com/en/service_terms.html

WeChat. (2016b). Privacy policy. Retrieved from http://www.wechat.com/en/privacy_policy.html

WeChat. (2016c). Terms of Use and Privacy Policy (Chinese language versions). Retrieved from http://weixin.qq.com/cgi-bin/readtemplate?uin=&stype=&promote=&fr=wechat.com&lang=zh_CN&ADTAG=&check=false&nav=faq&t=weixin_agreement&s=default

Whitman, R. (2015, Aug. 26). Study Finds Social Media Claiming Larger Share of Ad Budgets. *MediaPost.com*. Retrieved from http://www.mediapost.com/publications/article/257029/study-finds-social-media-claiming-larger-share-of.html

ADDITIONAL READING

Smythe, D. W. (1977). Communications: Blindspot of Western Marxism. *Canadian Journal of Political and Social Theory*, *1*(3), 1–27.

Svec, H. A. (2015). On Dallas Smythe's "Audience Commodity": An Interview with Lee McGuigan and Vincent Manzerolle. *TripleC*, *13*(1), 270–273.

van Dijck, J. (2013). *The Culture of Connectivity: A Critical History of Social Media*. Oxford: Oxford University Press. doi:10.1093/acprof:oso/9780199970773.001.0001

Yang, K., Zhang, C., & Tang, J. (2012). Internet Use and Governance in China. In Y. Chen & P. Chu (Eds.), *Governance and Cross-boundary Collaboration: Innovations and Advancing Tools* (pp. 305–324). Hershey, PA: IGI Global Publications. doi:10.4018/978-1-60960-753-1.ch017

Zhang, H., Lu, Y., Gupta, S., & Zhao, L. (2014). What Motivates Customers to Participate in Social Commerce? The Impact of Technological Environments and Virtual Customer Experiences. *Information & Management*, *51*(8), 1017–1030. doi:10.1016/j.im.2014.07.005

KEY TERMS AND DEFINITIONS

Attention Economy: A concept where media, in a modern competitive market, tries to gain and keep the attention of the audience to its channel, newspaper or online service.

Audience Commodification: A political economy concept related to media. Audiences contribute to media companies' profit-making process both as worker and buyer. But they gain no financial profit.

Government surveillance: A situation where the government observes the activities of an individual and group to collect information on them and in some cases it may affect their privacy.

Online Social Networking Services: Social platforms on the Internet that are created by building online profiles with valid email address. These are used to communicate and share contents with others.

Political Economy: The interdisciplinary study of the relationship between economy and politics.

Social Media Policies: Policies created and followed by the social media companies to conduct their activities and business.

User Data: All information and content collected and stored about a social media user by the government and/or companies of social media.

Chapter 15
Voyeurism in Social Networks and Changing the Perception of Privacy on the Example of Instagram

Serpil Kır
https://orcid.org/0000-0002-6653-6102
Hatay Mustafa Kemal University, Turkey

ABSTRACT

With the development of communication technologies and changing perceptions of privacy in Turkey, it has emerged to problematize as concept voyeurism. The basic element that framed the intimate place over the body is the place. In social networks, the reset function of the place transforms the private body into a public domain for consumption. The notion of voyeurism, which means watching, is also related to place as of origin. The pleasure of peeping the place belonging to others is also related to the pleasure of penetrating the boundaries of place. Social networks threaten privacy/space as a voyeur environment in the context of establishing this system of pleasure. In the context of social networks, place, and body, a conceptual framework will be discussed, as well as privacy and voyeurism. Also, the selected social network activities will be examined by Instagram's photo and video sharing content analysis method.

DOI: 10.4018/978-1-7998-1041-4.ch015

INTRODUCTION

With the rapid incorporation of computer and telecommunication networks into daily life, especially after the 1990s, it became impossible for individuals to remain indifferent to technology and developments in the world. With the constantly upgraded technology, especially the digital environment transformed into a point where all needs of an individual are met through an endless information acquisition, unlimited shopping, and unlimited socialization. The Internet and social media are constantly updated to meet the needs of the individual and in return expects the user to put something forward. The user responds to this expectation of social media with or without awareness.

Surrounded with all the opportunities offered by the internet in everywhere, the individual is in an activity race in many ways from socializing to shopping in order to always be *present* and *popular* on social media. In a time where curiosity is included to this activity race, the content of the privacy concept changes.

The sense of unlimited freedom in social media that stimulated the feeling *I can do anything* - especially in Instagram – poses a threat to that privacy. Instagram, which is a social media as a photo sharing platform, gives users the freedom to look at the photos of members. In addition to the introductory, educative, and longing feelings photos, the platform also host photos that require a certain age limit. Some photographs, which are accepted as weird, sin or shame in daily practice are considered normal in this virtual space.

In a study that will be conducted through Instagram where exhibitionism and voyeurism are legal and there is no diagnosis of psycho-pathological cases in unlimited conduction, the approaches to privacy should be reviewed and perhaps redefined. Because, the perception of privacy in social media and the perception of privacy shaped by real-life value judgments are no longer the same. In this study, the problem of privacy that has emerged on the axis of Instagram has been changed with shared photographs and live broadcasts and the extent to which the concept of voyeurism has started to be regarded as routine.

CONCEPT OF THE PRIVACY AND ITS LIMITATIONS

It is very difficult to make a concrete definition and to determine the limits of privacy, which is one of the most important defense mechanisms established by the individual against his/her environment. One of the important reasons for this difficulty is that the concept of privacy has a variable characteristic from individual to individual, from culture to culture and from year to year. Within these variables, the individual exists with the personal perception of privacy and uses this privacy as a kind of defense armor in many environments from social life to work, from family life to friendship circles and even to the virtual world that the individual has set up in the computer environment.

Attempting to explain the phenomenon of privacy, which shows variations in space and time by definition, there emerges another common point that literature emphasizes: the habitat (space) of the person. Because privacy is related to habitats. These habitats are open to others. For example, common life is the area known to many relatives. The private life is a hidden area where privacy and confidentiality are defined. "*The private space/life consists of the views of life kept to oneself and includes knowledge desired to be enclosed from the others*" (Yüksel, 2003, p. 189).

The individual realizes the need for socialization under the roof of privacy and within the limits. As a part of the need for socialization, a person sometimes needs to share her/his privacy with others. Privacy, thus, is one of the central concepts in terms of the sincerity and reliability of relationships and determines the dimension of the relationship in the individual's social domains in terms of the quality and depth (Dedeoglu, 2004).

What is essential and important in privacy, which is categorized as "*spatial privacy, information privacy, and individual privacy*" (Zhumagaziyeva & Koç, 2010, p. 114), is that the individual has the right to control the personal information and to determine who to reveal that information. The individual makes a decision about whom to incorporate into his/her relationships, including family life, and which style he/she will use some of the untouchable areas. The main conclusion reached at this point is that the concept of privacy refers to the space and this space is the determinant in the relationships of the individual. The individual immediately decides which communication style to use for a person, and who to include some of the private areas including family life into by immediately engaging the privacy phenomenon to the relationships around.

The private areas that the individual always needs and their protection with the privacy armor varies from individual to individual. The most common aspect of the privacy armor is the desire to be alone and the aspiration of isolation from the environment. Another common aspect is the desire to maintain a distance even from the closest relatives in order to limit their interference and supervision. The third aspect is the expression of one's demand to participate in public life without attracting attention (Yüksel, 2003, p. 182).

Thus, privacy is a right that individuals use not only to isolate themselves from other individuals but also to participate in social life within certain criteria. While evaluating privacy in the context of culture, Hilmi Yavuz proposed that the concept of privacy in Western culture was built on immunity, whereas that of eastern culture was built on invisibility (Yavuz, 2012, p. 17).

For Western culture, the privacy of the human body is guaranteed by its untouchability in the public sphere. In Eastern societies, privacy is "*hiding the intimate from the gaze of the others and keeping it closed to the others*" (Yavuz, 2012, p. 18). That is to say, privacy is not affiliated with the sense of touch, but associated with the sense of vision. Today, although people are in a great battle to protect their privacy, these areas of privacy can often be violated in terms of public interest and security. In particular, this intervention became more evident with a digital life that emerged in parallel with the development of technology, and all areas of the individual that could be considered intimate were under great threat, and even more serious, the individual has now come to the point of losing the struggle.

THE CONCEPT OF 'VOYEURISM'

Voyeurism is derived from the word "voir" which means the act of seeing and correspond to the Turkish slang terms of "röntgencilik" or "dikizcilik". In other words, on the contrary to the French origin of a pure glance or seeing, the word in is considered as a criminal element. Therefore, in explaining the concept of voyeurism, it is possible to see that the peeping and exhibitionism, which open doors to voyeurism, are used in the same circles.

Voyeurism is described as problematic in the health and legal literature. In the social context, this problematic situation has evolved into a non-problematic, insignificant and even encouraging situation, especially by social media. There is no gender discrimination in voyeurism, but it is a general belief that it is more common among men as they chose the female body as a secretly observing object at the point of arousing satisfaction.

The phenomenon of voyeurism is closely related to the urge to wonder. Because the characteristics of individuals that lead men to voyeurism are actually on the intimate thing that the other person is trying to hide. Curiosity is an instinctive behavior of any age and gender. But especially in men, when the subject is the intimate area of the opposite gender, the urge to wonder becomes more active. Men's sense of curiosity about watching the opposite gender, in particular, often results in the invasion of the privacy of the other. Therefore, voyeurism is very common and can be seen as a harmless behavior since it can be seen in many places at a moderate level, but this situation persists and proceed to a degree that disturb the other side, it can transform into a legal case.

When explaining the concept of voyeurism, it is necessary to mention the exhibitionism, which is defined as the act of opening up what should be left in private to others. Usually, there is voyeurism on one side of the medallion and exhibitionism on the other. It is the 'exposing' element that drags individuals to the act of looking and seeing naively with the will and desire.

CONCEPT OF EXHIBITIONISM

The exhibitionism, similar to voyeurism, is about private space. If we describe Voyeurism as the action of the observer, we can position the exhibitionism in the opposite pole as the action, object, or event that is in the interest of the observer.

While the observer performs a task attributed to himself/herself, s/he establishes a kind of sovereignty over the observed (Foucault, 2007, p. 23). This sovereignty is not only directed to the destruction of the sanctified space, but can also be directed towards the private individual space.

While voyeurism is based on secretly watching and observing, exhibitionism is based on opening the boundaries of private life to others, where confidentiality disappears without any shame. Since the two closely related term, Voyeurism and exhibitionism, are now accepted as moderately normal, especially on social networking platforms, the chronicity of these two conditions is an indication of a serious problem.

In an innocent sense, exhibitionism is a type of action that takes place around everyone in almost every situation and environment. However, health and law experts make new definitions for the recent deviation in the meaning for exhibitionism in the form of exposure and try to emphasize to the sexuality, sexual deviance, mental illness, and sexual crimes in the new definitions. It is not possible to set an age limit for a specific crime and disease, but experts state that it appeared before the age of 18 for exhibitionism and gradually began to decline after age 40. Just as in voyeurism, it is not possible to connect the disease to any gender, but men are in a more exhibitionistic position than women (http://elyadal.org).

As the literature considers exhibitionism as a disease, it is necessary to explain the underlying causes of this disease like any other disease. The most important reason is that the individual has an asocial stance in society. Unpretentious males and males over-dependent to their mothers have higher tendency to be exhibitionists (Blair & Lanyon, 1981).

SOCIAL MEDIA

Emerging from the developing communication technologies, social media platforms are new communication environments that are created with the idea of providing communication among people, marketing the products of the companies, and delivering them to individuals (Safko, 2010, p. 4). Considering the idea that every innovation is attributed to the economic and commercial motivation, social media is also a product with the commercial motivation as it is a new communication environment that has enumerated the needs of individuals via penetrating into the various aspects of personal life and interests of the individual and constantly updating itself in order to meet the needs. Updating itself continuously, social media accomplishes this in line with the developments in communication technologies.

The evolution of Web 1.0 to the Web 2.0 with the advancing technology and the shift from one-way information sharing to reciprocal and simultaneous information sharing is the latest update and evolution of social media.

The evolution of new communication technologies from the traditional ones implied laying of the foundation of social media and in this process, the individual communication tended to progress towards mass communication. "*Social media has enabled the development of virtual communication; thus, individual and mass communication began to change dimensions. Communication in social media can be personal or mass*" (Kırık, 2015, p. 169).

In 1997, the social media movement started with Sixdegrees. Sixdegrees, the pioneer of the movement, aimed to become a social media space for new relationships. When the dating site, Sixdegrees, was founded in 1997, it became a precursor to the digitalization of relationships, and in the following years it became a guide for social media platforms like Friendster and Facebook. In the same way, Sixdegrees stimulated the establishment of the other social media platforms in the new world order where flirting, loves and relationships are digitalized, while revived the question why not digitizing shopping, business life, and education. Especially since 2000, it has been a pioneer in the establishment of a social media platform that organizes every activity of the individual in response to this very question.

In Turkey, the term social media (sosyal medya) was first coined in 2008. After that, MySpace and Yonja.com entered our lives and social media awareness started to grow like avalanche. People were starting to share their personal information and intimate things, establishing a new virtual world for themselves. The social media boom took place in 2004 with Facebook. Facebook had a record high number of clicks and members became a channel that companies and institutions could not remain indifferent. Twitter and Instagram sharing sites followed Facebook and people involved in this new trend consciously or unconsciously.

In social media which transform in to a form of human communication where sharing and discussion is essential beyond the time and the space and is in a relentless need to update, people make appointments with each other, exchange information, buy and sell products, realize the virtual state of socialization, and most importantly they are under constant information bombardment on social media. People can no longer remain indifferent without writing anything on social media or even passively looking. Because social media has encompassed individuals as a living space, as a medium, as a publication, and as a sector.

MOBILE SURVEILLANCE PHENOMENON, "INSTAGRAM"

While the world has been changing with a focus around the information and communication technologies, the most important object of this change and evolution is the human. Developments in communication technologies have affected individuals intimately and the individual has lost the state of belonging to a place and becomes a part of common culture and common life due to these developments. Especially, changes in communication technologies have arisen a new identity obligation to individuals. This identity is a social media identity that offers the ability to be anywhere at any time. The individual joining in social media with this identity participates in different social platforms and blogs according to the personal needs and styles and spends a part of the time under this identity.

Different platforms within social media have become a complement to the emotions that individuals cannot find in real life. At the same time these platforms can be socializing, educative and informative for individuals. One of these platforms is Instagram.

Instagram is a mobile application (iOS and Android) that allows users to share photos that are thought to be visually appealing over the network (Salomon, 2013, p. 408). The reference statement "visual appeal" makes it clear that the use of Instagram is based on the urge for appreciation and recognition.

When Instagram was launched on October 6, 2010 after the extensive efforts of Mike Krieger and Kevin Systrom, the Systrom described Instagram project as an aim to send photos from from sunny countries to the people living in the countries with the winter months. (http://ekonomi.milliyet.com.tr/). But then, it deviated from the idea of sunny photographs and individuals began to use Instagram for various purposes, including sharing what they eat, showing where they go, and even going further to revealing their privacy.

When two young entrepreneurs along with a crew of 13 people exhibited Instagram in the world of apps on October 6, 2010, Instagram was chosen as the best IPhone application of the week and people began to recognize this application with their virtual identities. As of January 2017, live video broadcast feature such as Periscope has been added to Instagram. With the live broadcasts, the users have the ability to be visible not only to his followers but also to other users in the Discover tab.

This mobile application was downloaded by 1 million people in three months and attracted the attention of other social media platforms in a short period of time. In 2012, Facebook incorporated Instagram with a figure of 1 billion dollars and Instagram identities that users registered are now under the control of Facebook.

METHODOLOGY OF THE STUDY

Purpose and Scope

The developments that emerged from new communication technologies have led individuals to live in a mobile life. Social media platforms are the most important actors of these mobilized environments. With the help of these mobilized environments, the individual now forgets the concepts of leisure and free time and tries to satisfy himself in these environments.

In social media, that have been penetrated into the everyday lives, people shop, date, observe others, and exhibit themselves. As a result of this situation, the concept of privacy is destroyed by the instincts to look into and share with others, particularly through the pictures and images shared on social media platform Instagram.

For this reason, the aim of this study was to investigate the extent to which people exhibit their privacy on Instagram, which is based entirely on the urge for appreciation and recognition. For this purpose, the research includes the investigation of Instagram photo sharing application that functions on the voyeurism with ever increasing number of users (exceeding 1 billion in 2019) (https://tr.sputniknews.com).

Method

In order to reveal the aims and practices of exhibitionism and voyeurism as a privacy problem of social media through the example of Instagram, the conceptual framework of the study was created by literature search method and the research in the current study was conducted by Content Analysis method, which is one of the quantitative research methods.

The emergence of the content analysis method is directly attributed to the research conducted at the beginning of the 20th century measuring the effects of mass media. Data for the content is obtained by objective methods that allows to make inferences from the data.

Limitations

The conceptual framework of the study is limited to the concepts of Privacy, Voyeurism, Social Media and Instagram, which is a social sharing site and has an important place in the research. Content analysis method used in the research part of the study is limited to Instagram photo sharing application.

RESEARCH

Analysis and Evaluation

The concept of privacy basically co-existed with humanity. In particular, the concept of space, which is a protection of the individual's physical existence, is the most important determinant of the privacy/space. The privacy phenomenon that exists with humanity is a fundamental and legal necessity that is shaped by this concept of space and originates from the instinct of protecting the individual from others.

However, computer and telecommunication systems, which gained momentum especially with the developments in the industrial revolution and the processes of transition to the information society, have carried societies to another dimension in intellectual and vital aspects and the most important attacks in this dimension have been directed to the private area. The changing intellectual and vital forms in the information-oriented social structures, have offered different environments in relation to the life of the individual and the individual has always endeavored to exist in these different environments.

It would be appropriate to call these environments a reflection of the uniformity of living spaces in the globalizing world with the information society while they are rapidly consumed and replaced by another uniformity. Likewise, it would be appropriate to denote that social media as one of the most important practices of these environments.

The effort of the individual to exist in this mobile and virtual world has caused a shift in the definition of privacy and thus resulted the necessity of developing a new definition and boundary especially for the social media.

Nothing persists in the life of the individual for a long as it used to before. New messages, information, images constantly flow to the individual's life through social media and make the person insensitive. In the midst of this depersonalization, the individual is in the position of a founder or mediator of this depersonalization systematic. In social media, which is the most appropriate basis of a life of demonstration, individuals become the most important part of this demonstration and they can open their privacy to others or virtual public space in order to be always at the forefront of this depersonalization systematic.

The individual is curious by nature and wants to be recognized. However, while trying to satisfy these instincts of appreciation and curiosity through new identities (virtual identity), different channels are used according to the consumption practices of social media.

Instagram is a platform where the routine is always excluded and is a virtual environment satisfying curiosity and recognition needs of the individual. While the individual satisfies these impulses on photography, the person is also a medium for satisfying others' impulses. In this application, where admiration is a prerequisite, the individual moves away from the routine of photography and space in order to get more 'like' and reveals special moment and space photographs that no one knows/sees. In order to stay up-to-date, every day users -especially celebrities- share photos of their families, homes, bedrooms, and even intimate places on their bodies. In other words, the intimate is now drifting to the point of routine in social media.

In Instagram, where everything becomes routine through photographs, Exhibitionism and Voyeurism are now becoming routine. Due to the social media environments, 'confidentiality' in the definition of voyeurism and the 'disease' attributed to the concept of exhibitionism are now eliminated. Contrary to their definition in the literature, Exhibitionism and Voyeurism in Instagram are no longer considered as crime or disease. Both exhibitionists and voyeuristic are happy in this world.

One of the most important factors that make Instagram a popular application is that celebrities (football players, cinema artists, singers) love this application and that millions of people respond to their curiosity through photos on Instagram.

In figure 1, world-renowned American actor and filmmaker Channing Matthew Tatum shared his wife Jenna Dewan Tatum's photo from Instagram. After attending the Golden Globe Awards, Channing Tatum shared a photo of his wife sleeping naked that night and the photo had around 500,000 likes and over 6000 comments. The image, which was shared in 2017, is one of the best examples of the destruction of the privacy of the place through the body. After this destructive sharing, not only celebrities but

Figure 1. Presentation of celebrities' special living areas in Instagram

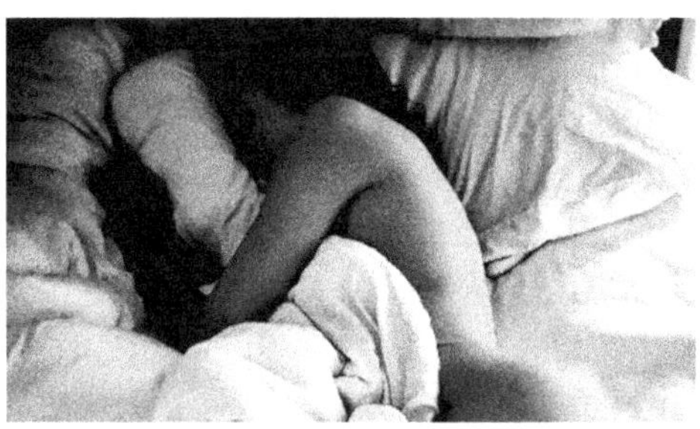

also ordinary individuals continue to share such moments. Considering that the content of the privacy varies in different societies and different times and individual characteristics is important in the concept of privacy that can be encountered in the private domain including the body, various parts of the body, sexuality, family, romantic relationships, and home. Therefore, this example is very important as it is one of the first sharing of 'private' and so widespread to the extent that it can be reported in newspapers while reducing the sharing of private space in an individual scale.

It is the place and body exhibited in Instagram which is also used for the satisfaction of the pleasure of exhibiting and recognition. This impulse to expose surrounded the individual in every aspect. The content of the privacy varies according to society, time and individual characteristics. Apart from the body, various parts of the body, political preferences, and religion (belief) are generally considered as private spaces. The role of religion in determining the content and boundaries of the private area should also be taken into account. In other words, privacy is not an area that the individual determines for himself or has a complete autonomy. This is also an area with metaphysical foundations dominated by divine determinism (Bağlı, 2011, pp. 185-187). As can be seen in Figure 2, even prayers are exposed on social media where peepers are expected to say 'amen'.

Instagram is a language of photographs and acts as a virtual photo album open to the public where the users often ignore to whom they share without even realizing it. With the development of new communication technologies and convergence phenomenon, contemporary individuals who have gone through a mobilized life phase also use mobile-surveillance as one of the standards offered by their mobility (Dolgun, 2008, pp. 133-135).

RESULTS

Surveillance is as old as human history. Surveillance is carried out in different ways in individuals and societies that need this practice from past to present and its aim is explained as social security and discipline. With the idea of the control and discipline of workers along with the capitalist system, the surveillance had been used as a means of power and evolved through different stages as medium, if not as a goal. Especially with the information society, these tools have been used in the field of computer technologies. However, the only fact that does not change is the sole governing power of surveillance practice.

The 1980s marked the beginning of a period in which computers were used for surveillance, and technology began to be recognized in the area. Rapidly evolving technology has transformed surveillance into a cultural activity that has led to the emergence of different surveillance practices and tools. Especially those who have power have developed new monitoring methods for the security and discipline of humanity in surveillance technologies employing computers effectively. The best examples of these are the cameras in certain places such as ATM cameras and street cameras. In other words, surveillance is increasingly under the hegemony of those who hold the power in parallel with the development of technology and is a secret and effective weapon.

With the modernization of society, tools such as telephone, computer, fax, etc. have been entered in the homes as an indicator of modern life as well as a sign of cultural and social change. With these tools, the individual started to have new identities (virtual identity) and began to change consumption practices through these tools. One of these changing practices is the phenomenon of surveillance.

Figure 2. Presentation of private living areas in Instagram

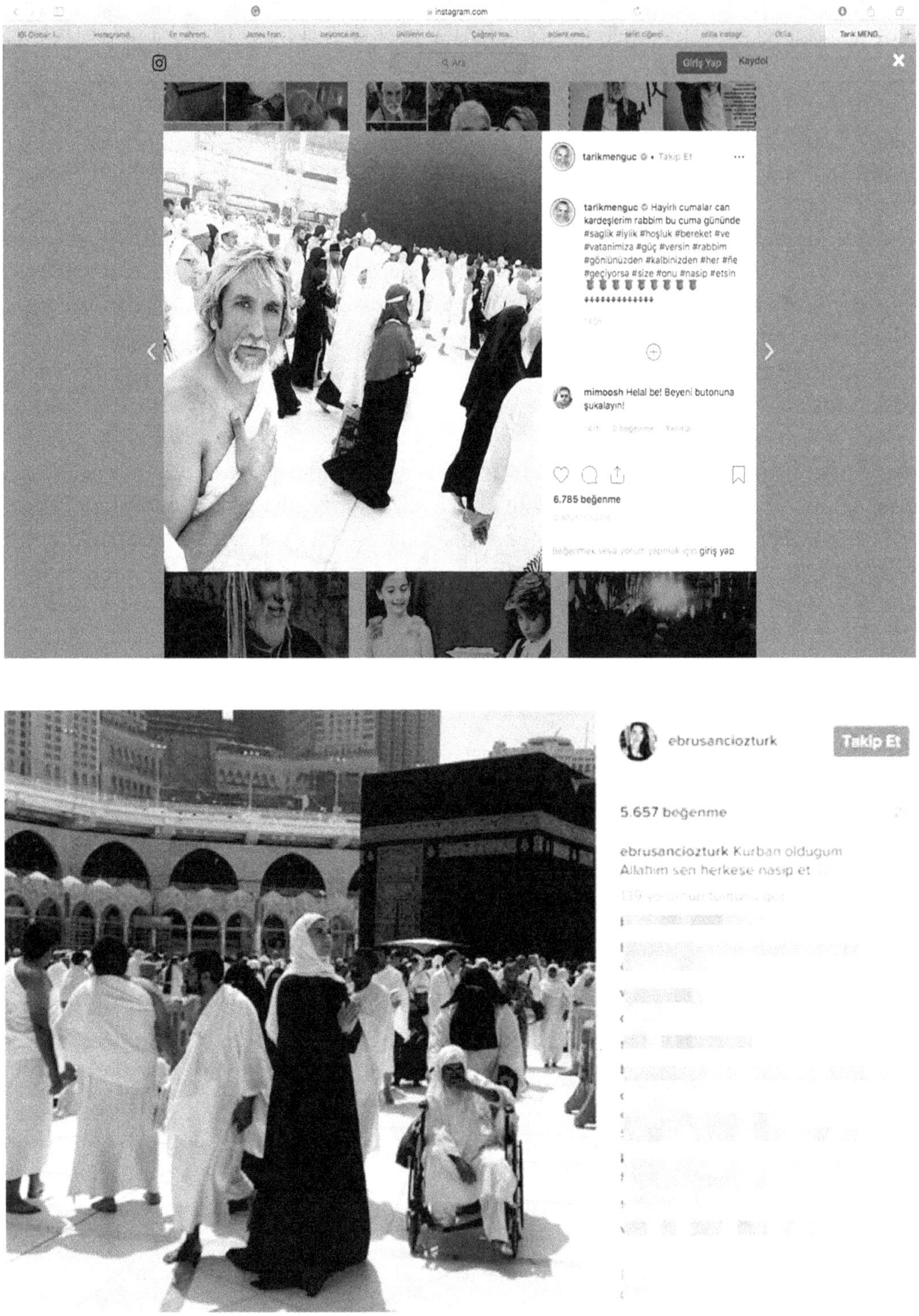

Becoming a global and integrated phenomenon, surveillance culture has started to manifest itself in almost every house with the help of the internet. The world is shrinking with the internet and the medium becoming indispensable and attractive with its easy and inexpensive access. What made the internet so attractive is undoubtedly the emerge of virtual social environments, blogs, and social networking sites.

The different applications emerge in the social media in a daily basis steal users' free time and people are trying to be active in this medium without realizing the cost.

With the mobile phones, individuals who are introduced to the concept of portable internet are constantly interacting with their virtual environment and live a mobilized life with their computers in their office and with the mobile phones outside. The individual who is increasingly accustomed to his virtual identity tries to exist with what cannot be realized in the real life.

Under the pressure of curiosity and recognition impulses, the individual is aiming to be ahead of others in the universe with virtual identities and this race comes with the cost of opening private space to the public sphere.

Various applications in social media are available to the individual. In addition to satisfying the informatics hunger, these applications obliged the individual to serve in the social media. Instagram social sharing application examined in this study shows both the degree of this obligation and that it can be useful while eliminating the informational hunger. However, it is clearly demonstrated that exhibitionism and voyeurism, which are focus of the present study, are offered to the consumption of the individual who is expected to serve social media.

In Instagram, where instinct of curiosity and recognition are at the forefront, users share their photos and expect to be recognized (liked) by the others. More practical for realizing the goals of celebrities, Instagram has started to populate its own celebrities. "The more 'tagged' I am, the more I am recognized" mentality implied in this application urged the users to go outside the routine with the 'tag' shows.

The user who does not want to be ordinary in Instagram tries to show this by sharing extraordinary photos. Photos that are not ordinary mostly imply sharing the privacy. The user, who does not mind sharing private life with the photographs, does not see any harm in destroying the sanctified space for the intimate. In Instagram where the space is exposed, a new culture is created where the observer does not encounter a problem in this a new culture: voyeuristic culture.

In Instagram, happiness and sadness are expressed with the dull photos, and these expressions receive a dull photo in celebrations and consolation as a response. In Instagram Live, another platform where happiness and sorrows are displayed, the situation becomes more fluid and exhibitionism and voyeurism are experienced in an interactive and instant way. In this application, the exhibited, the exposed object and the observer are online.

The charm of the individual space is the cornerstone of this practice. The place of the space in the privacy of Instagram and Instagram Live broadcasts are destroyed, one's house is everyone's house, one's room is considered everyone's room. With the Instagram, you can enter the bedrooms, share the happiest moments, search for partners in distress as well as search for partners for privacy. Instagram is a platform in which exposition and voyeurism are not regarded as problems, and understanding and reciprocity are the principles normalizing this situation. These principles, which is not applied at such a high level in the routine life practices, has caused an erosion in the point and punishment of the voyeurism and exhibitionism and meaning in the legal and health literature, and thus has caused disappearance of privacy and secret space.

In particular, the concept of space that blesses the privacy of an individual has almost disappeared through these two applications and the importance of the secret/place/space where the individual does not include anyone without the will is now lost. People have become very eager to exhibit in social media what they denote 'personal' secrets in their daily life that is closed to others' interference.

While the current century is shaped by the power of vision and envision, new communication technologies build "*Homo videns*" (man who sees) instead of "*Homo sapiens*" (man who thinks)(Sartori, 2006, p. 11). Compare to *Homo sapiens*, *Homo videns* is a profile that can access information sources very easily and inexpensively but seeks the solution of informational ignorance on the image.

In today's societies *Homo videns* is evidently a trend in which new communication technologies are built on goals and means of observation. These applications, such as Instagram, are the result of new communication technologies' aims and tools, as well as reflecting the situation of *Homo sapiens* in social and personal relationships. However, the act of seeing is used for other purposes rather than learning in applications such as Instagram, and the limitlessness leads to the observation transforming into voyeurism.

Instagram profile is also considered as a virtual space for user. It will not be wrong to describe membership as the new identities of the users, involving as the new house, and the privacy settings in these applications as the door and window sliders to ensure the security of the house in the applications. However, it is not a problem to leave these doors open on social media where others can enter without knocking the door or go to sleep in the bedroom without the asking permission.

Social media is no longer disturbing, no matter how or by whom it is viewed, on the contrary observation makes it observed more valuable. Therefore, the individual constantly seeks answers for 'How can I always be more valuable in this way?' Instagram allows people to accomplish their dream of becoming a celebrity while requesting them to put their privacy, secrets, decency, and 'uniqueness' on the table and expect some real-world social values to be ignored.

Socialization and communication activities, which are some of the reasons of individuals' existence in the virtual environment, take place via photos on Instagram. With shared photos and publications in which location and participants of the activity along with the choice of food and garments are revealed, the individual is deliberately or indecisively pushes the limits of privacy. Now, the sharing based on observation and exposure of the bedrooms turns private areas into public spaces.

With the social networking sites and applications, along with the sense of curiosity, users can develop a habit of following the individuals whom s/he may never actually have the opportunity to meet in real life. The user in this sense is unwittingly dragged into voyeurism through Instagram, and after some time regards this as an ordinary event. However, in the literature, voyeurism is seen as a disease if it lasts longer than 6 months. Considering that the number of regular users of Instagram application today reaches to a billion, it is very common that voyeurism is routinely performed in virtual space. Nonetheless, users are attracting each other into voyeurism and exhibitionism in these practices rather than considering it as a disease.

Instagram, a photo sharing site that can be easily accessed and allow sharing by people of all ages, has advantages and disadvantages compared to its intended use. It can be a very advantageous application in the context of visual learning particularly about a place planned to be visited or in the context of longings. However, these applications, which are open to use in different intentions, are also disadvantageous since it can capture regular users in name of popularity and socialization, and urges everything to be presented and shared.

REFERENCES

Bağlı, M. (2011). *Modern bilinç ve mahremiyet.* İstanbul, Turkey: Yarın.

Blair, C., & Lanyon, R. I. (1981). Exhibitionism: Etiology and treatment. *Psychological Bulletin, 89*(3), 439–463. doi:10.1037/0033-2909.89.3.439 PMID:7255626

Dedeoğlu, G. (2004). Gözetleme, mahremiyet ve insan onuru. *Turkey Informatics Association Journal, 89*, 35–36.

Dolgun, U. (2008). *Şeffaf hapishane yahut gözetim toplumu: küreselleşen dünyada gözetim, toplumsal denetim ve iktidar ilişkileri.* İstanbul, Turkey: Ötüken Neşriyat.

Foucault, M. (2007). *İktidarın gözü* (Çev: İ.Ergüden) İstanbul, Turkey: Ayrıntı. Retrieved from http://www.elyadal.org/pivolka/22/parafili.html

Kırık, M. (2015). Sivil toplumun sınırlandırılamayan sosyal medya sorunsalı. (Ed: M. Kırık, A. Çetinkaya, Ö. E. Şahin). Bilişim. S:161-183. İstanbul, Turkey: Hiperlink.

Milliyet. (2012). Şipşak 1 Milyar Dolar. Retrieved from Http://Ekonomi.Milliyet.Com.Tr/Sipsak-1-MilyarDolar/Ekonomi/Ekonomidetay/11.04.2012 /1526691/ Default.Htm.

Safko, L. (2010). *The social media bible tactics, tools and strategies for business success.* Hoboken, NJ: John Wiley & Sons.

Salomon, D. (2013). Moving on from facebook using ınstagram to connect with undergraduates and engage in teaching and learning. *Collage Researchlibraries News, 74*(8), 408–412. doi:10.5860/crln.74.8.8991

Sartori, G. (2006). *Görmenin iktidarı. (Çev: Gül Batuş, Bahar Ulukan).* İstanbul, Turkey: Karakutu.

Sputniknews. (n.d.). Retrieved from Https://tr.sputniknews.com/yasam/201812051036489809-instagram-kullanici-sayisi-turkiye/

Yavuz, H. (2012). *Budalalığın keşfi.* İstanbul, Turkey: Timaş.

Yüksel, M. (2003). Mahremiyet hakkı ve sosyo-tarihsel gelişimi. Ankara University Faculty of Political Sciences Journal, 58(1), 182-213.

Yüksel, M. (2003). Modernleşme ve mahremiyet. *Journal of Culture and Communication, 1*, 75–108.

Zhumagaziyeva, A., & Koç, F. (2010). Kültürel etkileşimde kadınların giysi alışkanlıklarındaki değişimin mahremiyet açısından incelenmesi (Kazakistan ve Türkiye Örneği). *E-Journal of New World Sciences Academy, 5*(2), 113–131.

ADDITIONAL READING

Akgül, S. K. ve Pazarbaşı, B., (2018), Küresel Ağlar Odağında Kültür, Kimlik ve Mekan Tartışmaları, Hiperyayın, İstanbul.

Bakıroğlu, C. T. (2013). *Sosyalleşme ve Kimlik İnşası Ekseninde Sosyal Paylaşım Ağları, Akademik Bilişim, XV.* Akademik Bilişim Konferansı Bildirileri.

Niedzviecki, H. (2010). *Dikizleme Günlüğü, Çev.: Gündüz, G.* İstanbul: Ayrıntı.

KEY TERMS AND DEFINITIONS

Exhibitionism: İs the act of exposing in a public or semi-public context those parts of one's body that are not normally exposed – for example, the breasts, genitals or buttocks. The practice may arise from a desire or compulsion to expose themselves in such a manner to groups of friends or acquaintances, or to strangers for their amusement or sexual satisfaction or to shock the bystander. Exposing oneself only to an intimate partner is normally not regarded as exhibitionism. In law, the act of exhibitionism may be called indecent exposure, "exposing one's person", or other expressions.

Instagram: Instagram is a photo and video-sharing social networking service owned by Facebook, Inc. The app allows users to upload photos and videos to the service, which can be edited with various filters, and organized with tags and location information. An account's posts can be shared publicly or with pre-approved followers. Users can browse other users' content by tags and locations, and view trending content. Users can "like" photos, and follow other users to add their content to a feed.

Privacy: Privacy is the ability of an individual or group to seclude themselves, or information about themselves, and thereby express themselves selectively. The boundaries and content of what is considered private differ among cultures and individuals, but share common themes. When something is private to a person, it usually means that something is inherently special or sensitive to them. The domain of privacy partially overlaps with security (confidentiality), which can include the concepts of appropriate use, as well as protection of information. Privacy may also take the form of bodily integrity. Privacy may be voluntarily sacrificed, normally in exchange for perceived benefits and very often with specific dangers and losses, although this is a very strategic view of human relationships.

Social Media: Social media is an interactive computer-mediated technology that facilitates the creation and sharing of information, ideas, career interests and other forms of expression via virtual communities and networks.

Surveillance: Can be viewed as a violation of privacy. A graph of the relationships between users on social networking sites. Social network analysis enables governments to gather detailed information about peoples' friends, family, and other contacts. Since much of this information is voluntarily made public by the users themselves, it is often considered to be a form of open-source intelligence. One common form of surveillance is to create maps of social networks based on data from social networking sites such as Facebook, Instagram, Twitter. These social network "maps" are then data mined to extract useful information such as personal interests, friendships & affiliations, wants, beliefs, thoughts, and activities.

Voyeurism: Voyeurism is the sexual interest in or practice of spying on people engaged in intimate behaviors, such as undressing, sexual activity, or other actions usually considered to be of a private nature. The term comes from the French for which means "to see". However, that term is usually applied to a human who observes somebody secretly and, generally, not in a public space.

Chapter 16
Carnivalesque Theory and Social Networks:
A Qualitative Research on Twitter Accounts in Turkey

Sefer Kalaman
Yozgat Bozok University, Turkey

Mikail Batu
https://orcid.org/0000-0002-6791-0098
Ege University, Turkey

ABSTRACT

Carnivalesque theory has been used as a model and a structure in the works carried out in many fields such as communication, literature, and sociology. In fact, Carnivalesque appears in many environments/areas, particularly in the social networks, which are the manifestation of social life. This chapter examines social networks in the context of carnivalesque theory to reveal facts of carnivalesque in Twitter. Content analysis technique was used in the research. Research data came from 10 Twitter accounts which have a maximum number of followers in Turkey. These data were analyzed and examined in terms of grotesque, dialogism, carnival laughter, upside-down world, marketplace, and marketplace speech belonging to the carnivalesque theory. According to the findings, the structure of Twitter, which is one of the most popular social networks in Turkey, is largely similar to the structure of the carnival and features of carnivalesque theory.

DOI: 10.4018/978-1-7998-1041-4.ch016

INTRODUCTION

Human beings, who started living in the big communes, have the birth and development of many innovations in social, political, artistic and technological aspects since they have the common mind, the ability to work together and the ability to think and invent individually. These innovations have led to the transformation of the individual and society in many respects. The internet, which is the most recent example of these innovations, has led to the transformation of everyday life in more ways and numbers than ever before although it is included human life into more than half a century ago.

The social networks within the Internet are the main actors of this transformation at the point of social relations. Social networks have influenced and changed the way people communicate, how they belong to a community, and how they display their selves. People are now socialized through this virtual network, belonging to the virtual communities in this network and express their ego with created virtual identities in these virtual networks.

These networks resemble a carnival because of their characteristics. People are doing that will never do in everyday life in this network. People say things they can never say and people act like they can never be this virtual network. This kind of virtual environment exists as their second life.

Social networks have a great deal of the characteristics of the carnivalesque theory of Michail Bakhtin. This digital network is a digital environment of carnival where makes people come together, form communities, act for a common purpose, oppose ins, derive a common language, produce a humor, laugh, change their bodies, reverse the existing order and authority just like the carnivalesque.

CARNIVALESQUE THEORY

Carnival

The settled life action that one of the most important turning points in the life of mankind trace to "Göbekli Tepe" that B.C. 10000 years in the Şanlıurfa province in Turkey. In time, the people coming together and forming large communities, and the formation of cities happened in B.C. 3900. The first samples of this city is now Turkey, Egypt, Iraq, are of such countries. At the time, these cities became the pioneers of civilization, production, education, art and many other fields of life. In Egypt, one of the most important of these civilizations, the whole habitat is designed with a hierarchical order and is divided into different categories (slave, ruling class, workers, farmers, craftsmen) according to their status. The social activities of the people belonging to these classes take place in their class in the general and their relationship with other classes continues on the basis of superior-subordinate relationship. However, in some social celebrations, such as carnival or festivities, these classes coexist, and at the time of the carnival, these distinctions are relatively uncertail/transparent compared to other times.

This is the first festival of history recorded Feyyum in Egypt, one of the first cities of history interrupted routine of everyday life that determined by strict rules and restrictions, is at ease Pharaoh and minorities have power, contains the most difficult conditions for the rest of the people. These festivities outside the day-to-day operations and rules have a free time feature and on other days, these festivities

Figure 1.

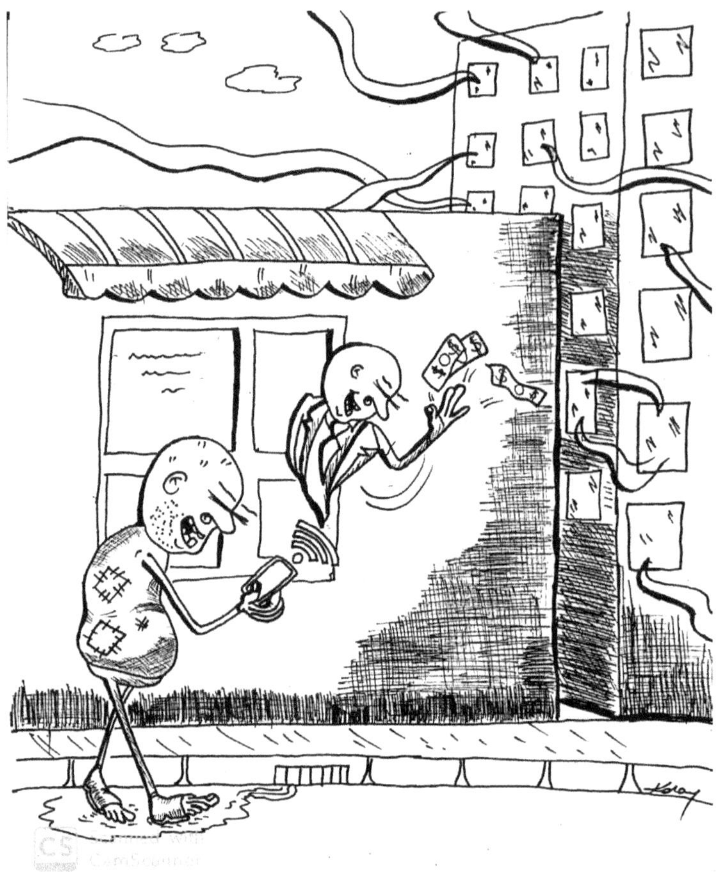

eliminate the necessities that shape the life of the people. In the given banquets, it is replaced by the abundance, people can enter and see places where they are not allowed to enter in the normal time and they can be found in the same environment as the Pharaoh and the minority they serve in the whole year (Yurdakul, 2006: 20). This structure, which was the first example of the carnival, showed itself in the Middle Ages. Carnival culture was born as a new life space in the oppressive, rigid and restricted environment of the Middle Ages and has been strictly adopted by people. In other words, the carnival became a second life space in the Middle Ages, built by people on laughter. The carnival, which differs from culture to culture and from society to society, has come to life as a celebration, entertainment or festival that Christians gathered in communities before they entered the Great Lent.

Carnival is a festival life. The festival contains odd features of all the shows and funny rituals of the Middle Ages. All features and applications of real and ideal life at the carnival are suspended. Special type of communication that will never be possible in everyday life shows up during the carnival. In this type of communication, everything can be said clearly, all kinds of satire, mockery, place, humiliation (this may be a king or a noble), to ignore the rules of etiquette, and to perform the desired gesture and mimic forms (Bakhtin, 1984: 8- 10).

Bakhtin alleges these carnivals as an anarchic and emancipatory time period when it is razed the political, legal, and ideological authority of the church and the state. In this period of time, beliefs and rules are used as a mockery fact, and a clean and clear beginning is allowed to create new ideas (Kıpçak, 2016: 110). The suspension of the hierarchical structure and the fear, respect and sublime concepts that develop in relation to this hierarchical structure, the disappearance of the distance between individuals and the beginning of a more intimate relationship, the monotony of everyday life to replace the strange and unusual, sacrilege, obscenity, satirize holy, select the metaphorical king and queen and overthrow them, are some of the carnival's features (Brandist, 2011: 208-209).

From this point of view, there must be two lives of people living in the Middle Ages. The first is the real life that is formal, ordinary and be submit to ins and eupatrid. The second is the second life, the carnival life, where they live their true selves, behave freely from oppression and authority, transcend bodily and linguistic boundaries, and be turn distinction of ins-people, wielder-governed, noble- commons.

The person who transforms himself into a carnival wants to subvert all the moral and social rules imposed by the official life. The characteristics of the rebellious behavior; it provides an emotional discharge of absurdity, contrast, extravagance and illegality. H owever, if such behavior is separated from the daily life by the usual timetable, it also begins to adapt the social order. Carnival is an area of freedom under the surveillance of the power and within the limits of power. This liberating and at the same time adaptive order of the carnival has its own characteristics and determinative functions. When carnival participants turn into carnival itself, they fulfill these functions and violate the rules of daily and official times. While they all exist in each other, these functions, which form autonomous areas on their own, are carnival itself (Kıpçak, 2016: 116-117). During the carnival, the power of the king, kingdom, nobles and autocracy weakens. However, when the carnival is over and people return to their normal life, these power centers continue to manage more powerful than before.

Mikhail Bakhtin and Carnivalesque Theory

Born in 1895, Mikhail Bakhtin was a theorist in the human sciences, linguistics and literature. He makes mark in history with his work in the Soviet Union between 1920-1975. Bakhtin's work has been recognized by many scientific disciplines, scientists and researchers when he died more than the period he lived. However, Bakhtin's work has been used by different researchers as a concept, theory or inspiration in different disciplines. He included dialogue, dialogic and carnival concepts in own works. Another concept that Bakhtin discusses in depth is carnivalesque. Carnivalesque is a theory which based on carnival, characteristic structure of carnival and carnival features.

Bakhtin said that "In its history, carnivalesque is both a historical phenomenon and a literature trend". In this sense, carnivalesque has the characteristics that evaluate the multiplicity of life in the theoretical field and in this context can be applied to the language or culture area (Sözen, 2009: 66). It is necessary to explain some essential concepts in terms of the structure, characteristics and framework of the carnivalesque theory. These concepts, which form the basis of the carnivalesque theory, can be listed as grotesque, dialogism, carnival laughter, upside-down world, marketplace and marketplace speech.

In the carnival studies of Bakhtin, the grotesque form is mentioned quite a lot. Hegel said that grotesque has three feature in the context of the archaic Indian forms. According to this, it is possible to list the grotesque characteristics as the mergering of different natural areas, immeasurable and exaggerated dimensions and proliferation of different members and organs (hands, feet and eyes of Indian gods) of the human body. In particular, the destruction of the body, which is the basis of the grotesque form, is one of the indispensable elements of the carnival squares. In particular, the destruction of the body, which is the basis of the grotesque form, is one of the indispensable elements of the carnival squares. It is a common behavior to compare phenomenon individuals (king, religious leaders, nobles) to dead people or overthrow them (Uygun ve Akbulut, 2018: 78; Bakhtin, 1984: 44). However, people experience a metamorphosis by shedding their own assets and bodies in carnival with outfit, mask and exaggerated costumes. This situation, in the words of Bakhtin (1984: 48), means that both the man and the woman are freed from the idealized body structure imposed by the society and the body is free from its standard structure.

Another important concept of the carnivalesque theory, except for grotesque, is dialogism. Bakhtin includes highly of the concept of dialogism in his books and clarifies how dialogism has come to life in the carnival environment. Bakhtin expresses the common language used in carnival with the concepts of dialogue, dialogic and dialogism. Bakhtin also reveals how this language differs from the official language of prohibition in normal life. According to Bakhtin (1981: 426; 2001: 20), dialogism is a characteristic epistemological type of a world dominated by multilingualism. In this language world, the obligation of people to establish dialogue is a guarantee that communication will not be a monologue. Even the simplest daily conversation is basically dialogic. This multiplicity, which forms the basis of dialogism, forms the basis of a free and creative communication environment. In this context, the free expression of different ideas and the emergence of a discussion environment on these differences become one of the basic characteristics of dialogic communication (Kıpçak, 2016: 125).

An irreplaceable element of this communication is the laughing act. In other words; laughing does not just mean the loosening of the formal environment. The laughing act takes on a renewal mission at the carnival. As a form of carnival laughter, parodies destroy the subordinate and upper relations, and ensure to be reinterpreted the objects of everyday life. Laughter also allows for the creation of a creative environment. Hierarchical order transforms everyday life into a colorless, cold and diseased environment. In this context, laughing takes the role of a therapeutic role on the society by destroying the hierarchical structure and language (Gülüm, 2016: 131; Kıpçak, 2016: 119-120). Laughing refers to the fact that people are suspended the dominant culture and authority (Turan, 2008: 16). This smile is a revolt in its essence, this is to denigrate ins and it is a shout of freedom. This situation is to reverse the existing order, habits and life.

Life is turned upside-down during the carnival. Identities are turned upside-down. Tasks, occupations, moral norms, sexes are turned upside-down by reversing the established hierarchical order. This upside-down state is a description of a utopian world. When the carnival ends and life returns to its old order, it is replaced by a flat, oppressive and boring world. While the possible new world order is being abandoned, things are restored and the order continues (Kıpçak, 2016: 130).

The first concrete step in the upside-down world order, which is at the head of the Carnival environment, is undoubtedly depose rituals that the king to be declared a clown and the clown to be declared a king. This ritual is done in an order and accompanied by ceremonies. However, this is a behavior that is unique to the carnival time frame. Even if the carnival appears that authority have lost a little power, it will not dare demolish it. Because although the carnival environment depicts a free formation, it obtains permission from the authority to be realized and to be presented to the public (Uygun and Akbulut, 2018: 78).

Carnival is a non-discriminatory performance between the performers and the audience. At the Carnival, everyone is an active participant. In the carnival act everyone comes together and unite. The carnival is not watched, or even more strident, it is not even an execution; participants live in the carnival. Because carnival life is a life out of the usual way, it is a life that has been reversed. Carnival is the reversed face of the world (Derin, 2017: 328). In the same way, daily, formal and restricted language has turned into a structure called marketplace speech.

The well-known language of the marketplace has a structure in which there is blasphemy, slang and underbred communication. Another living area where this situation is experienced is carnivals. Carnival is an environment in which the oppression, slang, humiliation and ridicule occur as in the same marketplace where the repressive regime, formal and official language can be taken out. Here, everyday life can be turned upside-down, people can swear and mock with each other as well as the noble group, and their existence may be overthrown during the carnival period (Bakhtin, 1984: 17).

The place where this action takes place is the carnival square, in other words the marketplace. Carnival, which is one of the most prominent cultural and popular gathering events of early Modern Europe, also hosted the marketplace where trade developed. In the marketplace, merchants, sellers, touts exhibit their products. In the marketplace where a wide variety of products exist, the phenomenon of shopping in the atmosphere of the carnival like feasts also come to forefront. As a result, carnivals, which temporarily liberate their attendees, also is allow for a commercial exchange field that enable to settle the capitalist system. While people are temporarily moving away from the dogmatic seriousness of hierarchical structures at the carnival, they also find themselves in the midst of the capitalist logic and power mechanisms of the established system with the drunkenness of laughter. Crowds gathered together with the carnival transform this rich and fertile environment into an advertising environment. People who advertise their products or themselves are attracted to the public's attention by using all kinds of humorous language in the city square and the streets. In this context, it will not be wrong to say that the marketplace is the place where the announcements and advertising strategies are implemented in the carnivalesque sense (Kıpçak, 2016: 13-133).

The entertainment on the marketplace as well as the carnival has also become legal. This legalization took place by force. However, these entertainments have been partial and have led to conflicts and new prohibitions. During the whole Middle Ages, church and the state were forced to make more or less concessions and satisfy the marketplace. The small islets of the time, which are allowed to leave the official routine under the laughing sheath only and are strictly limited by the feast days, are spread all year. Obstacles have been removed provided there is nothing but laughing (Bakhtin, 2001: 110).

The elements such as marketplace, marketplace speech, grotesque, dialogism, upside-down world and carnival laughter are the main features that constitute the carnivalesque theory. Although carnivalesque theory is a theory used in literature, it is adaptable to many fields of science and art because of its characteristics. One of these areas is undoubtedly the field of communication. Because of its structure, the concepts of carnival and communication are two elements which are already very much related to each other. For this reason, the carnivalesque theory, which can be adapted to many communication environments, can also be adapted for social networks, one of the most important communication environments of the era.

SOCIAL NETWORKS AND CARNIVALESQUE THEORY, TWITTER AS A CARNIVALESQUE FIELD

The traditional media, which has been created through mass media such as radio, television and newspapers, have been replaced by digital media. A new media environment has been created with the internet technology that provides an interactive participation to the individual. In this platform, which is also called as social media, individuals have the opportunity to tell their thoughts and their attitudes towards any situation freely, to express, support or criticize even an event in another center of the world (Uygun ve Akbulut, 2018: 76-77). Considering the elements of number of users, generated content and popularity, this virtual environment, which is used for different purposes, is the most important environment for social networks. Social networks are a digital communication environment where incorporates different communication tools, users can enter anytime and anywhere they want, users can create accounts individually, communicate with other users and generate content.

Boyd and Ellison (2007) emphasize three features to define social networks that product of web 2 technology. First, social networks allow users to create an open or semi-open profile within a system with a limited number of boundaries. Second, social networks provide a list that users share links with others and access to this list. Third, social networks allow users registered in this network to have access to others' links.

In other words, social networks are a virtual communication/life area where people get together and share their feelings and thoughts, react to ins or people and institutions who have power, create communities for a specific purpose, generate humor, mock ins with this humor, distance from distress of everyday life, oppose the social hierarchical order and reverse the existing social structure, can freely exhibit their own self and create a space of freedom.

Social networks are similar to a carnival because of these features. Today, people's living space is now surrounded by computer networks. In the digital world that lives 24 hours a day, 7 days a week, social networks cover a large part of people's lives. The carnivalesque theory, which Bakhtin put forward years ago, has also been adapted to the social networks of today (Kıpçak, 2016: 107). Because the nature of the carnival is related to popular culture. There are carnival if there are locals. Although it is directed towards transformation into mass culture with capitalism, people are popular culture are people in basis of popular culture. Popular culture provides a free environment and endless pleasure like carnival. Popular culture is spreading more quickly thanks to mass media. New media tools that replace the traditional media order are effective tools in this formation. Considering all this, it is seen that the new media and social networks have become carnival (Uygun and Akbulut, 2018: 79).

In a study by Mclean and Wallace (2013: 1530) on the carnivalesque theory and blogs, which are one of the social networks, it was revealed that social networks have a large part of the features of the carnivalesque theory. The actions of people in this social networks such as laughter, parody, used masks and costumes, break off daily formal life and transform different life, used marketplace speech, mock about ins are very similar to the carnivalesque theory. In addition, in the study of Uygun and Akbulut (2018: 88), in the context of social media researches, the carnivalesque theory of Mihail Bahtin is frequently mentioned. Because of the features it contains and its unique style, social networks and carnivalesque are overlapping. From a general perspective, the upside-down world, tout, body destruction, grotesque images, features of humor and laughter are found in Instagram, one of the most popular social networks.

Carnivalesque theory and features of carnivalesque are in social networks such as blogs, Instagram and Facebook. At the same time, the structural features of the carnivalesque theory are largely similar Twitter which is one of the most widely used environments of social networks. The accounts created on Twitter, the shares made on these accounts and the comments made on the shares correspond dimension of virtual environment such as marketplace, marketplace speech, grotesque, dialogism, upside-down world and carnival laughter which belong to carnivalesque theory.

Twitter is a social network established by Jack Dorsey and his colleagues in 2016 in San Francisco. The general purpose of Twitter provides to share opportunity to people in a web environment, like that feelings, thoughts, experiences, things in daily life. In other words, Twitter is a public sharing network that is open to everyone with internet access. The main feature that distinguishes Twitter from other social networks is that users can express themselves with a maximum of 140 characters, but there is no limit to the number or frequency of messages. Therefore, it shows a practical structure for writing and reading (Altunay, 2010: 36). As a result of its features, Twitter can be used by different people, states, institutions and organizations for different purposes.

For example, Twitter is a common place where protesters meet against Iranian elections, a public channel which gave the Zapatistas a right to speak and shared their thoughts with the whole world, the first mass media that Barack Obama has announced his presidential candidacy, a network of Haiti earthquake victims. In this context, it can be stated that new media and social networks carry their own democracy to the most powerful economies from the most repressive regimes (Altunay, 2010: 36).

According to Kıpçak (2016: 155), Twitter has parallels with the carnival environment with the above mentioned structural features and the usage patterns that arise from these structural features. The main determinants of the carnival can also be seen on Twitter and Twitter is becoming an online carnival. Throughout history, certain species may reappear when they meet the appropriate conditions. This is due to the fact that the species is updated according to time, but it has its core qualities. Carnivals can be repeated in a completely commercial way, far away from the traditional atmosphere. However, developing technologies and new media offerings; It creates a digital carnival that forms its presence on networks, independent of time and space. Twitter has a carnival qualities with humor, political satire, concealment of identities, advertising strategies and unique language.

METHODOLOGY

Method

The aim of the study is to examine social networks in the context of carnivalesque theory and to reveal the carnivalesque factors in Twitter, one of the most important social networks. In this study, a case study design which is one of the qualitative research methods is used. Case study that why and how to answer questions, is used for cases where the link between the content and the event is not clear and net (Sığrı, 2018: 161). In the case analysis, it is possible to examine in-depth details of a subject as well as to compare the details of many subjects. In this way, the issues that have not been previously considered

Figure 2. Examples of grotesque

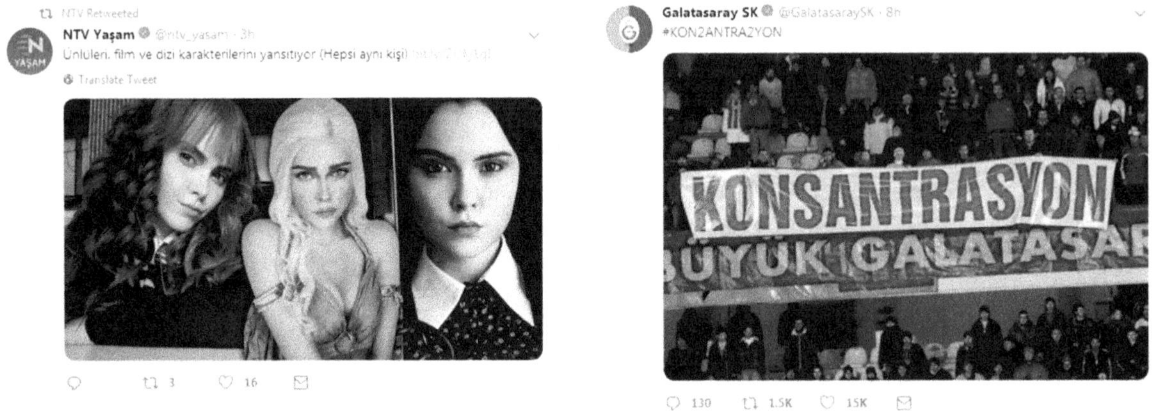

on the subject can be transferred to the reader. Content analysis technique was used in the analysis of the data obtained with case study. Content analysis is a qualitative data analysis technique that aims to reach the concepts and relationships that can explain the collected data. In the first stage, the data summarized and interpreted in the descriptive analysis are subjected to a deeper process in content analysis (Sığrı, 2018: 280).

Population and Sampling

The population of the study is Twitter accounts which the most followed in Turkey. 10 Twitter accounts which have a maximum number of followers in Turkey, represent sample of study. This Twitter accounts (Cem Yılmaz, Recep Tayyip Erdoğan, Abdullah Gül, Galatasaray SK, NTV, Okan Bayülgen, Ata Demirer, Cüneyt Özdemir, T.C. Cumhurbaşkanlığı ve Fenerbahce SK) is the first 10 accounts with the most followers in Turkey (Socialbakers, 2019). In this study, the shares sent from these accounts are discussed. Data was analyzed and examined in terms of dialogism, upside-down world, carnival laughter, grotesque, marketplace and marketplace speech belonging to the carnivalesque theory.

SOLUTIONS AND RECOMMENDATIONS

Shares of 10 Twitter accounts which have a maximum number of followers in Turkey were analyzed and examined in terms of carnival laughter, dialogism, grotesque, upside-down world, marketplace and marketplace speech belonging to the carnivalesque theory.

Grotesque: With the development of new communication technologies, new applications are increasing and with these applications, it becomes easier to perform different operations on the existing images, change the image in any way and to remove new meanings from the image. In carnivalesque theory, grotesque imagery is deformed by different processes on the body, or new forms or images can

Figure 3. Examples of dialogism

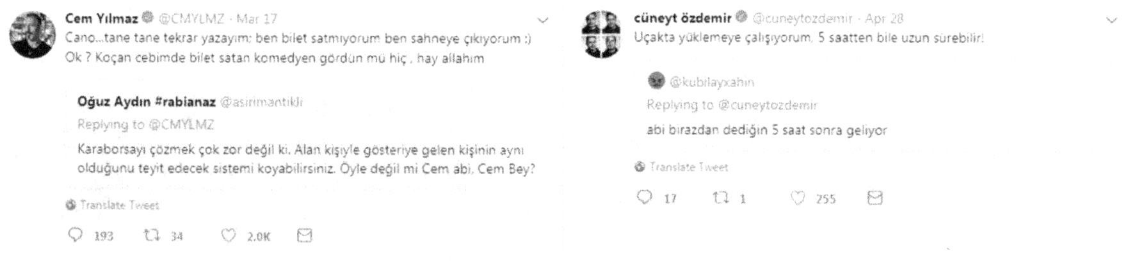

be revealed which can be derived from new meanings. There is an entertainment-oriented production in which people in different borders accepted by the society in general terms can be displaced with a cynical content, the images of the powerful ones are destroyed and the irregularity, satire and irony are highlighted. In this respect, it is possible to share changed, converted images or directly modified/converted/produced characters/information on Twitter. Thus, a social criticism is the basis for social entertainment and the meanings are presented. Below are examples for the grotesque subtitle.

In the first tweet, the character Hermione Jean Granger in Harry Potter has three different images. Sharing three different images of the same woman maters for grotesque content. In the second image in the first tweets, the bodily features of the person were changed and a completely different character was revealed. There are few similarities between the three photographs. In the second example, there is a text which can be explained with the sub-titles of satire and weird in grotesque title, and which hang in the match played by Galatasaray vs Konyaspor. In the article, the team name of Konyaspor was changed. The word "Konyaspor" was replaced by the word "konsantrasyon". The team of Galatasaray carried a banner photo containing this word on their Twitter account and shared it with all their followers in their social networks. In the two examples seen above, there are important contents for the grotesque title (out of existing reality) such as distortion, conversion in the messages shared on the Twitter account. Thanks to Twitter's opportunity of photo sharing, these photos were shared with all followers, and retweeters of the followers were also allowed to see people who were not followers.

Dialogism: There is dialogue in the essence of communication. Dialogue refers to the interaction and influence of two people in the communication process. Hence, there is mutual conversation in the dialogue, but the conversation does not have to be an agreement. In some cases, the end of the dialogue can turn into conflict. In the carnivalesque theory, people who have a dialogue can have different ideas. Dialogic communication is an interaction based on interaction and demonstrating the concrete outcome. This communication is based on reciprocity from the nature of explicity and dialogue. One of the social media, which is considered as the new media, Twitter also allows for dialogic communication, which allows for a carnival interaction. Below are two examples for dialogic communication.

The first tweet above is an important example of dialogic communication within Twitter's opportunity. Due to Twitter's standard number of words, Twitter has short and concise dialogues. However, the established dialogue can also have a acrimonious content within the framework of the carnival speech. In the first example above, a follower of the famous comedian Cem Yılmaz wrote his opinion about

Figure 4. Examples of carnival laughter

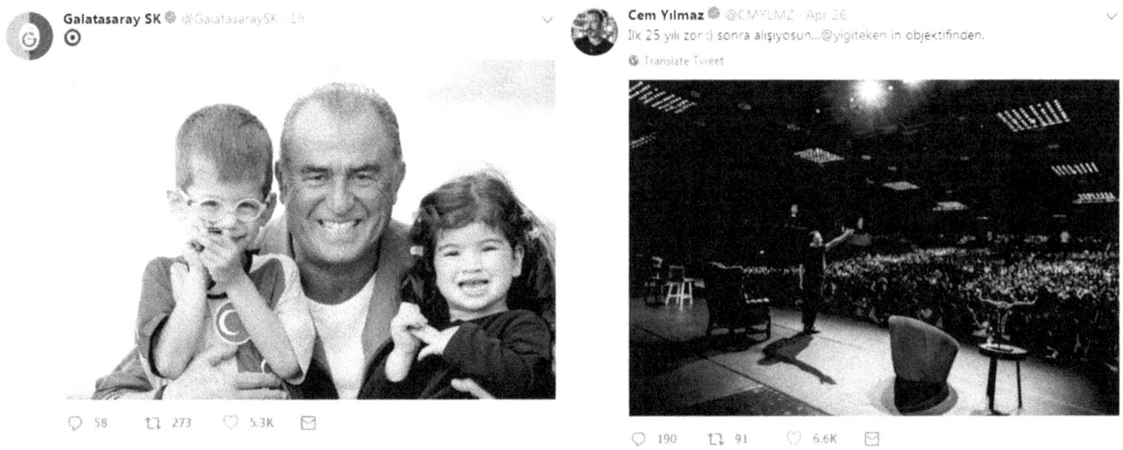

the show tickets on the black market: "It's not hard to solve the black market problem. You can put the system to confirm that the person who bought the ticket was the same as the person who came to the show. Isn't that right, Cem Abi, Cem Bey?". However, Cem Yılmaz responded in a humorous way that this was not about him and therefore he was not the solution point: "Honey ... let me rewrite it; I don't sell tickets I'm going on stage. OK? Did you ever see a comedian who sold tickets in his/her pocket? Oh my Gosh!". In the second example, after a well-known journalist Cüneyt Özdemir shared "shortly after" that his statement in a previous tweet. Then a follower responded in a critical way: "You're saying soon, but it's coming in five hours". Then he explains the situation. In the two examples above, there was a direct interaction between the people who had a Twitter account and their followers. Twitter's application was used, and it was answered directly in a short time and the responses were retweeted by other users. Thus, they communicated directly with their followers within the framework of dialogic communication, tried to explain problems them, and by making this interaction clear to other followers, they have prevented new problems in the subjects mentioned.

Carnival Laughter: Humor and humor can be evaluated in laughter, is a form of expression that encountered in different areas of social life. The laughter, which is an individual and social behavior, can be found in almost every area of humor. Laughing action has an important and central point in the carnival period. Carnival participants put their thoughts into behavior how they can reflect the different codes of laughter in many cases. In social networks, laughter behavior can be implied with photographs, as well as with different emojis, word games or abbreviations. In this direction, the participant can perform the exaggerated laugh on Twitter like that the carnival, and create a deep content on the humorous subject. Participants can do these how using the words that will emphasize the same emojis, letters or laughter. Below are two examples of Twitter being a carnivalesque laughter environment.

In the first tweeting photo, Galatasaray coach Fatih Terim poses with a spectacled boy wearing a Galatasaray shirt and a little girl with a downsendrome. However, the photograph can reflect the social sadness inside, but instead, it has been conveyed exaggerated laughter like carnival. Not only Fatih Terim, it is possible to see the same smile on children's faces. This photograph is a transfer of emotion that ignoring the existing situation and use the laughter.

Figure 5. Examples of upside-down world

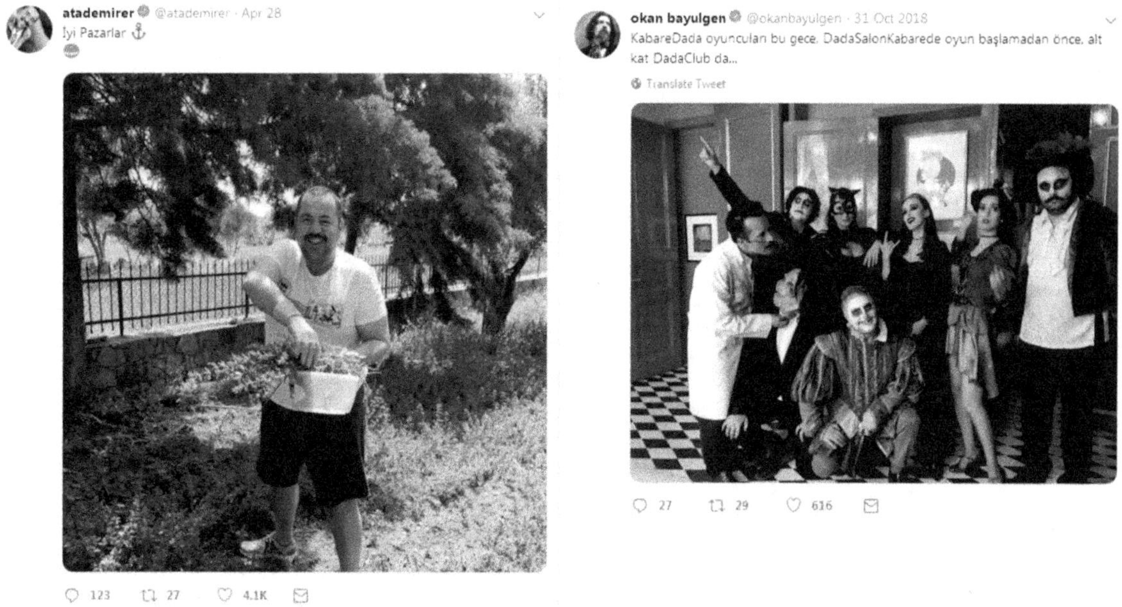

In the second example, there is a photograph from the famous comedian Cem Yılmaz's show and a sentence he wrote himself. Cem Yılmaz wrote "The first 25 years are getting hard, but then you get used to" on a photograph taken from the program. The sentence is a carnival expression, mocked for years and deepening it with smile emojis.

In the examples shown above, there is a smile about the situation and laughing as if it is mocked in the framework of the carnivalesque theory. In this respect, samples were retweeted and appreciated by hundreds of followers. The extent to which the current situation can be ridiculed is clearly seen in Cem Yılmaz's message.

Upside-Down World: In the social life, there is a system accepted by the majority. This order takes place in a system in which the lives of people from different backgrounds are carried out in a certain framework. In carnivals, a new and critical world is established by going beyond the accepted and stereotyped order in social life. In this new world, words are changed, clothing is exaggerated and worn out of the standard. Through masks and colorful costumes, it is disguised and new identities are created. In this way, the hierarchy in real life is destroyed, everyone is equalized, and social anger, rage, joy and revolt are expressed. In this direction, the existing order is turned upside-down. Twitter enables this upside-down world to be applied in terms of its application features and it offers an environment of behavior that will turn the general life in the system to all people with certain technologies. Below are examples for the world that has been turned upside-down on Twitter.

In the first tweeting photo above, the famous comedian Ata Demirer shared a photo of herbs in her garden like an ordinary citizen. Ata Demirer's dress and the garden in the background in the photo are different from the known Ata Demirer image. The photograph is far from the person who is represented under bright lights in the media or scenes and represent the opposite of its existing image.

Figure 6. Examples of marketplace speech

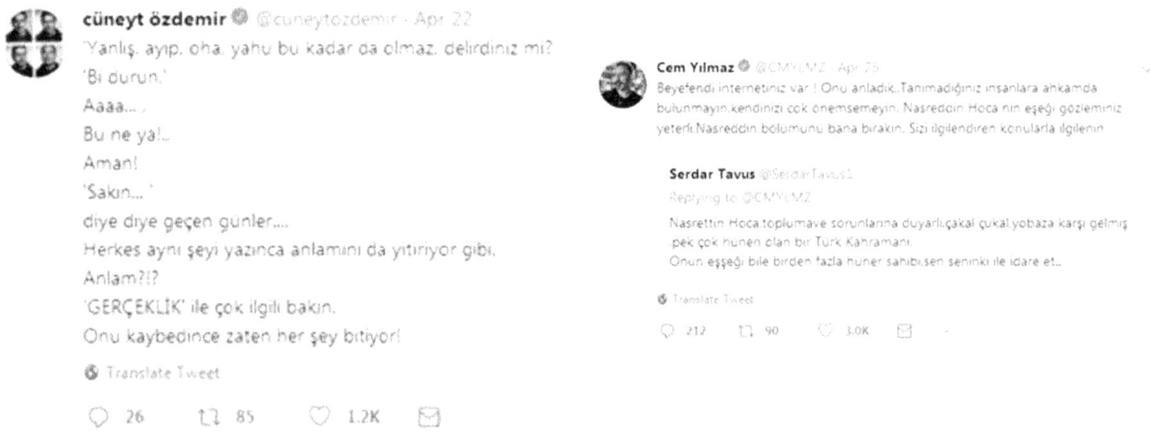

Okan Bayülgen who is the famous showman, shared a photo. Kabare Dada actors is in this photo. When looking at the actors, their clothes, their makeup and their postures are seen to be far beyond ordinary life on a carnivalesque basis. In the examples, via Twitter, was moved out of the world and was created upside-down world image by both disguise and going beyond the visible image.

Marketplace Speech: Public markets have a unique structure and language. The people can speak as they wish there, create their jokes by referring to the agenda and use a language that is natural, free from anyone, culture based and approach to culture humorously. Slang words, special sayings, abbreviations and implications can be found in this language. Through this intimate and direct language, the free environment is supported and a laugh-based festival is created. Thus, certain norms that are imposed on the people by some groups are avoided. Nowadays, marketplace speech can be reflected in social networks that users think are free. In Twitter, it is possible for users to communicate with political leaders and artists in everyday language without a problem. In some cases, Twitter users can express themselves by using local dialects and interesting dialogues can be experienced. The following tweets show examples of how marketplace speech is used in the context of carnivalesque theory.

In the first tweets above, the message of the famous journalist Cüneyt Özdemir is communicating with his followers. In the language of the message, it is observed that the words used in the public and the exclamation words are chosen in daily life. This content directly reflects the marketplace speech of carnivalesque theory.

In the second share, there is a dialogue with a follower of the famous comedian Cem Yılmaz. When we look at the language of dialogue, it is seen that daily Turkish language is used, in addition, a cynical language is used, sarcasm is made and Nasreddin Hoca, which has an important place in Turkish culture, is the subject of speech. His follower said “Nasreddin Hoca is a Turkish hero who is sensitive to society and its problems, has a lot of skills and against evil people. His donkey is more talent than us, you handle it with yours.”. Cem yılmaz in response to this tweet: “Gentleman, you have a internet! You have a gentleman internet! We understand that. Do not obey people you do not know, do not care about yourself. It is enough observation of the Nasreddin Hodja’s donkey. Leave me the part of Nasreddin. Take care of your interests.”. The whole of these statements directly reflects the subheading of the marketplace speech of the carnivalesque theory.

Figure 7. Examples of marketplace

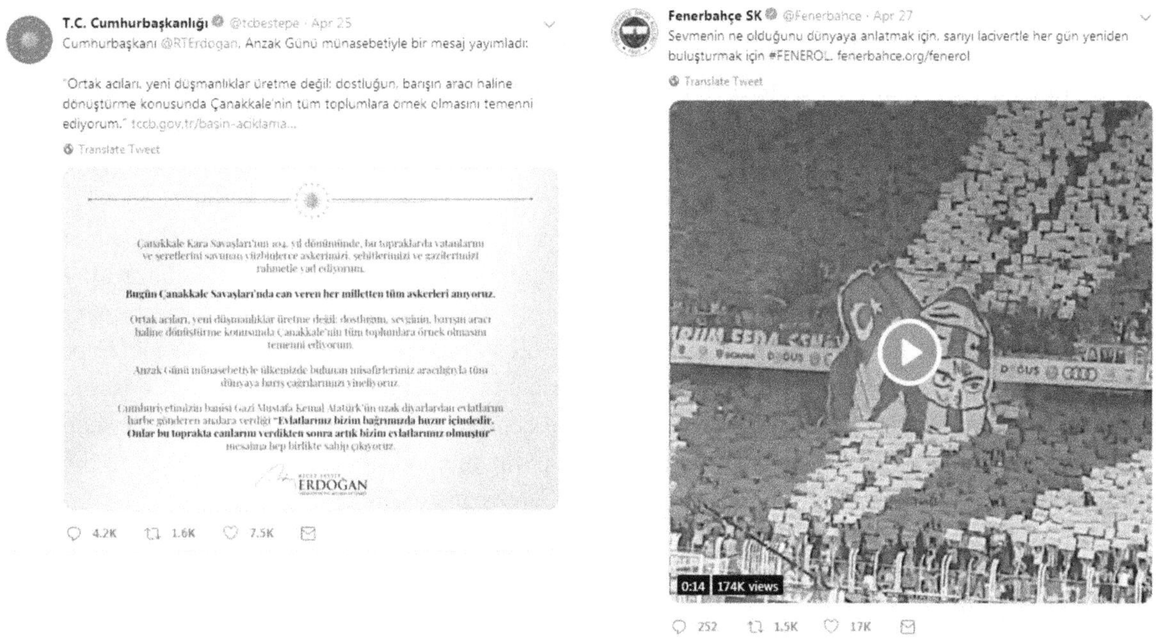

In the two above examples, there is a dialogic-based communication. Looking at the language used, a content is used in the street in daily life. Thanks to Twitter, the message was quickly responded and a dialogue took place. The words chosen in the speech will be used by people from every group in daily life and can be an example of the marketplace speech.

Marketplace: Today, as in the cities, there were marketplaces of villages and towns in the 20th century. Marketplaces were the places where there were many sought-after products, witnessed places of encounter, promotions were made, and there were places for today's advertising billboard. The marketplace is parallel to the carnival rituals with the language used, its purpose, color and design. Promotions or campaigns are brought to the front of the showcase, the official discourses are replaced by the expressions that the people are free and comfortable in this environment. Twitter is a square where people feel comfortable, away from the official and promotional campaigns and also meet people or institutions from different fields. Below are the shares on the marketplace in the context of the carnivalesque theory.

The first tweet in the photo above, Presidency Republic of Turkey published a document to mark Anzac Day in April 25. The message stated that it was remembered all soldiers who died from all nations in the Dardanelles War. This issue that great importance for the history of Turkey, has announced to everyone on Twitter with the signature of the President Recep Tayyip Erdoğan.

In the second example, a share was made on the Turkish football team Fenerbahçe's Twitter account. In the share, there are videos of "Fener OL" campaigns launched on April 4 by Ali Koç to remove Fenerbahçe from the economic crisis. Fenerbahçe Twitter account, which is shared by "Fener OL" campaign information, is presented to followers and more support is requested. In the two examples, thousands of people were reached via Twitter. Thus, large audiences were given the desired message through Twitter. In these messages, it was both expressed a national feeling for a painful event and requested support for a campaign by a team.

FUTURE RESEARCH DIRECTIONS

In this study, the similarity between the carnivalesque theory of Michail Bakhtin and the social networks was examined. To reveal this similarity, Twitter, which is one of the most important social networks, was analyzed by using qualitative content analysis technique. To obtain the data, 10 Twitter accounts which the most followers in Turkey were examined and similarities were found between these accounts and the carnivalesque theory.

In this study, although carnivalesque theory is discussed in social networks which is a branch of communication field, it is a theory which can be applied in almost all sub-dimensions of communication field (cinema, television, journalism etc.). Carnivalesque theory, however, is also a very appropriate theory for use in the fields such as literature, sociology, psychology, fine arts and political science. Especially first of all the communication field, the most appropriate research technique to be used in research in the fields as part of carnivalesque theory is qualitative content analysis. In addition, questionnaire and interview (semi-structured, unstructured) techniques are the methods that can be used to learn the emotions, thoughts and behaviors of target groups in these areas and to obtain data.

CONCLUSION

The carnivalesque theory of Michail Bakhtin is a literary theory which contains the basic features of carnival. Carnival laughter, dialogism, upside-down world, grotesque, marketplace and marketplace speech is the basic features of the carnivalesque theory. Although the theory of carnivalesque is a theory which is put forward in the field of literature, it is adaptable to many fields of science and art because of its characteristics. One of these areas is undoubtedly the area of communication. In fact, the concepts of carnival and communication are two elements that are already very much related to each other. For this reason, the carnivalesque theory, which can be adapted to many communication environments, can also be adapted for social networks, one of the most important communication environments. From this point of view, a study has been done to reveal the relationship and similarities between the carnivalesque theory and Twitter one of the social networks.

In this study, a case study design which is one of the qualitative research methods is used. In the study, with a maximum number of Turkey 10 Twitter follower examples of the account identified and analyzed. In the study, examples that obtained from 10 Twitter accounts with a maximum number of followers in Turkey, identified and analyzed. It is possible to see grotesque examples although not all of the pages examined have grotesque sharing. It was found that there was disguise and change words in the shares examined in the sub-title grotesque. It is thought that this situation is made due to the relevant shares and it has no special reason. It was found out that the shared messages that digress the fact and a distortion/changing was made. It was found that the followers did not react differently to the others. In terms of the dialogism in which mutual communication took place, it was seen that dialogue was established with followers in almost all accounts. This situation shows that there is interaction in the title of dialogism. n addition, it was determined that retweets were made under the shares and some headers were labeled with the hashtag sign. In this way, it was realized that the shares were seen on other people's pages and people from different sectors were able to reach the shares. In dialogism, it can be said that not only two people but also more people are affected by the existing dialogue.

The carnival laughter expresses a reaction with gestures and mimics in carnivalesque theory. This response does not have to be spoken or vocal. Similar cases were observed in the shares analyzed. There was a smile to mock the situation in the sharing. In almost every sub-title of the carnivalesque theory, it is possible to talk about the upside-down world. In the examined Twitter pages, examples of the upside-down world where the existing system was reversed, there is no difference in level between people and the routine order has been changed, was reached. In these examples, it has been seen that it is out of the general world, it was created new world by disguise, it was shared in opposite direction of image.

Marketplace speech is another topic in which oral expression becomes important for carnivalesque theory. In this title, the language of daily life is returned and the street mouth used by the people is preferred. Such samples were also found in the examined pages. These examples are expected to be seen. Because Twitter is a social network used by individuals belonging to different social groups. In this network, it is possible to come across every expression like carnival environment. On the examined pages, examples about the marketplace title were also seen. In these examples, it was determined that the account holders sent messages to thousands of people without any limits and interacted with these messages. Therefore, in the framework of carnivalesque theory, it was found equivalent marketplace expression.

As can be seen from the examples, carnivalesque shares subsist in Twitter which is one of the most popular social networks in Turkey. These shares represent carnivalistic details in real life in the context of carnivalesque theory. In this context, Twitter, which is one of the most important social networks, has a great similarity with the carnivalesque theory of Michail Bakhtin.

REFERENCES

Altunay, M. C. (2010). Gündelik Yaşam ve Sosyal Paylaşım Ağları: Twitter ya da "Pıt Pıt Net". *Galatasaray Üniversitesi İletişim Dergisi*, *12*, 31–56.

Bakhtin, M. M. (1981). *The Dialogic Imagination*. Austin, TX: University of Texas Press.

Bakhtin, M. M. (1984). *Rabelais and His World*. Bloomington, IN: Indiana University Press.

Bakhtin, M. M. (1986). *Speech Genres and Other Late Essays*. Austin, TX: University of Texas Press.

Bakhtin, M. M. (2001). *Karnavaldan Romana*. İstanbul, Turkey: Ayrıntı Yayınları.

Boyd, D. M., & Ellison, N. B. (2007). Social Network Sites: Definition, History, and Scholarship. *Journal of Computer-Mediated Communication*, *13*(1), 210–230. doi:10.1111/j.1083-6101.2007.00393.x

Brandist, C. (2011). *Bahtin ve Çevresi, Felsefe, Kültür ve Politika*. Ankara, Turkey: Doğu Batı Yayınları.

Derin, Ö. (2017). İtici Bir Güç Olarak Oluşsal Şiddet. *Felsefe ve Sosyal Bilimler Dergisi*, *24*, 317–335.

Gülüm, E. (2016). Karnavalesk Bir İmgelemin Taşıyıcısı Olarak Bektaşi Mizahı. *Milli Folklor*, *28*(112), 130–141.

Kıpçak, N. S. (2016). *Yeni Karnaval Olarak Yeni Medya: Karnavalesk Nitelikleri ile Twitter*. İstanbul, Turkey: Marmara Üniversitesi Sosyal Bilimler Enstitüsü Radyo TV Sinema Anabilim Dalı, Doktora Tezi.

Mclean, P. B., & Wallace, D. (2013). Blogging the Unspeakable: Racial Politics, Bakhtin, and the Carnivalesque. *International Journal of Communication*, *7*, 1518–1537.

Sığrı, Ü. (2018). *Nitel Araştırma Yöntemleri*. İstanbul, Turkey: Beta Yayınları.

Socialbakers (2019). *Twitter statistics for Turkey*. Retrieved from https://www.socialbakers.com/statistics/twitter/profiles/turkey/

Sözen, M. F. (2009). Bakhtin'in Romanda "Karnavalesk" Kavramı ve Sinema. *Akdeniz Sanat Hakemli Dergi*, *2*(4), 65–86.

Turan, M. (2008). Kaos Teorisi: Bauman ve Bakhtin. *Ankara Üniversitesi Dil ve Tarih-Coğrafya Fakültesi Felsefe Bölümü Dergisi*, *19*(1), 45–66.

Uygun, E., & Akbulut, D. (2018). Karnavalesk Kuramı ve Instagram Ortamına Yansımaları. *Yeni Medya Elektronik Dergi*, *2*(2), 73–89. doi:10.17932/IAU.EJNM.25480200.2018.2/2.73-89

Yıldız, T. (2014). Diyaloji Diyalektiğe Karşı. *Psikoloji Çalışmaları Dergisi*, *34*(1), 79–85.

Yurdakul, S. (2006). *Bir Ürün ve Toplumsal Bütünleşme Aracı Olarak Modern Kent Karnavalları ve Türkiye'deki Örneklerde Görülen Uygulama Farklılıkları*. İstanbul, Turkey: Marmara Üniversitesi Sosyal Bilimler Enstitüsü, İletişim Bilimleri Anabilim Dalı, Yüksek Lisans Tezi.

ADDITIONAL READING

Bakhtin, M. M. (2004). *Dostoyevski Poetikasının Sorunları*. İstanbul: Metis Yayınları.

Castells, M. (2010). *The Rise of the Network Society*. Malden, MA: Wiley-Blackwell.

Cross, M. (2011). *Bloggerati, Twitterati: How Blogs and Twitter Are Transforming Popular Culture*. Santa Barbara, Denver, Oxford: Praeger.

Karimova, G. (2010). Interpretive Methodology From Literary Critcism: Carnivalesque Analysis of Popular Culture: Jackass, Southpark and Everyday Culture. *Studies in Popular Culture*, *33*(1), 37–51.

Lindahl, C. (1996). Bakhtin's Carnival Laughter and the Cajun Country Mardi Gras. *Folklore*, *107*(1-2), 57–70. doi:10.1080/0015587X.1996.9715915

Rheingold, H. (1994). *The Virtual Community: Homesteading on the Electronic Frontier*. New York: Harper Perennial.

St John, G. (2008). Protestival: Global Days of Action and Carnivalized Politics in the Present. *Social Movement Studies*, *7*(2), 167–190. doi:10.1080/14742830802283550

Theall, D. F. (1999). The Carnivalesque, the Internet and Control of Content: Satirizing Knowledge, Power and Control. *Continuum (Perth)*, *13*(2), 153–164. doi:10.1080/10304319909365789

Tufekci, Z. (2008). Can You See Me Now? Audience and Disclosure Regulation in Online Social Network Sites. *Bulletin of Science, Technology & Society, Vol*, *28*(: 1), 20–36. doi:10.1177/0270467607311484

Vaneigem, R. (2012). *The Revolution of Everyday Life*. Oakland: PM Press.

KEY TERMS AND DEFINITIONS

Mikhail Bakhtin: Was a Russian philosopher, literary critic, semiotician[4] and scholar who worked on literary theory, ethics, and the philosophy of language. His writings, on a variety of subjects, inspired scholars working in a number of different traditions (Marxism, semiotics, structuralism, religious criticism) and in disciplines as diverse as literary criticism, history, philosophy, sociology, anthropology and psychology. Although Bakhtin was active in the debates on aesthetics and literature that took place in the Soviet Union.

Carnival: The carnival, whose origin is based on the Middle Ages and which varies from culture to culture, from society to society, is a celebration, entertainment or festival that Christians gathered in communities before Great Lent.

Carnivalesque: The carnivalesque, which Mikhail Bakhtin formulated from the concept of carnival, is a theory based on the basic characteristics (grotesque, dialogism, carnival laughter, upside-down world, marketplace and marketplace speech) of the Middle Age carnivals and argues that people communicate with a second identity in a community by getting rid of their everyday identities.

Content Analysis: Content analysis is a technique that provides an objective and systematic description and analysis of text content such as news, advertising texts, interview, television programs.

Qualitative Research: is primarily exploratory research. It is used to gain an understanding of underlying reasons, opinions, and motivations. It provides insights into the problem or helps to develop ideas or hypotheses for potential quantitative research.

Social Networks: Social networking is a virtual communication environment where people come together with different purposes such as information, entertainment and socialization, communicate independently of time and space, create and display their own profile and access other users' profiles.

Twitter: İs an American online news and social networking service on which users post and interact with messages known as "tweets".

Compilation of References

Abdulahi, A., Samadi, B., & Gharleghi, B. (2014). A study on the negative effects of social networking sites such as Facebook among Asia Pacific university scholars in Malaysia. *International Journal of Business and Social Science*, *5*(10).

Abutaleb, Y. & Bhattacharjee, N. (2015, Aug. 13). Facebook Struggles to Sell Advertising in India. *Reuters*. Retrieved from http://in.reuters.com/article/facebook-india-ads-idINKCN0QH0DY20150812

Adler, A. (2013). *Understanding Human Nature (Psychology Revivals)*. Abingdon, UK: Routledge. doi:10.4324/9780203438831

Adquickly Blog. (2017, December 21). Importance of Visual communication in Digital Media [Web Blog Post]. Retrieved from https://www.adquicky.com/blogs/importance-visual-communication-digital-media/s

Advertorial. (2019, 02 27). *Bu Yaza Damgasını Vuracak En Yeni Estetik Ve Güzellik Trendleri!* Retrieved from http://www.hurriyet.com.tr/bu-yaza-damgasini-vuracak-en-yeni-estetik-ve-guzellik-trendleri-41131904

AFP. (2015, March 2). 5 of China's 10 wealthiest to take part in key political meetings. *Business Insider*. Retrieved from http://www.businessinsider.com/afp-chinas-wealthiest-to-take-part-in-key-political-meetings-2015-3?IR=T&

Ahi, G. (2017). Gerçeklik ötesi (post-truth) ve popülizm. Retrieved from https://digitalage.com.tr/makale/gerceklik-otesi-post-truth-populizm/

Akarsu, B. (1988). *Felsefe Terimleri Sözlüğü*. İstanbul, Turkey: İnkılap Kitabevi.

Akgül, S. K. ve Pazarbaşı, B. (2018). Küresel Ağlar Odağında Kültür, Kimlik ve Mekan Tartışmaları, Hiperyayın, İstanbul.

Akıncı Vural, B. (2005). *Kurum Kültürü*. İstanbul, Turkey: İletişim Yayınları.

Akmeşe, Z. ve Deniz, K. (2017). Stalk, Benliğin İzini Sürmek. *Yeni Düşünceler*, *8*, 23–32.

Aktuğlu, I. K. (2004). *Marka Yönetimi*. İstanbul, Turkey: İletişim Yayınları.

Akyazı, A. (2019). Mahremiyetin Dönüşümü: Ünlülerin Instagram Paylaşımları Üzerine Bir Araştırma. Gaziantep University Journal of Social Sciences, 18(1).

Akyol, A. Ç. (2011). *Reklam Mesajlarında Bilgilendirici İçerik: Dergi Reklamları Üzerine Bir İçerik Analizi*, (Unpublished doctoral thesis), Selçuk University/Social Sciences Institute, Konya, Turkey.

Alexander, F. (1948). Development of the ego psychology. In S. Lorand (Ed.), *Psycho Analysis Today*. London, UK: George Allen & Unwin.

Alpan, G. (2008). Ders kitaplarındak imetin tasarımı. *Türk Eğitim Bilimleri Dergisi*, *6*(1), 107–134.

Alpan, G. (2008). Görsel Okuryazarlik Ve Öğretim Teknolojisi. *Yüzüncü Yıl Üniversitesi Eğitim Fakültesi Dergisi*, *5*(2), 74–102.

Alpman, P. S. (2016). Necip milletin millet-i hakime hassasiyeti ve yerli-yurtlu gayr-ı milliler. *Toplum ve Kuram*, *11*(Spring), 13–38.

Al-Rawi, A. (2017). Audience Preferences of News Stories on Social Media[. *The Journal of Social Media in Society*, *6*(2), 343–367.

Altunay, M. C. (2010). Gündelik Yaşam ve Sosyal Paylaşım Ağları: Twitter ya da "Pıt Pıt Net". *Galatasaray Üniversitesi İletişim Dergisi*, *12*, 31–56.

Ambroseg, H., & Harris, P. (2010). *Görsel Grafik Tasarım Sözlüğü*. İstanbul, Turkey: LiteratürYayınları.

An, J., Quercia, D., Cha, M., Gummadi, K., & Crowcroft, J. (2013). Traditional media seen from social media. In *Proceedings of the 5th Annual ACM Web Science Conference* (pp. 11-14). New York, NY: ACM. 10.1145/2464464.2464492

Appel, H., Gerlach, A. L., & Crusius, J. (2016). The interplay between Facebook use, social comparison, envy, and depression. *Current Opinion in Psychology*, *9*, 44–49. doi:10.1016/j.copsyc.2015.10.006

Armstrong, H. (2010). *Grafik Tasarım Kuramı*. İstanbul, Turkey: EspasYayınları.

Aronson, E. (1980). *The social animal* (3rd ed.). San Francisco, CA: Freeman & Co.

Ashley, C., & Tuten, T. (2015). Creative strategies in social media marketing: An exploratory study of branded social content and consumer engagement. *Psychology and Marketing*, *32*(1), 15–27.

Atabek, Gülseren Şendur ve Ümit Atabek. (2007). Medya Metinlerini Çözümlemek, Ankara, Turkey: Siyasal Kitabevi.

Atikkan, Z., & Tunç, A. (2011). *Blog'dan Al Haberi: Haber Blogları, Demokrasi ve Gazeteciliğin Geleceği Üzerine*. İstanbul, Turkey: Yapı Kredi Publications.

Aydın, B. (2013). Sosyal Medya Mecralarında Mahremiyet Anlayışının Dönüşümü. İstanbul Arel Üniversitesi İletişim Çalışmaları Dergisi, 3(5), 131-146.

Aygül, E. (2013). Yeni medyada nefret söyleminin üretimi: bir toplumsal paylaşım ağı olarak Facebook örneği. (*Unpublicated Master Thesis)*, Gazi University, The Institute of Social Sciences, Ankara/Turkey.

Aytaç, A. M. & Demirkent, D. (2017). Ahmet Murat Aytaç: "hakikatin kendinden menkul bir enerjisi yoktur; hakikati savunmak gerekir". *Ayrıntı Dergi Yalan Özel Sayısı*. Retrieved from http://ayrintidergi.com.tr/ahmet-murat-aytac-hakikatin-kendinden-menkul-bir-enerjisi-yoktur-hakikati-savunmak-gerekir/

Aytaç, A. M. & Demirkent, D. (2017). Hakikatin kendinden menkul bir enerjisi yoktur, hakikati savunmak gerekir. *Ayrıntı Dergi*, Retrieved from http://ayrintidergi.com.tr/ahmet-murat-aytac-hakikatin-kendinden-menkul-bir-enerjisi-yoktur-hakikati-savunmak-gerekir/

Aytaç, A. M. (2017). Bir Türk'ü nereden tanırsınız. *Gazete Duvar*. Retrieved from https://www.gazeteduvar.com.tr/yazarlar/2017/11/13/bir-turku-nereden-tanirsiniz/

Bağlı, M. (2011). *Modern Bilinç ve Mahremiyet*. İstanbul, Turkey: Yarın Yayınevi.

Bağlı, M. (2011). *Modern bilinç ve mahremiyet*. İstanbul, Turkey: Yarın.

Bajaj, A., & Samuel, B. (2018). Beyond beauty: Design symmetry and brand personality. *Journal of Consumer Psychology*, *28*(1), 77–98. doi:10.1002/jcpy.1009

Bakan, Ö. (2005). *Kurumsal İmaj*. Konya, Turkey: Tablet Yayınları.

Baker, U. (2017). *Beyin ekran*. İstanbul, Turkey: Birikim.

Baker, U. (2017). *Beyin Ekran*. İstanbul, Turkey: İletişim Publications.

Baker, U. (2018). *Kanaatlerden İmajlara: Duygular Sosyolojisine Doğru*. İstanbul, Turkey: Birikim Publications.

Bakhtin, M. M. (1981). *The Dialogic Imagination*. Austin, TX: University of Texas Press.

Bakhtin, M. M. (1984). *Rabelais and His World*. Bloomington, IN: Indiana University Press.

Bakhtin, M. M. (1986). *Speech Genres and Other Late Essays*. Austin, TX: University of Texas Press.

Bakhtin, M. M. (2001). *Karnavaldan Romana*. İstanbul, Turkey: Ayrıntı Yayınları.

Bakıroğlu, C. T. (2013). *Sosyalleşme ve Kimlik İnşası Ekseninde Sosyal Paylaşım Ağları, Akademik Bilişim, XV*. Akademik Bilişim Konferansı Bildirileri.

Balcı, E. V. (2018). Sanal Dünyave Dijital Oyun Kültürü, Yeni Medyada Yeni Yaklaşımlar, Ed. Rengim Sine, Gülşah Sarı, Konya: Literatürk Yayınları, pp.11-43.

Balmer, J., & Greyser, S. (2003). Managing the Multiple Identities of the Corporation. In J. Balmer & S. Greyser (Eds.), *Revealing The Corporation* (pp. 15–30). London, UK: Routledge. doi:10.4324/9780203422786

Barkuş, F., & Koç, M. (2019). Dijital mahremiyet kavramı ve ilgili çalışmalar üzerine bir derleme. Bilim, Eğitim [BEST Dergi]. *Sanat ve Teknoloji Dergisi*, *3*(1), 35–44.

Barnard, M. (2002). *Sanat, Tasarım ve Görsel Kültür. (Güliz Korkmaz, Çev.)*. Ankara, Turkey: Ütopya Yayınları.

Barroso, P. M. (2017). The semiosis of sacred space. Versus – Quaderni di Studi Semiotici, (125), 343-359.

Barthes, R. (1994). The Semiotic Challenge. (Trans. R. Howard). Berkeley: The University of California Press.

Barthes, R. (1991). *Mythologies* (A. Lavers, Trans.). New York, NY: The Noonday Press.

Bartholmé, R., & Melewar, T. (2011). Remodelling the corporate visual identity construct: A reference to the sensory and auditory dimension. *Corporate Communications*, *16*(1), 53–64. doi:10.1108/13563281111100971

Batmaz, V., (1981). Popüler Kültür Üzerine Değişik Kuramsal Yaklaşımlar, İletişim 1981/1, Ankara, Turkey: AGTGA Gazetecilik ve Halkla İlişkiler Yüksek Okulu Yayını No:2, 163-192.

Baudrillard, J. (1997). *Simulacra and Simulation* (S. F. Glaser, Trans.). Michigan: The University of Michigan Press.

Baudrillard, J. (1998). *Simulakrlar ve Simülasyon* (O. Adanır, Trans.). İzmir: Dokuz Eylül University Press.

Baudrillard, J. (1998). *Simulakrlar ve Simülasyon. Oğuz Adanır (Trans.)*. İzmir, Turkey: Dokuz Eylül Publications.

Bauman, Z. (2016). *Postmodern Etik. Alev Türker (Trans.)*. İstanbul, Turkey: Ayrıntı Publications.

Bauman, Z. (2017). *Kimlik*. Ankara, Turkey: Heretik.

Bauman, Z., & Lyon, D. (2013). *Akışkan gözetim. Çev., Elçin Yılmaz*. İstanbul, Turkey: Ayrıntı Yayınları.

Bayad, A., (2016). Erving Goffman'ın Benlik Kavramı ve İnsan Doğası Yaklaşımı, Psikoloji Çalışmaları / Studies in Psychology, 36-1 (81-93).

Bayraktutan, G., Binark, M., Çomu, T., İslamoğlu, G., Doğu, B., & Aydemir, A.T. (2013). Web 1.0'dan Web 2.0'a Barış ve Demokrasi Partisi. İletişim ve Diplomasi Dergisi, 1, 31-54.

BBC. (2018, Dec. 19). Facebook's data-sharing deals exposed. Retrieved from https://www.bbc.com/news/technology-46618582

Becer, E. (2006). *İletişim ve Grafik Tasarım*. Ankara, Turkey: Dost Kitabevi.

Beigi, G. & Liu, H. (2018). Privacy in social media: Identification, mitigation and applications. *arXiv preprint arXiv:1808.02191*.

Benney, J. (2014). The Aesthetics of Chinese Microblogging: State and Market Control of Weibo. *Asiascape: Digital Asia*, *1*(3), 169–200. doi:10.1163/22142312-12340011

Berger, J. (1990). *Görme Biçimleri, Çev.; Salman, Y.* İstanbul, Turkey: Metis.

Berger, J. (2018). *Bir Fotoğrafı Anlamak. Geoff Dyer (Prep.), Beril Eyüboğlu (Trans.)*. İstanbul, Turkey: Metis Publications.

Beverungen, A., Böhm, S., & Land, C. (2015). Free Labour, Social Media, Management: Challenging Marxist Organization Studies. *Organization Studies*, *36*(4), 473–489. doi:10.1177/0170840614561568

Bignell, J. (2002). *Media Semiotics*. Manchester, UK: Manchester University Press.

Bilgin, N. (2007). *Aşina Kitaplar*. İzmir, Turkey.

Binark, M. (2001). Kadının Sesi Radyo Programı ve Kimliği Konumlandırma Stratejisi. In *Toplumbilim, Sayı:14*. Ankara, Turkey: Bağlam Yayınları.

Binark, M., & Gencel Bek, M. (2007). *Eleştirel medya okuryazarlığı*. İstanbul, Turkey: Kalkedon.

Blair, C., & Lanyon, R. I. (1981). Exhibitionism: Etiology and treatment. *Psychological Bulletin*, *89*(3), 439–463. doi:10.1037/0033-2909.89.3.439 PMID:7255626

Blue Communications. (2019, January 20). *The role of visual communications in today's digital age*. Retrieved from https://www.blue-comms.com/the-role-of-visual-communications-in-todays-digital-age/

Bocock, R. (1997). *Tüketim, Çev., Kutluk, İ.* Ankara, Turkey: Dost Yay.

Bolay, S. H. (2009). *Felsefe Doktrinleri Ve Terimleri Sözlüğü*. Ankara, Turkey: Nobel Yayın Dağıtım.

Bosch, A., Elving, W., & de Jong, M. (2006). The impact of organisational characteristics on corporate visual identity. *European Journal of Marketing*, *40*(7/8), 870–885. doi:10.1108/03090560610670034

Bosch, V. (2005). How corporate visual identity supports reputation. *Corporate Communications*, *10*(2), 108–116. doi:10.1108/13563280510596925

Boyd, D. M., & Ellison, N. B. (2007). Social network sites: Definition, history, and scholarship. *Journal of Computer-Mediated Communication*, *13*(1), 210–230. doi:10.1111/j.1083-6101.2007.00393.x

Brandist, C. (2011). *Bahtin ve Çevresi, Felsefe, Kültür ve Politika*. Ankara, Turkey: Doğu Batı Yayınları.

Bülbül, M. (2017). *İmgesel İletişim*. İstanbul, Turkey: ÇizgiKitabevi.

Burn, A., & Durran, J. (2007). *Media Literacy in Schools: Practice, Production and Progression*. London, UK: Paul Chapman Publishing.

Byod, D. (2014). *The Social Live of Networked Teens*. New Haven, CT: Yale University Press.

Çağan, K. (2003). *Popüler Kültür ve Sanat*. Ankara, Turkey: Altınküre.

Çağlayan, S., Korkmaz, M., & Öktem, G. (2014). Sanatta görsel algının literatür açısından değerlendirilmesi. *Eğitim ve Öğretim Araştırmaları Dergisi*, *3*(1), 160–173.

Campbell, C. (2013). *Günah keçisi, başkalarını suçlamanın tarihi. Gizem Kastamonulu (Trans.)*. İstanbul, Turkey: Ayrıntı.

Caplan, G. (1964). *Principles of preventive psychiatry*. New York, NY: Basic Books.

Castells, M. (2004). *The network society: a cross-cultural perspective*. Cheltenham, UK: Edward Elgar Publishing. doi:10.4337/9781845421663

Chaykowski, K. (2015, Nov. 4). Facebook Beats Third-Quarter Earnings, Revenue Estimates, Shares Rise. *Forbes*. Retrieved from https://www.forbes.com/sites/kathleenchaykowski/2015/11/04/facebook-beats-third-quarter-earnings-revenue-estimates-shares-rise/#37a0771a37a0

Chen, G. (2016). Quick Pass Churns Payment Market. Retrieved from http://en.ce.cn/Insight/201604/18/t20160418_10589502.shtml

Chepken, C. (2012). *Telecommuting in the developing world: a case of the day-labour market* (Doctoral dissertation, University of Cape Town).

Chepken, C. K., Blake, E. H., & Marsden, G. (2011, September). Software design for informal setups: Centring the benefits. In *proceedings of the 14th Southern Africa Telecommunication Networks and Applications Conference*.

China Internet Watch. (2016). WeChat Payment Reached over 300K Retail Stores. Retrieved from http://www.chinainternetwatch.com/17437/wechat-payment-reached-over-300k-retail-stores

Chou, H. L., Liu, Y. L., & Chou, C. (2019). Privacy behavior profiles of underage Facebook users. *Computers & Education*, *128*, 473–485. doi:10.1016/j.compedu.2018.08.019

Christophers, B. (2010). *Envisioning Media Power: On Capital and Geographies of Television*. Lanham, MD: Lexington Books.

Clow, K. E., & Baack, D. (2017). *Integrated advertising, promotion and marketing communication. (8th edit.)*. Harlow, UK: Pearson Education Limited.

Cohen, H. (2019). Social media definitions. Available at https://heidicohen.com/social-media-definition/

Connolly, J. (2007). The New World Order: Greek Rhetoric in Rome. In I. Worthington (Ed.), *A Companion to Greek Rhetoric* (pp. 139–165). Oxford, UK: Blackwell Publishing. doi:10.1002/9780470997161.ch11

Cornelissen, J. (2004). *Corporate Communications Theory and Practice*. London, UK: Sage.

Creemers, R. (2016, May 24-25). Disrupting the Chinese State: New Actors and New Factors. Paper Presented at the *Conference on Digital Disruption in Asia: Methods and Issues*, University of Leiden, The Netherlands.

Curran, J. (1997). Medya ve Demokrasi, 139-197. In Medya, Kültür, Siyaset, (Eds: Süleyman İrvan), Ark Yayınları, Ankara, Turkey.

Cuthbertson, A. (2019, March 22). Facebook Admits Storing Millions of User Passwords in Plain Text Files for Years. *Independent.* Retrieved from https://www.independent.co.uk/life-style/gadgets-and-tech/news/facebook-passwords-plain-text-instagram-admission-a8833941.html?utm_medium=Social&utm_source=Facebook&fbclid=IwAR35iRRkYT9ezasULRIaY0YrWdJqUicx9TEWmKvUiBD8KKwS_Ws2dgL58XQ#Echobox=1553185822

Danforth, N. (2008). Ideology and Pragmatism in Turkish Foreign Policy: From Atatürk to The AKP". Turkish Policy Quarterly, 7(3), Washington DC, Spring.

David, H. (1997). *Postmodernliğin Durumu. (Sungur Savran, Çev).* İstanbul, Turkey: Metis Yayınları.

De Lencastre, P., & Côrte-Real, A. (2010). One, two, three: A practical brand anatomy. *Journal of Brand Management, 17*(6), 399–412. doi:10.1057/bm.2010.1

Deagon, B. (2015, March 9). Facebook Making Progress with Major Brand Advertisers. *Investor's Business Daily.* Retrieved from http://www.investors.com/facebook-targets-national-brand-advertisers/

Debord, G. (1995). *The Society of the Spectacle* (D. Nicholson-Smith, Trans.). New York, NY: Zone Books.

Dedeoğlu, G. (2004). Gözetleme, mahremiyet ve insan onuru. *Turkey Informatics Association Journal, 89*, 35–36.

Deleuze, G. (1997). *Cinema 2 – The Time-Image* (H. Tomlinson & R. Galeta, Trans.). Minneapolis, MN: University of Minnesota Press.

Demirkent, D. (2017). Siyasal toplumun kurucu yalanlarıyla ne yapacağız? *Ayrıntı Dergi Yalan Özel Sayısı.* Retrieved from http://ayrintidergi.com.tr/siyasal-toplumun-kurucu-yalanlariyla-ne-yapacagiz/ adresinden ulaşıldı.

Demirtaş, H. A. (2003). Sosyal Kimlik Kuramı, Temel Kavram ve Varsayımlar. İletişim Araştırmaları, 1(1), 123-144.

Derin, Ö. (2017). İtici Bir Güç Olarak Oluşsal Şiddet. *Felsefe ve Sosyal Bilimler Dergisi, 24*, 317–335.

Deuze, M. (2003). The Web And İts Journalisms: Considering The Consequences Of Different Types Of News media Online. *New Media & Society, 5*(2), 203–230. doi:10.1177/1461444803005002004

Dijck, J. (2013). *The culture of connectivity: A critical history of social media.* New York, NY: Oxford University Press. doi:10.1093/acprof:oso/9780199970773.001.0001

Dinev, T., Xu, H., & Smith, H. J. (2009, January). Information privacy values, beliefs and attitudes: An empirical analysis of web 2.0 privacy. In *2009 42nd Hawaii International Conference on System Sciences* (pp. 1-10). Piscataway, NJ: IEEE.

Dolgun, U. (2008). *Şeffaf hapishane yahut gözetim toplumu: küreselleşen dünyada gözetim, toplumsal denetim ve iktidar ilişkileri.* İstanbul, Turkey: Ötüken Neşriyat.

Donath, J. S. (1998). Identity and deception in the virtual community. In P. Kollock & M. Smith (Eds.), *Community in Cyber Spaces.* London, UK: Routledge.

Duncum, P. (2002). Visual Culture Art Education: Why, What and How? *Journal Art and Design Education, 21*(1), 14–23. doi:10.1111/1468-5949.00292

Dyer, G. (1982). *Advertising as Communication*. London, UK: Routledge. doi:10.4324/9780203328132

Eco, U. (1976). *A Theory of Semiotics*. Bloomington, IN: Indiana University Press. doi:10.1007/978-1-349-15849-2

Eco, U. (1984). *Semiotics and the Philosophy of Language*. London, UK: MacMillan. doi:10.1007/978-1-349-17338-9

Eco, U. (2001). *A Estrutura Ausente. (Trans, Pérola de Carvalho from the original La Struttura Assente)*. São Paulo, Brazil: Editora Perspectiva.

Elden, M. (2009). *Reklam ve Reklamcılık*. İstanbul, Turkey: Say Yayınları.

Elden, M., & Bakır, U. (2010). *Reklam Çekicilikleri: Cinsellik, Mizah, Korku*. İstanbul, Turkey: İletişim Publications.

Elden, M., & Okat Özdem, Ö. (2015). *Reklamda Görsel Tasarım-Yaratıcılık ve Sanat*. İstanbul, Turkey: Say Yayınları.

Elden, M., & Özdem, Ö. O. (2015). *Reklamda Görsel Tasarım - Yaratıcılık ve Sanat*. İstanbul, Turkey: Say Publications.

Elden, M., & Yeygel, S. (2006). *Kurumsal Reklamın Anlattıkları*. İstanbul, Turkey: Beta.

Eltantawy, N., & Wiest, J. B. (2011). Social media in the Egyptian Revolution: Reconsidering resource mobilization theory. *International Journal of Communication*, (5), 1207–1224.

Erdoğan, İ. ve Alemdar K. (2005). Popüler Kültür ve İletişim, Erk, Ankara.

Erdoğan, İ. (2004). Popüler Kültürün Ne Olduğu Üzerine, Bilim ve Aklın Aydınlığında Eğitim Dergisi: Popüler Kültür ve Gençlik. *Sayı*, *57*, 7–19.

Erengül, B. (1997). *Kültür Sihirbazları - Rekabet Üstünlüğü Sağlayan Yönetim*. İstanbul, Turkey: Evrim Yayınevi.

Erkul, R. E. (2009). Sosyal medya araçlarının (web 2.0) kamu hizmetleri ve uygulamalarında kullanılabilirliği. Türkiye Bilişim Derneği-Bilişim Dergisi (116).

Ertan Keskin, Z. (2004). Türkiye'de Haber İncelemelerinde Van Dijk Yöntemi. In *Dursun, Ç. (Der.), Haber-Hakikat ve İktidar İlişkisi*. Ankara, Turkey: Elips Yayınları.

Espinar, E., Frau, C., González, M., & Martínez, R. (2006). *Introducción a la Sociología de la Comunicación*. Alicante, Spain: Publicaciones Universidad de Alicante.

Ethridge, D. E. (2004). *Research methodology in applied economics*. New York, NY: John Wiley & Sons.

Facebook says 50 million user accounts affected by major security breach. (2017, September 28). Retrieved from: https://www.thehindu.com/sci-tech/technology/internet/facebook-reveals-security-breach-affecting-50-million-user-accounts/article25074368.ece

Facebook. (2014). *Facebook Annual Report*. Retrieved from https://materials.proxyvote.com/Approved/30303M/20150413/AR_245461/#/8/

Facebook. (2015a). Facebook Reports Third Quarter 2015 Results. Retrieved from http://investor.fb.com/releasedetail.cfm?ReleaseID=940609

Facebook. (2015b). Data Policy. Retrieved from https://www.facebook.com/full_data_use_policy

Facebook. (2015c). Data Policy. Retrieved from https://www.facebook.com/privacy/explanation

Facebook. (2016a). Facebook Reports Fourth Quarter and Full Year 2015 Results. Retrieved from http://investor.fb.com/releasedetail.cfm?ReleaseID=952040

Facebook. (2016b). Government request report. Retrieved from https://govtrequests.facebook.com/#

Facebook. (2016c). Publishing Information about Government Requests to Facebook. Retrieved from https://govtrequests.facebook.com/about/

Facebook. (2016d). Three Million Business Stories. What's Yours? Retrieved from https://www.facebook.com/business/news/3-million-advertisers

Facebook. (2019). AdChoices. Retrieved from https://www.facebook.com/help/568137493302217

Fahnestock, J., & Secor, M. (2004). *A rhetoric of argument: a text and reader*. New York, NY: McGraw-Hill.

Fleisher, L. & Fairless, T. (2015, May 15). Belgian Watchdog Raps Facebook for Treating Personal Data 'With Contempt'. *The Wall Street Journal*. Retrieved from https://www.wsj.com/articles/belgian-watchdog-slams-facebooks-privacy-controls-1431685985

Floch, J.-M. (2000). *Visual Identities*. London, UK: Continuum.

Foroudi, P., Melewar, T., & Gupta, S. (2017). Corporate Logo: History, Definition, and Components. *International Studies of Management & Organization*, *47*(2), 176–196. doi:10.1080/00208825.2017.1256166

Forstenzer, J. (2018). Something has cracked: post-truth politics and Richard Rorty's postmodernist Bourgeois liberalism. In Tony Saich (Ed.), *Ash Center Occasional Papers Series*, Cambridge, MA: Harvard Kennedy School Publication.

Foucault, M. (2007). *İktidarın gözü* (Çev: İ.Ergüden) İstanbul, Turkey: Ayrıntı. Retrieved from http://www.elyadal.org/pivolka/22/parafili.html

Fougère, M., & Moulettes, A. (2007). The Construction of the Modern West and the Backward Rest: Studying the Discourse of Hofstede's Culture's Consequences. *Journal of Multicultural Discourses*, *2*(1), 1–19.

Fox, M. (2018). Fake News: Lies spread faster on social media than truth does. Available at https://www.nbcnews.com/health/health-news/fake-news-lies-spread-faster-social-media-truth-does-n854896

Fox-Brewster, T. (2015, Oct. 6). 'Landmark' Decision Threatens Facebook Use Of European Personal Data. *Forbes.com*. Retrieved from https://www.forbes.com/sites/thomasbrewster/2015/10/06/safe-harbour-invalid/#1d98763a1d98

Freund, G. (2016). *Fotoğraf ve Toplum. Şule Demirkol (Trans.)*. İstanbul, Turkey: Sel Puplications.

Fu, J. (2016, Jan. 26). Optimism Over China's Economy Won Out at Davos. *China Daily*. Retrieved from http://europe.chinadaily.com.cn/opinion/2016-01/26/content_23244212.htm

Fuchs, C. (2017). Social media: A critical introduction. Thousand Oaks, CA: Sage.

Fuchs, C. (2012a). Critique of the Political Economy of Web 2.0 Surveillance. In C. Fuchs, K. Boersma, A. Albrechtslund, & M. Sandoval (Eds.), *Internet and Surveillance: The Challenges of Web 2.0 and Social Media* (pp. 31–70). New York, NY: Routledge.

Fuchs, C. (2012b). Dallas Smythe Today - The Audience Commodity, the Digital Labour Debate, Marxist Political Economy and Critical Theory. Prolegomena to a Digital Labour Theory of Value. *TripleC*, *10*(2), 692–740. doi:10.31269/triplec.v10i2.443

Fuchs, C. (2012c). The Political Economy of Privacy on Facebook. *Television & New Media*, *13*(2), 139–159. doi:10.1177/1527476411415699

Fuchs, C. (2014). Social media and the public sphere. *TripleC*, *12*(1), 57–101. doi:10.31269/triplec.v12i1.552

Fuchs, C. (2016). Baidu, Weibo and Renren: The Global Political Economy of Social Media in China. *Asian Journal of Communication*, *26*(1), 14–41. doi:10.1080/01292986.2015.1041537

Fuchs, C. (2017). Google kapitalizmi. In F. Aydoğan (Ed.), *Yeni Medya Kuramları* (pp. 71–83). İstanbul, Turkey: Der Publications.

Galician, M.-L., & Merskin, D. L. (2007). *Critical Thinking About Sex, Love, and Romance in the Mass Media – Media Literacy Applications*. New Jersey: Lawrence Erlbaum Associates, Publishers. doi:10.4324/9781410614667

Garber, L. L., & Hyatt, E. M. (2008). Color as a Tool for Visual Persuasion. In L. M. Scott & R. Batra (Eds.), *Persuasive Imagery - A Consumer Response Perspective* (pp. 313–336). New Jersey: Lawrence Erlbaum Associates.

Garcia, M. R. (1997). *Contemporary Newspaper Design*. Upper Saddle River, NJ: Prentice Hall.

Garfield, S. (2012). *Tam Benim Tipim/Bir Font Kitabı, Çev: Sabri Gürses*. İstanbul, Turkey: Domingo Yayınları.

George, D. & Mallery, M. (2010). SPSS for Windows Step by Step: A Simple Guide and Reference, 17.0 update (10a ed.) Boston, MA: Pearson.

Giddens, A. (2006). *Sociology*. Cambridge, UK: Polity Press.

Giddens, A. (2012). *Sosyoloji*. İstanbul, Turkey: Kırmızı Yay.

Girard, R. (2005). *Günah keçisi. Işık Ergüden (Trans.)*. İstanbul, Turkey: Kanat Kitap.

Glaveanu, V. P. (2017). Psychology in the post-truth era. *Europe's Journal of Psychology*, *13*(3), 375–377. doi:10.5964/ejop.v13i3.1509 PMID:28904590

Global Digital Report. (2018). Erişim Adresi https://digitalreport.wearesocial.com/

Goel, V. (2015, April 23). Facebook Reports Quarterly Results Dominated by Shift to Mobile and Video. *The New York Times*. Retrieved from http://www.nytimes.com/2015/04/23/technology/facebook-q1-earnings.html?_r=0

Goffman, E. (2014). *Damga: örselenmiş kimliğin idare edilişi üzerine notlar. Ş. Geniş at all (Trans.)*. Ankara, Turkey: Heretik.

Goffman, E. (2018). *Gündelik Yaşamda Benliğin Sunumu*. İstanbul, Turkey: Metis.

Göker, G. & Keskin, S. (2015). haber medyası ve mülteciler: suriyeli mültecilerin türk yazılı basınındaki temsili. İletişim Kuram ve Araştırma Dergisi, 46, 229-256.

Goldhaber, M. H. (1997). The Attention Economy and the Net. *First Monday*, *2*(4). Retrieved from http://journals.uic.edu/ojs/index.php/fm/article/view/519/440

Göle, N. (1992). Modern mahrem: medeniyet ve örtünme. Ankara, Turkey: Metis yayınları.

Gombrich, E. H. (1972). The Visual Image. *Scientific American*, *227*(3), 82–97. doi:10.1038cientificamerican0972-82 PMID:4114778

Goransson, K., & Anna-Sara, F. (2018). Towards visual strategic communications: An innovative interdisciplinary perspective on visual dimensions within the strategic communications field. *Journal of Communication Management*, *22*(1), 46–66. doi:10.1108/JCOM-12-2016-0098

Gottfried, M. (2016, Jan. 27). Facebook: It's Hard to Argue with Results Like These; The Social Network's Results Show it Hasn't Missed a Step. *The Wall Street Journal Online*. Retrieved from https://www.wsj.com/articles/facebook-its-hard-to-argue-with-results-like-these-1453936786

Gregg, M. (2011). *Work's Intimacy*. Cambridge, UK: Polity.

Güçlü, A., Uzun, E., Uzun, S., & Hüsrev Yolsal, Ü. (2003). *Felsefe sözlüğü*. Ankara, Turkey: Bilim ve Sanat Yayınları.

Gülüm, E. (2016). Karnavalesk Bir İmgelemin Taşıyıcısı Olarak Bektaşi Mizahı. *Milli Folklor*, *28*(112), 130–141.

Gündüz, A., Ertong Attar, G., & Altun, A. (2018). Üniversite Öğrencilerinin Instagram'da Benlik Sunumları. *DTCF Dergisi*, *58*(2), 1862–1895. doi:10.33171/dtcfjournal.2018.58.2.32

Gunitsky, S. (2015). Corrupting the Cyber-Commons: Social Media as a Tool of Autocratic Stability. *Perspectives on Politics*, *13*(1), 42–54. Retrieved from http://individual.utoronto.ca/seva/corrupting_cybercommons.pdf doi:10.1017/S1537592714003120

Gürel, E., & Alem, J. (2014). *Ürün Yerleştirme*. Ankara, Turkey: Nobel Yayıncılık.

Gutnick v. Dow Jones & Company Inc. [2001] VSC 305

Hair, J. F., Jr., Tomas, G., Hult, M., Ringle, C. M., & Sarstedt, M. (2014). A Primer on Partial Least Squares Structural Equation Modeling (PLS-SEM), Thousand Oaks, CA: Sage.

Hall, S. (1997). The Work of Representation. In S. Hall (Ed.), Representations. Cultural Representations and Signifying (pp. 13–74). London: Sage Publications.

Hall, S. (1999). İdeolojinin yeniden keşfi: medya çalışmalarında baskı altında tutulanın geri dönüşü. In M. Küçük (Ed.), *Medya, İktidar, ideoloji* (pp. 77–126). Ankara, Turkey: Bilim ve Sanat Publications.

Harsin, J. (2018). Post-truth and critical communication. In *Oxford Research Encyclopedia of Communication* (pp. 1–36). Oxford, UK: Oxford University Press. doi:10.1093/acrefore/9780190228613.013.757

Heath, R. (2012). *Seducing the Unconscious: The Psychology of Emotional Influence in Advertising*. Chichester, UK: Wiley Blackwell. doi:10.1002/9781119967637

Hendricks, D. (2014, March 4). *Socialnomics.com*. Retrieved from: https://socialnomics.net/2014/03/04/the-shocking-truth-about-social-networking-crime/

Hill, C. A. (2008). The psychology of rhetorical images. In C. A. Hill & M. Helmers (Eds.), *Defining Visual Rhetorics* (pp. 25–40). New Jersey: Lawrence Erlbaum Associates Publishers.

Hofstede, G. (2011). Dimensionalizing cultures: The Hofstede model in context. *Online Readings in Psychology and Culture*, *2*(1), 1–26. doi:10.9707/2307-0919.1014

Holsanova, J. (2014). In the eye of the beholder: Visual communication from a recipient perspective. In D. Machin (Ed.), *Visual Communication* (pp. 331–355). Berlin, Germany: Mouton De Gruyter.

Holtzschue, L. (2009). *Rengi Anlamak, Çev: Fuat Akdenizli*. İzmir, Turkey: Duvar Yayınları.

https://en.oxforddictionaries.com/word-of-the-year/word-of-the-year-2016

https://tr.sputniknews.com/ortadogu/201810231035793591-dunyanin-en-etkili-musluman-lideri-erdogan/

https://www.aa.com.tr/tr/gunun-basliklari/cumhurbaskani-erdogan-eger-racon-kesilecekse-bu-raconu-bizzat-kendim-keserim/888565

https://www.facebook.com/anticiler/

https://www.facebook.com/ErdoganaSevdalilar/

https://www.facebook.com/groups/1582158098546652/

Huey, L. S. & Yazdanıfard, R. (2014). How Instagram can be used as a tool in social networking marketing. *ResearchGate*. Retrieved from http://www.researchgate.net

Imai, T. (2017). The collapse of faith in policy and the establishment. My Vision, 31. Retrieved from http://www.nira.or.jp/pdf/e_myvision31_A.pdf

İnal, A. (1996). *Haberi Okumak*. İstanbul, Turkey: Temuçin Yayınları.

Işık, U. & Oz, K. A. (2014). Çöp Yığınlarında Haber Aramak: İnternet Gazeteciliği Üzerine Bir Çalışma. *Humanities Science*, *9*(2), 27–43. doi:10.12739/NWSA.2014.9.2.4C0178

Jabeur, N., Zeadally, S., & Sayed, B. (2013). Mobile social networking applications. *Communications of the ACM*, *56*(3), 71–79. doi:10.1145/2428556.2428573

Jaenichen, C. (2017). Visual Communication and Cognition in Everyday Decision-Making. *IEEE Computer Graphics and Applications*, *37*(6), 10–18. doi:10.1109/MCG.2017.4031060 PMID:29140778

Jain, B. (2018 September 6). Google, Facebook, Twitter to self-censor political content. *The Times of India*. Retrieved from: http://timesofindia.indiatimes.com/articleshow/65693822.cms?utm_source=contentofinterest&utm_medium=text&utm_campaign=cppst

Jenkins, H. (2006). *Convergence Culture: Where Old and New Media Collide*. New York, NY: New York University Press.

Jenkins, H. (2017). Medya yöndeşmesinin kültürel mantığı. In F. Aydoğan (Ed.), *Yeni Medya Kuramları* (pp. 33–45). İstanbul, Turkey: Der Yayınları.

Jenkins, H., Purushotma, R., Weigel, M., Clinton, K., & Robinson, A. (2009). *Confronting the challenges of participatory culture: media education for the 21st century*. London, UK: MIT. doi:10.7551/mitpress/8435.001.0001

Jenkins, R. (2016). *Bir Kavramın Anatomisi Sosyal Kimlik*. İstanbul, Turkey: Everest Yay.

Jupp, V. (n.d.). The Sage Dictionary of Social Research Methods. London, UK: Sage.

Kahraman, M. (2014). *Sosyal Medya 2.0*. İstanbul, Turkey: Media Cat Kitapları.

Kaplan, A. M. (2012). If you love something, let it go mobile: Mobile marketing and mobile social media 4x4. *Business Horizons*, *55*(2), 129–139. doi:10.1016/j.bushor.2011.10.009

Kaplan, A. M., & Haenlein, M. (2010). Users of the world, unite! The challenges and opportunities of Social Media. *Business Horizons*, *53*(1), 59–68. doi:10.1016/j.bushor.2009.09.003

Karaatlı, M. (2010). Verilerin Düzenlenmesi ve Gösterimi içinde SPSS Uygulamalı Çok Değişkenli İstatistik Teknikleri, 5. Baskı, Ed. Şeref Kalaycı, Asil Yayın Dağıtım, Ankara, ss. 3-47.

Karaduman, N. (2017). Popüler Kültürün Oluşmasında ve Aktarılmasında Sosyal Medyanın Rolü, Erciyes Üniversitesi Sosyal Bilimler Enstitüsü Dergisi XLIII, 2017/2, 7-27.

Karaduman, S. (2010). Modernizmden Postmodernizme Kimliğin Yapısal Dönüşümü. *Journal of Yasar University*, *17*(5), 2886–2899.

Karagöz, K. (2013). Yeni Medya Çağında Dönüşen Toplumsal Hareketler ve Dijital Aktivizm Hareketleri. İletişim ve Diplomasi Dergisi, (1),131-157.

Karagöz, K. (2018). Post-truth çağında yayıncılığın geleceği. *TRT Akademi*, *3*(6), 678–708.

Karataş, Ş., & Binark, M. (2016). Yeni medyada yaratıcı kültür: Troller ve ürünleri caps'ler. *TRT Akademi*, *1*(2), 427–448.

Kartarı, A. (2017). Nitel Düşünce ve Etnografi: Etnografik Yönteme Düşünsel Bir Yaklaşım. Moment Dergi, 4(1), 207-220. Retrieved from http://dergipark.gov.tr/moment/issue/36383/411586

Kavuran, T., & Özpolat, K. (2016). Görsel İletişim Aracı Olan Dergilerin Tasarlanma Süreci. *Fırat Üniversitesi Sosyal Bilimler Dergisi*, *26*(2), 267–275.

Kawamoto, K. (2003). Digital Journalism: Emerging Media and the Changing Horizons of Journalism. In K. Kawamoto (Ed.), Digital Journalism: Emerging Media and the Changing Horizons of Journalism, pp. 1–29. Lanham, MD: Rowman & Littlefield Publishers.

Kaya, Z. (2018). Yeni medyaya geçiş sürecinde ajans haberlerinin yapısal dönüşümü (Anadolu Ajansı Örneğiyle). (Unpublished doctoral thesis). Atatürk University İnstitute of Social Science, The Department of Basic Communication Science, Erzurum.

Kayış, A. (2010). Güvenilirlik Analizi içinde SPSS Uygulamalı Çok Değişkenli İstatistik Teknikleri, 5. Baskı, Ed. Şeref Kalaycı, Asil Yayın Dağıtım, Ankara, ss. 404-419.

Kearney, R. (2012). *Yabancılar, tanrılar ve canavarlar. Barış Özkul (Trans.).* İstanbul, Turkey: Metis.

Kellner, D. (2001). Popüler Kültür ve Postmodern Kimliklerin İnşası, Çev: Gülcan Seçkin, Doğu Batı, Sayı: 15.

Kemp, S. (2015). *Digital, Social and Mobile in 2015.* Retrieved from http://wearesocial.com/uk/special-reports/digital-social-mobile-worldwide-2015

Kendra, C. S. (2017, January 20). *Ministry of Electronics and Information Technology, Government of India.* Retrieved from https://www.cyberswachhtakendra.gov.in/

Kerlinger, F. N. (1973). Foundations of Behavioral Research. 2nd edition. New York, NY: Holt, Rinehart, & Winston.

Kerns, C. (2014). *Trendology: building an advantage through data-driven real-time marketing*. New York, NY: Palgrave MacMillan. doi:10.1057/9781137479563

Keskin, S. (2017). Dini ötekilik ve iletişimsel pratikler: din değiştirenler üzerine bir araştırma. (*Unpublicated Master Thesis)*, Fırat University, Intitute of Social Sciences, The Department of Communication Sciences, Elazığ/Turkey.

Keskin, S. (2018). Reklam Gerçeğinin Dijital Failleri Olarak "Efekt Kimlikler": Turkcell'in Emocanları Üzerinden Bir Kimlik Okuması. *Global Media Journal TR Edition.*, *8*(16), 328–353.

Keskin, S., & Baltacı, F. (2018). Sinemanın Toplumsallaşması ve Sosyal Temsili: Facebook'un 'Kadife Olmayan' Perdesi. In M. G. Genel (Ed.), *İletişim Çağında Dijital Kültür* (pp. 163–196). İstanbul, Turkey: Eğitim Yayınevi.

Ketenci, H. F., & Bilgili, C. (2006). *Yongaların 10.000 Yıllık Gizemli Dansı: Görsel İletişim ve Grafik Tasarım*. İstanbul, Turkey: Beta.

Keyes, R. (2004). *Post-Truth Era: Dishonesty and Deception in Contemporary Life*. New York: St. Martin's Press.

Keyes, R. (2004). *The Post-Truth Era: Dishonesty and Deception in Contemporary Life*. New York, NY: St. Martin's Press.

Key, W. B. (1973). *Subliminal Seduction*. Upper Saddle River, NJ: Prentice-Hall.

Key, W. B. (1976). *Media Sexploitation*. Upper Saddle River, NJ: Prentice Hall.

Kietzmann, J. H., Hermkens, K., McCarthy, I. P., & Silvestre, B. S. (2011). Social media? Get serious! Understanding the functional building blocks of social media. *Business Horizons*, *54*(3), 241–251. doi:10.1016/j.bushor.2011.01.005

Kim, J. Y., Giurcanu, M., & Fernandes, J. (2017). Documenting the Emergence of Grassroots Politics on Facebook: The Florida Case. *The Journal of Social Media in Society*, *6*(1), 5–41.

King, G., Pan, J., & Roberts, M. E. (2013). How Censorship in China Allows Government Criticism but Silences Collective Expression. *The American Political Science Review*, *107*(2), 326–343. doi:10.1017/S0003055413000014

Kıpçak, N. S. (2016). *Yeni Karnaval Olarak Yeni Medya: Karnavalesk Nitelikleri ile Twitter*. İstanbul, Turkey: Marmara Üniversitesi Sosyal Bilimler Enstitüsü Radyo TV Sinema Anabilim Dalı, Doktora Tezi.

Kırık, M. (2015). Sivil toplumun sınırlandırılamayan sosyal medya sorunsalı. (Ed: M. Kırık, A. Çetinkaya, Ö. E. Şahin). Bilişim. S:161-183. İstanbul, Turkey: Hiperlink.

Kırık, A. M., & Saltık, R. (2017). Sosyal medyanın dijital mizahı: İnternet meme/caps. *Atatürk Journal of Communication*, *12*, 99–118.

Kitchenham, B., Brereton, O. P., Budgen, D., Turner, M., Bailey, J., & Linkman, S. (2009). Systematic literature reviews in software engineering–a systematic literature review. *Information and Software Technology*, *51*(1), 7–15. doi:10.1016/j.infsof.2008.09.009

Kocabaş, F., & Elden, M. (1997). *Reklam ve Yaratıcı Strateji: Konumlandırma ve Star Stratejisinin Analizi*. İstanbul, Turkey: Yayınevi Yayıncılık.

Korkmaz, İ. (2013). Facebook ve mahremiyet: Görmek ve gözetle (n) mek. Yalova Üniversitesi Sosyal Bilimler Dergisi, 3(5).

Kozinets, R. V. (2002). The Field Behind the Screen: Using Netnography for Marketing Research in Online Communities. *JMR, Journal of Marketing Research*, *39*(February), 61–72. doi:10.1509/jmkr.39.1.61.18935

Kumar, C. (2017, June 22). One cybercrime in India every 10 minutes. *The Times of India*. Retrieved from https://timesofindia.indiatimes.com/india/one-cybercrime-in-india-every-10-minutes/articleshow/59707605.cms

Kumar, S. & Shah, N. (2018). False information on web and social media: A survey. *arXiv preprint arXiv:1804.08559*.

Kurt, M. (2002). Görsel-Uzaysal Yeteneklerin Bileşenleri. *Klinik Psikiyatri*, *5*(2), 120–125.

Kuss, D. J., & Griffiths, M. D. (2011). Online social networking and addiction—A review of the psychological literature. *International Journal of Environmental Research and Public Health*, *8*(9), 3528–3552. doi:10.3390/ijerph8093528 PMID:22016701

Langer, R., & Beckman, S. C. (2005). Sensitive Research Topics: Netnography Revisited. *Qualitative Market Research*, *8*(2), 189–203. doi:10.1108/13522750510592454

Leiss, W., Kline, S., Jhally, S., & Botterill, J. (2005). *Social Communication in Advertising – Consumption in the Mediated Marketplace*. New York, NY: Routledge.

Leslie Becker-Phelps. (2016). Social Media Fosters Insecurity: How to Overcome It. Available at https://www.psychologytoday.com/us/blog/making-change/201603/social-media-fosters-insecurity-how-overcome-it

Lester, P. M. (2006). *Syntactic Theory of Visual Communication*. Retrieved from http://paulmartinlester.info/writings/viscomtheory.html

Li, C. (2009). The Chinese Communist Party: Recruiting and Controlling the New Elites. *Journal of Current Chinese Affairs*, *38*(3), 13–33. doi:10.1177/186810260903800302

Lyon, D. (2013). *Gözetim çalışmaları*. İstanbul, Turkey: Kalkedon.

Ma, S. & Cao, Y. (2015, Dec. 18). Experts Call to Keep Data Safe. *China Daily*. Retrieved from http://www.chinadaily.com.cn/business/tech/2015-12/18/content_22741317

Macy, B., & Thompson, T. (2011). *The Power of Real-Time Social Media Marketing*. New York, NY: McGraw Hill.

Madden, M. & Rainie, L. (2015). *Americans' Attitudes about Privacy, Security and Surveillance*. Retrieved from http://www.pewinternet.org/files/2015/05/Privacy-and-Security-Attitudes-5.19.15_FINAL.pdf

Madden, M., Lenhart, A., Cortesi, S., Gasser, U., Duggan, M., Smith, A., & Beaton, M. (2013). Teens, social media, and privacy. *Pew Research Center*, *21*, 2–86.

Malmelin, N. (2010). What is Advertising Literacy? Exploring the Dimensions of Advertising Literacy. *Journal of Visual Literacy*, *29*(2), 129–142.

Manovich, L. (1997). Automation of Sight from Photography to Computer. Retrieved from http://manovich.net/index.php/projects/automation-of-sight-from-photography-to-computer-vision

Manovich, L. (1998) Database as A Symbolic Form. Retrieved from http://manovich.net/index.php/projects/database-as-a-symbolic-form

Manovich, L. (1998). *Database as A Symbolic Form*. Retrieved from http: //manovich.net/index.php/projects/database-as-a-symbolic-form

Manovich, L. (2006a). After Effect or Velvet Revolution (Part-1). Retrieved from http://manovich.net/index.php/projects/after-effects-part-1

Manovich, L. (2006a). *After Effect, or Velvet Revolution (Part 2)*. Retrieved from http: //manovich.net/index.php/projects/after-effects-part-2

Manovich, L. (2006b). After Effect or Velvet Revolution (Part-2). Retrieved from http://manovich.net/index.php/projects/after-effects-part-2

Manovich, L. (2006c). Import/Export: Design Workflow and Contemporary Aesthetics. Retrieved from http://manovich.net/content/04-projects/051-import-export/48_article_2006.pdf

Manovich, L. (2007). Understanding Hybrid Media. Retrieved from http://manovich.net/index.php/projects/understanding-hybrid-media

Manovich, L. (2010). What is Visualization. Retrieved from http://manovich.net/index.php/projects/what-is-visualization

Manovich, L. (2012). Media after software. Retrieved from http://manovich.net/index.php/projects/article-2012

Manovich, L. (2012). Media After Software. Retrieved from http://manovich.net/index.php/projects/article-2012

Manovich, L. (2016a). Instagrammism and Contemporary Cultural Identity. Retrieved from http://manovich.net/index.php/projects/notes-on-instagrammism-and-mechanisms-of-contemporary-cultural-identity

Manovich, L. (2016b). What Makes Photo Cultures Different? Retrieved from http://manovich.net/index.php/projects/what-makes-photo-cultures-different

Manovich, L. (2017). Automating Aesthetics: Artificial Intelligence and Image Culture. Retrieved from http://manovich.net/index.php/projects/automating-aesthetics-artificial-intelligence-and-image-culture

Manovich, L. (1999). Avant-Garde as Software. In M. Revolutions (Ed.), *S. Kovats*. Frankfurt, Germany: Campus Verlag.

Manovich, L. (2001). *The Language of New Media*. Cambridge, MA: MIT Press.

Manovich, L. (2006b). *What Is New Media? In* R. Hassan & J. Thomas, (Eds.), *The New Media Theory Reader*. Berkshire: Open University Press.

ManovichL. (2011). Inside Photoshop. Retrieved From http://manovich.net/index.php/projects/inside-photoshop

Mansell, R. vd., (2015), The International Encyclopedia of Digital Communication and Society. UK: Wiley-Blackwell.

Mao, J. (2014). Social media for learning: A mixed methods study on high school students' technology affordances and perspectives. *Computers in Human Behavior*, *33*, 213–223. doi:10.1016/j.chb.2014.01.002

Mattelart, A., & Siegelaub, S. (1979). *Communication and Class Struggle – 1. Capitalism, Imperialism*. New York, NY: International General/International Mass Media Research Center.

Maxwell, L. (2016). Donald Trump's Campaign of feelings. Retrieved from http://contemporarycondition.blogspot.com/2016/07/donald-trumps-campaign-of-feeling.html

Mayfield, A. (2008). What is Social Media. iCrossing. e-book. Retrieved from http://www.icrossing.com/sites/default/files/what-is-socialmedia-uk.pdf

McCombs, M. E., & Shaw, D. L. (1972). The agenda setting function of mass media. *Public Opinion Quarterly*, *36*(2), 176–187. doi:10.1086/267990

McCroskey, J. C. (2015). *An introduction to rhetorical communication*. New York, NY: Routledge.

Mclean, P. B., & Wallace, D. (2013). Blogging the Unspeakable: Racial Politics, Bakhtin, and the Carnivalesque. *International Journal of Communication*, *7*, 1518–1537.

McLuhan, M. (1964). *Understanding media: The extensions of man*. London, UK: Routledge.

McLuhan, M. (1994). *Understanding Media – The Extensions of Man*. Cambridge, MA: The MIT Press.

McLuhan, M. (2005). *Yaradanımız medya: Medyanın etkileri üzerine bir keşif yolculuğu. Ü. Oskay (Çev.)*. İstanbul, Turkey: Merkez Kitapçılık.

McQuail, D. (1994). *Kitle İletişim Kuramı: Giriş. (Çev. Ahmet Haluk Yüksel)*. Eskişehir: Kibele Sanat Merkezi Yayınları.

Melewar, T. C. (2001). Measuring visual identity: A multi-construct study. *Corporate Communications*, *6*(1), 36–42. doi:10.1108/13563280110381206

Melewar, T. C. (2003). Determinants of the corporate identity construct: A review of the literature. *Journal of Marketing Communications*, *9*(4), 195–220. doi:10.1080/1352726032000119161

Melewar, T., & Karaosmanoğlu, E. (2006). Seven dimensions of corporate identity: A categorisation from the practitioners' perspectives. *European Journal of Marketing*, *40*(7/8), 846–869. doi:10.1108/03090560610670025

Melewar, T., & Saunders, J. (1998). Global corporate visual identity systems: Standardization, control and benefits. *International Marketing Review*, *15*(4), 291–308. doi:10.1108/02651339810227560

Meng, J. (2016a, April 19). Tencent's WeChat Launches New App for Enterprises. *China Daily*. Retrieved from http://www.chinadaily.com.cn/business/tech/2016-04/19/content_24649488.htm

Meng, J. (2016b, Feb. 16). Tencent Claims Surge in Digital Red Envelopes over New Year. *China Daily*. Retrieved from http://www.chinadaily.com.cn/business/tech/2016-02/16/content_23496768.htm

Mercin, L. (2017). Müze Eğitimi, Bilgilendirme Ve Tanitim Açisindan Görsel İletişim Tasarimi Ürünlerinin Önemi. *Milli Eğitim Dergisi*, *46*(214), 209–237.

Miller, C. C. (2015, May 15). How Facebook's Experiment Changes the News (and How It Doesn't). *The New York Times*. Retrieved from http://www.nytimes.com/2015/05/15/upshot/how-facebooks-experiment-changes-the-news-and-how-it-doesnt.html

Miller, D. (2012). Social networking sites. Digital anthropology, 156-161.

Miller, D., Costa, E., Haynes, N., McDonald, T., Nicolescu, R., Sinanan, J., & Wang, X. (2016). *How the world changed social media*. UK: UCL Press. doi:10.2307/j.ctt1g69z35

Milliyet. (2012). Şipşak 1 Milyar Dolar. Retrieved from Http://Ekonomi.Milliyet.Com.Tr/Sipsak-1-MilyarDolar/Ekonomi/Ekonomidetay/11.04.2012 /1526691/ Default.Htm.

Mirzoeff, N. (1998). What is Visual Culture? (Nicholas Mirzoeff, Ed.), The Visual Culture Reader. (s: 3-13). USA: Routledge.

Mitchell, W. J. T. (2005). *What do Pictures Want? – The Lives and Loves of Images*. Chicago, IL: The University of Chicago Press. doi:10.7208/chicago/9780226245904.001.0001

Mlicki, P. P., & Naomi, E. (1996). Being Different or Being Better? National Stereotypes and Identifications of Polish and Dutch Students. *European Journal of Social Psychology*.

Modreanu, S. (2017). The post-truth era? *HSS*, *3*, 7–9.

Monggilo, Z. M. Z. (2016, Jan. 26-28). Internet Freedom in Asia: Case of Internet Censorship in China. Paper Presented at the *International Conference on Social Politics, Universitas Muhammadiyah Yogyakarta*, Indonesia. 10.18196/jgp.2016.0026

Möngü, B. (2013). Postmodernizm ve Postmodern Kimlik Anlayışı, Atatürk Üniversitesi Sosyal Bilimler Enstitüsü Dergisi, 2/17.

Montgomery, K. C. (2015). Youth and Surveillance in the Facebook Era: Policy Interventions and Social Implications. *Telecommunications Policy*, *39*(9), 771–786. doi:10.1016/j.telpol.2014.12.006

Moreno, M. A., Jelenchick, L. A., Egan, K. G., Cox, E., Young, H., Gannon, K. E., & Becker, T. (2011). Feeling bad on Facebook: Depression disclosures by college students on a social networking site. *Depression and Anxiety*, *28*(6), 447–455. doi:10.1002/da.20805 PMID:21400639

Morrison, K. (2014). The growth of social media: from passing trend to international obsession. *Social Times 2014*. Available at https://www.adweek.com/digital/the-growth-of-social-media-from-trend-to-obsession-infographic/

Morva, O. (2014). In S. Çakır (Ed.), *Goffman'ın Dramaturjik Yaklaşımı ve Dijital Ortamda Kimlik Tasarımı: Sosyal Paylaşım Ağı Facebook Üzerine Bir İnceleme, Medya ve Tasarım* (pp. 231–255). İstanbul, Turkey: Urzeni.

Mutlu, E. (2004). *İletişim Sözlüğü*. Ankara, Turkey: Bilim ve Sanat Yayınları.

Nassaji, H. (2015). Qualitative and descriptive research: Data type versus data analysis. *Language Teaching Research*, *19*(2), 129–132. doi:10.1177/1362168815572747

Nations, D. (2018). What is social media. *What are Social Media*. Available at https://www.lifewire.com/what-is-social-media-explaining-the-big-trend-3486616

Neves, J. C. (2014). *Introdução à Ética Empresarial* [*Introduction to Business Ethics*]. Lisbon, Portugal: Principia.

New Testament/John. Retrieved from https://incil.info/kitap/Yuhanna/8

Ng, J. Q. (2015). Politics, Rumors, and Ambiguity: Tracking Censorship on WeChat's Public Accounts Platform. Retrieved from https://citizenlab.org/2015/07/tracking-censorship-on-wechat-public-accounts-platform/

Niedzviecki, H. (2010). *Dikizleme Günlüğü, Çev.: Gündüz, G.* İstanbul, Turkey: Ayrıntı.

Niedzviecki, H. (2010). *Dikizleme günlüğü. Gökçe Gündüç (Trans.)*. İstanbul, Turkey: Ayrıntı.

Niedzviecki, H. (2011). *Ben özelim; Bireylik Nasıl Yeni Konformizm Haline Geldi, Çev.; Erduman, S.* İstanbul, Turkey: Ayrıntı.

North Central University. (2017). Available at https://www.ncu.edu/blog/dangers-social-media-marriage-and-family#gref

NTV. (2010). Peygamber diyen Ak Partili'ye ihraç. Retrieved from https://www.ntv.com.tr/turkiye/peygamber-diyen-ak-partiliye-ihrac,tcyQEPpYpEKHylMB1Qixfw

Odden, L. (2012). *How to Attract and Engage More Customers by Integrating Seo, Social Media, and Content Marketing*. Hoboken, NJ: John Wiley and Sons.

Oğuz, G. Y. (2010). Güzellik Kadınlar İçin Nasıl Vaade Dönüşür: Kadın Dergilerindeki Kozmetik Reklamları Üzerine Bir İnceleme, Selçuk iletişim, C.: 6, S.: 3, (184-195).

Okay, A. (2005). *Kurum Kimliği*. İstanbul, Turkey: Mediacat.

Old Testament/Isaiah. Retrieved from https://incil.info/kitap/Yesaya/53

Old Testament/Leviticus. Retrieved from https://incil.info/kitap/Levililer/16

Onat, F. ve Alikılıç, Ö. A. (2008). Sosyal Ağ Sitelerinin Reklam ve Halkla İlişkiler Ortamları. *Journal of Yaşar University*, *3*(9), 1111–1143.

Öncel Taşkıran, N., & Bolat, N. (2013). Reklam Ve Algı İlişkisi: Reklam Metinlerinin Alımlanmasında Duyu Organlarının İşlevleri Hakkında Bir İnceleme. *Beykent Üniversitesi Sosyal Bilimler Dergisi*, *6*(1), 49–69.

Öncü Yıldız, M. (2012). Görsel Okuryazarlik Üzerine. *Marmara İletişim Dergisi*, *19*, 64–77.

Onursoy, S. (2017). Görsel Kültür ve Görsel Okuryazarlık. *Türk Kütüphaneciliği*, *31*(1), 47–54. doi:10.24146/tkd.2017.4

Open Society Institute. (2018). Common sense wanted resilience to 'post-truth' and its predictors in the new media literacy index 2018. Retrieved from http://osi.bg/downloads/File/2018/MediaLiteracyIndex2018_publishENG.pdf

Oskay, Ü. (2001). *İletişimin ABC'si*. İstanbul, Turkey: Simav Publications.

Özarslan, Z. (2003). *Söylem ve İdeoloji Mitoloji, Din, İdeoloji. Çev. Nurcan Ateş, Barış Çoban, Zeynep Özarslan*. İstanbul, Turkey: Su Yayınları.

Özcan, M. (2018). Öznenin Ölümü: Post-Truth Çağında Güvenlik ve Türkiye. INSAMER, (January), 1-11.

Oz, M. (2014). Sosyal Medya Kullanımı ve Mahremiyet Algısı: Facebook kullanıcılarının mahremiyet endişeleri ve farkındalıkları. *Journal of Yasar University*, *35*(9), 6099–6260.

Özmen, S., & Keskin, S. (2018). Sosyal medyada öz-temsil ve 'ötekiliğin' öteki boyutu: Karikateist toplumsalı üzerine inceleme. *IntJCSS*, *4*(2), 533–558.

Özmen, S., & Keskin, S. (2018). Sosyal medyada öz-temsil ve ötekiliğin 'öteki' boyutu: 'karikateist' toplumsalı üzerine inceleme. *IntJCSS*, *4*(2), 533–558.

Özüdoğru, Ş. (2014). Nitel Araştırmanın İletişim Araştırmalarında Rol ve Önemi Üzerine Bir Deneme, Global Media Journal, S. 4(8), 260-275.

Özyal, B. (2016). Tık Odaklı Habercilik: Tık Odaklı Haberciliğin Türk Dijital Gazetelerindeki Kullanım Biçimleri. *Global Media Journal, TR Edition*, *6*(12), 273–301.

Packard, V. (2007). *The Hidden Persuaders*. New York, NY: Ig Publishing.

Padilla, A. M., & Perez, W. (2003, February). Acculturation, Social Identity, and Social Cognition: A New Perspective. *Hispanic Journal of Behavioral Sciences*, *25*(1), 35–55. doi:10.1177/0739986303251694

Palma, A. (2017). 'When the future catches up with us, the past will no longer be valid' Descartes could be a yardstick. *Uno Magazine*, *27*, 17–19.

Pan, B. & Crotts, J. C. (2012). Theoretical models of social media, marketing implications, and future research directions. Social media in travel, Tourism and hospitality: Theory, practice and cases, 73-85.

Pan, B., & Crotts, J. (2012). Theoretical models of social media, marketing implications, and future research directions. In M. Sigala, E. Christou, & U. Gretzel (Eds.), *Social Media in Travel, Tourism and Hospitality: Theory, Practice and Cases* (pp. 73–86). Surrey, UK: Ashgate.

Pandit, S. (2013, February 7). Two held in J&K girl Rockband Case. *The Times of India.* Retrieved from http://timesofindia.indiatimes.com/india/Two-held-in-JK-girl-rock-band-case/articleshow/18375067

Parthasarathy, S. (2015, March 26). *The judgment that silenced Section 66-A.* Retrieved from http://www.thehindu.com/opinion/lead/the-judgment-that-silenced-section66a/article7032656.ece

Patton, M. Q. (2005). Qualitative research. Encyclopedia of statistics in behavioral science.

Pavlik, J. (1999). New Media and News: Implications for the Future of Journalism. *New Media & Society, 1*(1), 54–59. doi:10.1177/14614448990010010 09

Pektaş, H. (2001). İnternette Görsel Kirlenme, Ankara, Turkey. *Tübitak ve Bilim Dergisi, 400,* 72–75.

Peng, L.-H., & Hung, C.-C. (2016). The practice of corporate visual identity—A case study of Yunlin Gukeng Coffee Enterprise Co., Ltd. *2016 International Conference on Applied System Innovation (ICASI).* Piscataway, NJ: IEEE. 10.1109/ICASI.2016.7539844

Perez, B. (2016, April 4). China's Social Networks Set for Boom, As Advertising Spending Forecast to Jump 56 Per Cent to Us$5.33b This Year. *South China Morning Post.* Retrieved from http://www.scmp.com/tech/enterprises/article/1933558/chinas-social-networks-set-boom-advertising-spending-forecast-jump

Perrin, A. (2015). Social media usage. Pew Research Center, 52-68.

Perry, A., & Wisnom, D. III. (2003). *Markanın DNA'sı.* İstanbul, Turkey: Kapital Medya.

Pingo, Z. & Narayan, B. (2018, September). Privacy Literacy and the Everyday Use of Social Technologies. In *European Conference on Information Literacy* (pp. 33-49). Cham, Switzerland: Springer.

Poe, T. M. (2019). *İletişim Tarihi: Konuşmanın Evriminden İnternete Medya ve Toplum.* İstanbul, Turkey: Islık Yayınları.

Porter, M. (2007). *Rekabet Stratejisi* (Vol. 4). İstanbul, Turkey: Sistem Yayıncılık.

Potter, J. W. (1998). *Media Literacy.* London, UK: Sage.

Poynter, R. (2012). *İnternet ve Sosyal Medya Araştırmaları El Kitabı Pazar Araştırmaları İçin Araçlarve Teknikler.* İstanbul, Turkey: Optimist Yayınları.

Primack, B. A., Shensa, A., Sidani, J. E., Whaite, E. O., Lin, L., Rosen, D., ... Miller, E. (2017). Social media use and perceived social isolation among young adults in the US. *American Journal of Preventive Medicine, 53*(1), 1–8. doi:10.1016/j.amepre.2017.01.010 PMID:28279545

Proudfoot, J. G., Wilson, D., Valacich, J. S., & Byrd, M. D. (2018). Saving face on Facebook: Privacy concerns, social benefits, and impression management. *Behaviour & Information Technology, 37*(1), 16–37. doi:10.1080/0144929X.2017.1389988

Ranjit D. Udeshi v. State Of Maharashtra (1965) AIR 881, Section 24 of Indian Penal Code- Whoever does anything with the intention of causing wrongful gain to one person or wrongful loss to another person, is said to do that thing dishonestly.

Ray, T. (2015, Oct. 3). Rich Get Richer as Google and Facebook Dominate Web Ads. *Barron's Asia*. Retrieved from http://www.barrons.com/articles/rich-get-richer-as-google-and-facebook-dominate-web-ads-1443851396

Rickitt, R. (2000). *Special Effects: The History and Technique*. New York, NY: Billboard Books.

Riddle, J. (2017). All Too Easy: Spreading Information Through Social Media. Available at https://ualr.edu/social-change/2017/03/01/blog-riddle-social-media/

Riley, H. (2004). Perceptual modes, semiotic codes, social mores: A contribution towards a social semiotics of drawing. *Visual Communication*, *3*(3), 294–315. doi:10.1177/1470357204045784

Robin, C., McCoy, S., & Yáñez, D. (2017). WhatsApp. In G. Meiselwitz (Ed.), Lecture Notes in Computer Science: Vol. 10283. Social Computing and Social Media. Applications and Analytics. SCSM 2017. Berlin, Germany: Springer. doi:10.1007/978-3-319-58562-8_7

Robins, K. (2013). *İmaj Görmenin Kültür ve Politikası, (N. Türkoğlu, Çev)*. İstanbul, Turkey: Ayrıntı Yayınları.

Rohm, A., & Kaltcheva, V. D., & R. Milne, G. (2013). A mixed-method approach to examining brand-consumer interactions driven by social media. *Journal of Research in Interactive Marketing*, *7*(4), 295–311. doi:10.1108/JRIM-01-2013-0009

Rosson, P., & Brooks, M. (2004). M&As and Corporate Visual Identity: An Exploratory Study. *Corporate Reputation Review*, *7*(2), 181–194. doi:10.1057/palgrave.crr.1540219

Rowe, D. (1996). *Popüler Kültürler. (Çev. Mehmet Küçük)*. İstanbul, Turkey: Ayrıntı Yayınları.

Ruhela, V. S., & Parween, S. (2018). Effect of visual communication in tracking activity schedule among children with autism spectrum disorder. *Indian Journal of Health & Wellbeing*, *9*(5), 748–751.

Rutter, D., Stephenson, G., & Dewey, M. (1981). Visual communication and the content and style of conversation. *British Journal of Social Psychology*, *20*(1), 41–52. doi:10.1111/j.2044-8309.1981.tb00472.x PMID:7237005

Safko, L. (2010). *The social media bible tactics, tools and strategies for business success*. Hoboken, NJ: John Wiley & Sons.

Sain-Dieguez, V. (2015). Are you overlooking the most valuable real-time marketing strategy?. Retrieved from http://www.convinceandconvert.com/digital-marketing/are-you-overlooking-the-most-valuable-real-time-marketing-strategy/

Salomon, D. (2013). Moving on from facebook using ınstagram to connect with undergraduates and engage in teaching and learning. *Collage Researchlibraries News*, *74*(8), 408–412. doi:10.5860/crln.74.8.8991

Sanches, C. ve Restrepo, J. C. (2015). Strategic real-time marketing. Advances in the Area of Marketing and Business Communication, pp. 164-184.

Sanders, C. E., Field, T. M., Miguel, D., & Kaplan, M. (2000). The relationship of Internet use to depression and social isolation among adolescents. *Adolescence*, *35*(138), 237. PMID:11019768

Sandoval, M. (2012). A Critical Empirical Case Study of Consumer Surveillance on Web 2.0. In C. Fuchs, K. Boersma, A. Albrechtslund, & M. Sandoval (Eds.), *Internet and Surveillance: The Challenges of Web 2.0 and Social Media* (pp. 147–169). New York, NY: Routledge.

Sarı, G. (2018) Sosyal Medyanın Yeni Starları: Youtuberlarla Değişen Popülerlik, Yeni Medyada Yeni Yaklaşımlar, Ed. Rengim Sine, Gülşah Sarı, Konya, Turkey: Literatürk Yayınları, pp: 277-296.

Sarıkavak, N. (2009). *Çağdaş Tipografinin Temelleri*. Ankara, Turkey: Seçkin Yayıncılık.

Sartori, G. (1998). *Homo Videns: La Sociedad Teledirigida* [*Homo Videns: Teledirected Society*]. Buenos Aires, Argentina: Taurus.

Sartori, G. (2006). *Görmenin iktidarı. (Çev: Gül Batuş, Bahar Ulukan).* İstanbul, Turkey: Karakutu.

Satar, B. (2015). *Popüler Kültür ve Tekrarlanan İmajlar.* İstanbul, Turkey: Kozmos Yayınları.

Saussure, F. (1959). *Course in General Linguistics* (W. Baskin, Trans.). New York, NY: The Philosophical Library.

Schmitt, B., & Simonson, A. (2000). *Pazarlama Estetiği.* İstanbul, Turkey: Sistem Yayıncılık.

Schroeder, R. (1994). Cyberculture, cyborg post-modernism and the sociology of virtual reality Technologies. *Future, 26*(5), 519–528. doi:10.1016/0016-3287(94)90133-3

Schudson, M. (1999). *Popüler Kültürün Yeni Gerçekliği: Akademik Bilinçlilik ve Duyarlılık, Popüler Kültür ve İktidar, (Derleyen: Nazife Güngör).* Ankara, Turkey: Vadi Yayınları.

Scott, L. (1994). Images in advertising: The need for a theory of visual rhetoric. *The Journal of Consumer Research, 21*(2), 252–273. doi:10.1086/209396

Section 26 of Indian Penal Code- A person is said to do a thing fraudulently if he does that thing with intent to defraud but not otherwise.

Section 294 of Indian Penal Code- punishes obscene act or song with imprisonment of either description for a term which may extend to three months or with fine or both. Section 67 of IT Act provides punishment for such act.

Section 503 of Indian Penal Code defines Criminal Intimidation. Whosoever threatens another with any injury to his person, reputation or property, or to the person or reputation of any one in whom that person is interested, with intent to cause alarm to that person, or to cause that person to do any act which he is not legally bound to do, or to omit to do any act which that person is legally entitled to do, as the means of avoiding the execution of such threat, commits Criminal intimidation.

Section 66A of Information Technology Act, 2000 deals with, Punishment for sending offensive messages through communication service, etc. Any person who sends, by means of a computer resource or a communication device,—any information that is grossly offensive or has menacing character; or any information which he knows to be false, but for the purpose of causing annoyance, inconvenience, danger, obstruction, insult, injury, Criminal Intimidation, enmity, hatred or ill will, persistently by making use of such computer resource or a communication device, any electronic mail or electronic mail message for the purpose of causing annoyance or inconvenience or to deceive or to mislead the addressee or recipient about the origin of such messages, shall be punishable with imprisonment for a term which may extend to three years and with fine. Explanation.— For the purpose of this section, terms "electronic mail" and "electronic mail message" means a message or information created or transmitted or received on a computer, computer system, computer resource or communication device including attachments in text, images, audio, video and any other electronic record, which may be transmitted with the message.

Seetharaman, D. (2015, July 30). Facebook, Google Tighten Grip on Mobile Ads; Facebook's Quarterly Revenue Jumps 39% as it Lures Big Brand Advertisers and Smartphone Users. *The Wall Street Journal Online.* Retrieved from https://www.wsj.com/articles/facebook-revenue-rises-39-1438200350

Şeker, M. (2006). Sayfa Düzeni Ekollerinin Estetik ve İçeriğe Etkileri. *Selçuk İletişim Dergisi, 4*(2), 30–40.

Şener, G. (2013). *Sosyal Ağlarda Mahremiyet ve Yeni Mahremiyet Stratejileri, Yeni Medya Çalışmaları I. Ulusal Kongre Kitabı* (pp. 397–401). Eskişehir: Haz. Burak Özçetin, Gamze Göker, Günseli Bayraktutan, İdil Sayımer, Tuğrul Çomu.

Senft, T. M. (2008). *Camgirls, Celebrity & Community in The Age of Social Networks*. Retrieved from https://books.google.com.tr/books

Shabrina, F., & Winarsih, A. S. (2016, Jan. 26-28). Businesspeople Co-Optation In China's Communist Party Adaption. Paper Presented at the *International Conference on Social Politics, Universitas Muhammadiyah Yogyakarta*, Indonesia.

Shokri, A. & Dafoulas, G. (2016). A Quantitative analysis of the role of social networks in educational contexts.

Sığrı, Ü. (2018). *Nitel Araştırma Yöntemleri*. İstanbul, Turkey: Beta Yayınları.

Silverblatt, A., Smith, A., Miller, D., Smith, J., & Brown, N. (2014). *Media Literacy - Keys to Interpreting Media Messages*. Santa Barbara, California: Praeger.

Simoes, C., & Dibb, S. (2008). Illustrations of the internal management of corporate identity. In T. C. Melewar (Ed.), *Facets of Corporate Identity, Communication and Reputation* (pp. 66–80). Oxon, UK: Routledge. doi:10.4324/9780203931943.ch4

Şimşek, V. (2018). Post-truth ve yeni medya: sosyal medya grupları üzerinden bir inceleme. Global Media Journal Turkish Edition, 8 (16), 1-14.

Sine, R. (2017). *Alternatif Medya ve Haber Toplumsal Hareketler de Habercilik Pratikleri*. Konya, Turkey: Literatürk Yayınları.

Sloane, G. (2015, March 13). Facebook Buys a Company That Could Solve an Annoying Online Ad Problem. *Adweek*. Retrieved from http://www.adweek.com/news/technology/facebook-buys-company-could-solve-annoying-ad-problem-163475

Smith, G. (2007). Social software building blocks. Available at http://nform.com/ideas/social-software-building-blocks/

Smythe, D. W. (1981). *Dependency Road: Communications, Capitalism, Consciousness, and Canada*. Norwood, NJ: Ablex.

Socialbakers (2019). *Twitter statistics for Turkey*. Retrieved from https://www.socialbakers.com/statistics/twitter/profiles/turkey/

Sontag, S. (1993). *Fotoğraf Üzerine. (Trans.) Reha Akçakaya*. Istanbul, Turkey: Altıkırkbeş Publications.

Sontag, S. (1993). *Fotoğraf Üzerine. Reha Akçakaya (Trans.)*. İstanbul, Turkey: Altıkırkbeş Yayınları.

Sözen, E. (1999). *Söylem, Belirsizlik, Mücadele, Bilgi/ Güç ve Refleksivite*. İstanbul, Turkey: Paradigma Yayınları.

Sözen, M. F. (2009). Bakhtin'in Romanda "Karnavalesk" Kavramı ve Sinema. *Akdeniz Sanat Hakemli Dergi*, *2*(4), 65–86.

Spivak, G. C. (1988). Can the subaltern speak? In C. Nelson & L. Grossberg (Eds.), *Marxism and The Interpretation of Culture* (pp. 271–313). Basingstoke, UK: Macmillan Education. doi:10.1007/978-1-349-19059-1_20

Sputniknews. (n.d.). Retrieved from Https://tr.sputniknews.com/yasam/201812051036489809-instagram-kullanici-sayisi-turkiye/

Statista. (2016a). Facebook's Revenue and Net Income from 2007 to 2015 (in Million U.S. Dollars). Retrieved from http://www.statista.com/statistics/277229/facebooks-annual-revenue-and-net-income/

Statista. (2016b). Number of Monthly Active Facebook Users Worldwide as of 4th quarter 2015 (in Millions). Retrieved from http://www.statista.com/statistics/264810/number-of-monthly-active-facebook-users worldwide/

Statista. (2016c). Number of Monthly Active WeChat Users from 2nd Quarter 2010 to 4th Quarter 2015 (in Millions). Retrieved from http://www.statista.com/statistics/255778/number-of-active-wechat-messenger-accounts/

Statista. (2019, Nisan 9). Retrieved from https://www.statista.com/statistics/272014/global-social-networks-ranked-by-number-of-users/

Stelter, B. (2015, May 13). Why Facebook is Starting a New Partnership with 9 News Publishers. *CNN*. Retrieved from http://money.cnn.com/2015/05/13/media/facebook-instant-articles-news-industry/

Stewart, D. R. (2019). *Social Media and the Law*. London: Routledge publishing.

Storey, J. (2009). *Cultural Theory and Popular Culture* (5th ed.). Harlow, UK: Pearson.

Suiter, J. (2016). Post-truth politics. *Sage Journals*, *7*. doi:10.1177/2041905816680417

Sullivan, L. (2015, March 26). Google Takes Backseat to Facebook's Digital Display Ad Revenue. *MediaPost.com*. Retrieved from http://www.mediapost.com/publications/article/246520/google-takes-backseat-to-facebooks-digital-displa.html?utm_source=newsletter&utm_medium=email&utm_content=headline&utm_campaign=81438

Sumit, B. (2018, March 31). Surge in cyber-crime rate. *The Hindu*. Retrieved from. https://www.thehindu.com/news/cities/Visakhapatnam/surge-in-cyber-crimerate/article23395834.ece

Swartz, A. (2015, March 25). How Facebook is Courting Small Businesses and Their Advertising Dollars. *Silicon Valley/San Jose Business Journal Online*. Retrieved from http://www.bizjournals.com/sanjose/news/2015/03/25/how-facebook-is-courting-small-businesses-and.html

Tabachnick, B. G., Fidell, L. S., & Ullman, J. B. (2007). *Using multivariate statistics* (Vol. 5). Boston, MA: Pearson.

Tajfel, H. (1982). Social psychology of intergroup relations. *Annual Review of Psychology*, *33*(1), 1–39. doi:10.1146/annurev.ps.33.020182.000245

Taniguchi, M. (2017). Confronting to 'Post-truth Era'. My Vision, 31. Retrieved from http://www.nira.or.jp/pdf/e_myvision31.pdf

Taş, O. (1993). *Örnekleriyle Çağdaş Gazete Tasarımı*. Ankara, Turkey: Makro Yayın.

Tavin, K. (2009). Engaging Visuality: Developing a University Course on Visual Culture. *The International Journal of the Arts in Society*, *4*(3), 115–123. doi:10.18848/1833-1866/CGP/v04i03/35641

Teker, U. (2003). *Grafik Tasarımve Reklam*. İzmir, Turkey: Dokuzeylül Yayınları.

Tencent. (2015). Tencent Announces 2014 Fourth Quarter and Annual Results. Retrieved from http://www.tencent.com/en- us/content/ir/news/2015/attachments/20150318.pdf

Tencent. (2016a). Tencent Announces 2015 Fourth Quarter and Annual Results. Retrieved from http://www.tencent.com/en-us/content/ir/news/2016/attachments/20160317.pdf

Tencent. (2016b). *Tencent Announces 2016 First Quarter Results*. Retrieved from http://www.tencent.com/en-us/content/ir/news/2016/attachments/20160518.pdf

Tifferet, S. (2018). Gender differences in privacy tendencies on social network sites: A meta-analysis. *Computers in Human Behavior*.

Tiryakioğlu, F. (2012). *Sayfa Tasarımıve Gazeteler*. Ankara, Turkey: Detay Yayıncılık.

Tiryakioğlu, F., & Top, D. (2010). Sayfa Tasarımı ve Kurumsal Kimlik Oluşturma: Türkiye'deki Ulusal Gazetelerin Birinci Sayfaları Üzerine Bir Araştırma. *Selçuk İletişim Dergisi*, *6*(3), 137–146.

Today, I. (2017, August 11). Retrieved from: https://www.indiatoday.in/education-today/gk-current-affairs/story/nccc-cyber-india-1029203-2017-08-11

Tombul, I. (2018). In R. Sine & G. Sarı (Eds.), *Sosyal Medyada Mahremiyetin Dönüşümü ve Gençlerin Mahremiyete Bakışı. Yeni Medyaya Yeni Yaklaşımlar* (pp. 131–175). Konya: Literatürk Yayınları.

Toprak, A. vd (2014). Toplumsal paylaşım ağı Facebook: "görülüyorum öyleyse varım!". Kalkedon Yayınları.

Trottier, D. (2016). *Social Media as Surveillance: Rethinking Visibility in a Convergence World*. New York, NY: Routledge. doi:10.4324/9781315609508

Tse, E. (2015). *China's Disruptors*. New York, NY: Penguin.

Tungate, M. (2007). *Adland – A Global History of Advertising*. London, UK: Kogan Page.

Turan, M. (2008). Kaos Teorisi: Bauman ve Bakhtin. *Ankara Üniversitesi Dil ve Tarih-Coğrafya Fakültesi Felsefe Bölümü Dergisi*, *19*(1), 45–66.

Türkiye İnternet Kullanım ve Sosyal Medya İstatistikleri. (2019). Erişim Adresi https://dijilopedi.com/2019-turkiye-internet-kullanim-ve-sosyal-medya-istatistikleri/

Turner, G. (2004). *Understanding Celebrity*. London, UK: Sage. Retrieved from http://eclass.uoa.gr/modules/document/file.php/MEDIA118/celebrity+culture/Book_understanding+celebrity_graham+turner.pdf

Uçar. Tevfik Fikret. (2004). Görsel İletişimve Grafik Tasarım. İstanbul, Turkey: İnkılâp Yayınları.

Uçar, T. F. (2004). *Görsel İletişim ve Grafik Tasarım*. İstanbul, Turkey: İnkılap.

Ülkü, G. (2004). Söylem Çözümlemesinde Yöntem Sorunu ve Van Dijk Yöntemi. In *Dursun, Ç. (Der.), Haber-Hakikat ve İktidar İlişkisi*. Ankara, Turkey: Elips Yayınları.

Uluç, G., ve Yarcı, A. (2017), Sosyal Medya Kültürü, Dumlupınar Ünv., Sosyal Bilimler Dergisi, S.: 52, 88-102.

Underwood, L. (2016). Mobile to become dominant device by 2019. Available at https://wearesocial.com/uk/blog/2016/04/mobile-to-become-dominant-device-by-2019

US Securities and Exchange Commission. (2019, January 20). *Microcap Fraud*. Retrieved from https://www.investor.gov/investing-basics/avoiding-fraud/types-fraud/microcap-fraud

Uyan Dur, B. İ. (2015). Türk Görsel İletişim Tasarimi Ve Kültürel Değerlerle Bağlari. *Journal of International Social Research*, *8*(37), 443–453. doi:10.17719/jisr.20153710615

Uygun, E., & Akbulut, D. (2018). Karnavalesk Kuramı ve Instagram Ortamına Yansımaları. *Yeni Medya Elektronik Dergi*, *2*(2), 73–89. doi:10.17932/IAU.EJNM.25480200.2018.2/2.73-89

Valenzuela, S., Halpern, D., & Katz, J. E. (2014). Social network sites, marriage well-being and divorce: Survey and state-level evidence from the United States. *Computers in Human Behavior*, *36*, 94–101. doi:10.1016/j.chb.2014.03.034

Van Dick, J. (2006). *The Network Society Social Aspects of New Media*. London, UK: Sage.

Van Dijk, T. A. (2003). Söylem ve İdeoloji, Çokalanlı Bir Yaklaşım. In B. Çoban & Z. Özarslan (Eds.), Söylem ve İdeoloji: Mitoloji, Din, İdeoloji. İstanbul, Turkey: Su Yayınları, (Çev.: N. Ateş).

Van Dijk, T. (1998). Ideology. *Sage (Atlanta, Ga.)*.

Van Dijk, T. A. (1985). (In the Press). News Analysis [Mahwah, NJ: Lawrence Erlbaum Associates Publication.]. *Case Studies of International and National News*.

Van Dijk, T. A. (1999). *Söylemin Yapıları ve İktidarın Yapıları, Medya İktidar İdeoloji*. Mehmet Küçük, Ark.: Der. ve Çev.

Van Dijk, T. A. (2008). *Discourse and Power. Basingstoke, UK: Palgrave* Macmillan. doi:10.1007/978-1-137-07299-3

Van Riel, C. (1997). Research in corporate communication: An overview of an emerging field. *Management Communication Quarterly*, *11*(2), 288–309. doi:10.1177/0893318997112005

Vella, K., & Melewar, T. (2008). Corporate Identity. In T. C. Melewar (Ed.), *Facets of Corporate Identity, Communication, and Reputation*. Oxon, UK: Routledge.

Viera, A. J., & Garrett, J. M. (2005). Understanding interobserver agreement: The kappa statistic. *Family Medicine*, *37*(5), 360–363. PMID:15883903

Viswanathan, A. (2013, February 20). Reasonable Restrictions. *The Hindu*. Retrieved from: http://www.thehindu.com/opinion/lead/an-unreasonable-restriction/article4432360.ece

Walizer, M. H., & Wienir, P. L. (1978). *Research Methods and Analysis: Searching for Relationships*. New York, NY: Harper & Row.

Wasko, J. (2005). Studying the Political Economy of Media and Information. *Comunicação e Sociedade*, *7*(0), 25–48. doi:10.17231/comsoc.7(2005).1208

WeChat. (2016a). Terms of Service. Retrieved from http://www.wechat.com/en/service_terms.html

WeChat. (2016b). Privacy policy. Retrieved from http://www.wechat.com/en/privacy_policy.html

WeChat. (2016c). Terms of Use and Privacy Policy (Chinese language versions). Retrieved from http://weixin.qq.com/cgi-bin/readtemplate?uin=&stype=&promote=&fr=wechat.com&lang=zh_CN&ADTAG=&check=false&nav=faq&t=weixin_agreement&s=default

Weeks, J. (1998). *Farklılığın Değeri. Kimlik: Topluluk/Kimlik/Farklılık içinde. Der: J. Rutherford, Çev: D.Sağlamer*. İstanbul, Turkey: Sarmal Yayınları.

Wertime, K., & Fenwick, I. (2008). *Digimarketing: The Essential Guide to New Media and Digital Marketing. Hoboken, NJ:* John Wiley & Sons.

WhatsApp working with Reliance Jio to curb fake news menace. (2016, September 23). Retrieved from: https://www.thehindu.com/business/Industry/whatsapp-working-with-reliance-jio-to-curb-fake-news-menace/article25045384.ece

Wheeler, A. (2009). *Designing Brand Identity* (Vol. 3). Hoboken, NJ: John Wiley & Sons.

Whitman, R. (2015, Aug. 26). Study Finds Social Media Claiming Larger Share of Ad Budgets. *MediaPost.com*. Retrieved from http://www.mediapost.com/publications/article/257029/study-finds-social-media-claiming-larger-share-of.html

Wicke, J. (1988). *Advertising Fictions: Literature, Advertisement and Social Reading*. New York, NY: Columbia University Press.

Williams, D. L., Crittenden, V. L., Keo, T., & McCarty, P. (2012). The use of social media: An exploratory study of usage among digital natives. *Journal of Public Affairs*, *12*(2), 127–136. doi:10.1002/pa.1414

Williamson, J. (1978). *Decoding Advertisements: Ideology and Meaning in Advertising*. London, UK: Marion Boyars.

Williams, R. (1993). *Kültür* (S. Aydın, Trans.). Ankara, Turkey: İmge Kitabevi Yayınları.

Wyche, S. P., Schoenebeck, S. Y., & Forte, A. (2013, February). Facebook is a luxury: An exploratory study of social media use in rural Kenya. In *Proceedings of the 2013 conference on Computer supported cooperative work* (pp. 33-44). New York, NY: ACM. 10.1145/2441776.2441783

Yaban, N. T. (2012). Sanat Ve Görsel İletişimin Buluşma Noktasi: Ekslibris. *Batman Üniversitesi Yaşam Bilimleri Dergisi*, *1*(1), 973–984.

Yang, X., & Luo, J. (2017). Tracking illicit drug dealing and abuse on Instagram using multimodal analysis. *ACM Transactions on Intelligent Systems and Technology*, *8*(4), 58. doi:10.1145/3011871

Yanık, A. (2017). Popülizm (V): Post-Truth. *Birikim Dergi*. Retrieved from http://www.birikimdergisi.com/haftalik/8463/populizm-v-post-truth#.XJFJdCgzZPY

Yanık, A. (2017). Popülizm (V): Post-Truth. Retrieved from http://www.birikimdergisi.com/haftalik/8463/populizm-v-post-truth#.XM4S6I4zZPY

Yavuz, H. (2012). *Budalalığın keşfi*. İstanbul, Turkey: Timaş.

Yegen, C., & Yanık, H. (2015). Yeni Medya İle Değişen Tüketim Anlayışı: Kadınların İnstagram Üzerinden Alış-Veriş Pratiği. In T. Kara & E. Ozgen (Eds.), *Ağdaki Şüphe Bir Sosyal Medya Eleştirisi*. İstanbul, Turkey: Beta Yayınları.

Yeni Akit. (2018). *ABD'lilerin uykuları kaçtı! 'Erdoğan'ı Deccal olarak görüyorlar!'*. Retrieved from https://www.yeniakit.com.tr/haber/abdlilerin-uykulari-kacti-erdogani-deccal-olarak-goruyorlar-465167.html

Yeniçıktı, N. T. (2016). Halkla ilişkiler aracı olarak Instagram: Sosyal medya kullanan 50 şirket üzerine bir araştırma. *Selçuk İletişim*, *9*(2), 92–115.

Yıldırım, A. & Şimşek, H. (2013). *Sosyal Bilimlerde Nitel Araştırma Yöntemleri,* Ankara: Seçkin Publications. Retrieved from https://brandfinance.com/knowledge-centre/reports/brand-finance-turkey-100-2016/

Yıldız, T. (2014). Diyaloji Diyalektiğe Karşı. *Psikoloji Çalışmaları Dergisi*, *34*(1), 79–85.

Yilmaz, R. (2017). *Narrative Advertising Models and Conceptualization in the Digital Age*. Hershey, PA: IGI Global. doi:10.4018/978-1-5225-2373-4

Yoo, Y., & Alavi, M. (2001). Media and group cohesion: Relative influences on social presence, task participation, and group consensus. *Management Information Systems Quarterly*, *25*(3), 371–390. doi:10.2307/3250922

Yüksel, M. (2003). Mahremiyet hakkı ve sosyo-tarihsel gelişimi. Ankara University Faculty of Political Sciences Journal, 58(1), 182-213.

Yüksel, M. (2003). Mahremiyet Hakkı ve Sosyo-tarihsel Gelişimi. *Ankara Üniversitesi SBF Dergisi*, *58*(1), 181–213.

Yüksel, M. (2003). Modernleşme ve mahremiyet. *Journal of Culture and Communication*, *1*, 75–108.

Yüksel, M. (2009). Mahremiyet Hakkına ve Bireysel Özgürlüklere Felsefi Yaklaşımlar. *Ankara Üniversitesi SBF Dergisi*, *64*(01), 275–298. doi:10.1501/SBFder_0000002130

Yurdakul, S. (2006). *Bir Ürün ve Toplumsal Bütünleşme Aracı Olarak Modern Kent Karnavalları ve Türkiye'deki Örneklerde Görülen Uygulama Farklılıkları*. İstanbul, Turkey: Marmara Üniversitesi Sosyal Bilimler Enstitüsü, İletişim Bilimleri Anabilim Dalı, Yüksek Lisans Tezi.

Yurdigül, Y. & Yurdigül, A. (2014).Tv Haberlerinin Anlatı yapısının Oluşturulması Sürecinde Özel Efekt Teknolojileri: NTV ve CNN Türk Ana Haber Bültenleri üzerinden Bir İnceleme. İstanbul Üniversitesi İletişim Fakültesi Dergisi. 46, 123-148.

Yurdigül, Y., & Zinderen, İ. E. (2013). *Sinema ve Televizyonda Özel Efekt*. İstanbul, Turkey: Doğu Kitapevi.

Zarzalejos, J. A. (2017). Communication, journalism and fast-checking. *Uno Magazine*, *27*, 11–13.

Zengin, A. M., Zengin, G., & Altunbaş, H. (2015). Sosyal medya ve değişen mahremiyet "facebook mahremiyeti. Gümüşhane Üniversitesi İletişim Fakültesi Elektronik Dergisi, 3(2).

Zhumagaziyeva, A., & Koç, F. (2010). Kültürel etkileşimde kadınların giysi alışkanlıklarındaki değişimin mahremiyet açısından incelenmesi (Kazakistan ve Türkiye Örneği). *E-Journal of New World Sciences Academy*, *5*(2), 113–131.

Žižek, S. (2005). *Yamuk Bakmak. Tuncay Birkan (Trans.)*. İstanbul, Turkey: Metis Publications.

Related References

To continue our tradition of advancing information science and technology research, we have compiled a list of recommended IGI Global readings. These references will provide additional information and guidance to further enrich your knowledge and assist you with your own research and future publications.

Adesina, K., Ganiu, O., & R., O. S. (2018). Television as Vehicle for Community Development: A Study of Lotunlotun Programme on (B.C.O.S.) Television, Nigeria. In A. Salawu, & T. Owolabi (Eds.), *Exploring Journalism Practice and Perception in Developing Countries* (pp. 60-84). Hershey, PA: IGI Global. doi:10.4018/978-1-5225-3376-4.ch004

Adigun, G. O., Odunola, O. A., & Sobalaje, A. J. (2016). Role of Social Networking for Information Seeking in a Digital Library Environment. In A. Tella (Ed.), *Information Seeking Behavior and Challenges in Digital Libraries* (pp. 272–290). Hershey, PA: IGI Global. doi:10.4018/978-1-5225-0296-8.ch013

Ahmad, M. B., Pride, C., & Corsy, A. K. (2016). Free Speech, Press Freedom, and Democracy in Ghana: A Conceptual and Historical Overview. In L. Mukhongo & J. Macharia (Eds.), *Political Influence of the Media in Developing Countries* (pp. 59–73). Hershey, PA: IGI Global. doi:10.4018/978-1-4666-9613-6.ch005

Ahmad, R. H., & Pathan, A. K. (2017). A Study on M2M (Machine to Machine) System and Communication: Its Security, Threats, and Intrusion Detection System. In M. Ferrag & A. Ahmim (Eds.), *Security Solutions and Applied Cryptography in Smart Grid Communications* (pp. 179–214). Hershey, PA: IGI Global. doi:10.4018/978-1-5225-1829-7.ch010

Akanni, T. M. (2018). In Search of Women-Supportive Media for Sustainable Development in Nigeria. In A. Salawu & T. Owolabi (Eds.), *Exploring Journalism Practice and Perception in Developing Countries* (pp. 126–149). Hershey, PA: IGI Global. doi:10.4018/978-1-5225-3376-4.ch007

Akçay, D. (2017). The Role of Social Media in Shaping Marketing Strategies in the Airline Industry. In V. Benson, R. Tuninga, & G. Saridakis (Eds.), *Analyzing the Strategic Role of Social Networking in Firm Growth and Productivity* (pp. 214–233). Hershey, PA: IGI Global. doi:10.4018/978-1-5225-0559-4.ch012

Al-Rabayah, W. A. (2017). Social Media as Social Customer Relationship Management Tool: Case of Jordan Medical Directory. In W. Al-Rabayah, R. Khasawneh, R. Abu-shamaa, & I. Alsmadi (Eds.), *Strategic Uses of Social Media for Improved Customer Retention* (pp. 108–123). Hershey, PA: IGI Global. doi:10.4018/978-1-5225-1686-6.ch006

Almjeld, J. (2017). Getting "Girly" Online: The Case for Gendering Online Spaces. In E. Monske & K. Blair (Eds.), *Handbook of Research on Writing and Composing in the Age of MOOCs* (pp. 87–105). Hershey, PA: IGI Global. doi:10.4018/978-1-5225-1718-4.ch006

Altaş, A. (2017). Space as a Character in Narrative Advertising: A Qualitative Research on Country Promotion Works. In R. Yılmaz (Ed.), *Narrative Advertising Models and Conceptualization in the Digital Age* (pp. 303–319). Hershey, PA: IGI Global. doi:10.4018/978-1-5225-2373-4.ch017

Altıparmak, B. (2017). The Structural Transformation of Space in Turkish Television Commercials as a Narrative Component. In R. Yılmaz (Ed.), *Narrative Advertising Models and Conceptualization in the Digital Age* (pp. 153–166). Hershey, PA: IGI Global. doi:10.4018/978-1-5225-2373-4.ch009

An, Y., & Harvey, K. E. (2016). Public Relations and Mobile: Becoming Dialogic. In X. Xu (Ed.), *Handbook of Research on Human Social Interaction in the Age of Mobile Devices* (pp. 284–311). Hershey, PA: IGI Global. doi:10.4018/978-1-5225-0469-6.ch013

Assay, B. E. (2018). Regulatory Compliance, Ethical Behaviour, and Sustainable Growth in Nigeria's Telecommunications Industry. In I. Oncioiu (Ed.), *Ethics and Decision-Making for Sustainable Business Practices* (pp. 90–108). Hershey, PA: IGI Global. doi:10.4018/978-1-5225-3773-1.ch006

Averweg, U. R., & Leaning, M. (2018). The Qualities and Potential of Social Media. In M. Khosrow-Pour, D.B.A. (Ed.), Encyclopedia of Information Science and Technology, Fourth Edition (pp. 7106-7115). Hershey, PA: IGI Global. doi:10.4018/978-1-5225-2255-3.ch617

Azemi, Y., & Ozuem, W. (2016). Online Service Failure and Recovery Strategy: The Mediating Role of Social Media. In W. Ozuem & G. Bowen (Eds.), *Competitive Social Media Marketing Strategies* (pp. 112–135). Hershey, PA: IGI Global. doi:10.4018/978-1-4666-9776-8.ch006

Baarda, R. (2017). Digital Democracy in Authoritarian Russia: Opportunity for Participation, or Site of Kremlin Control? In R. Luppicini & R. Baarda (Eds.), *Digital Media Integration for Participatory Democracy* (pp. 87–100). Hershey, PA: IGI Global. doi:10.4018/978-1-5225-2463-2.ch005

Bacallao-Pino, L. M. (2016). Radical Political Communication and Social Media: The Case of the Mexican #YoSoy132. In T. Deželan & I. Vobič (Eds.), *R)evolutionizing Political Communication through Social Media* (pp. 56–74). Hershey, PA: IGI Global. doi:10.4018/978-1-4666-9879-6.ch004

Baggio, B. G. (2016). Why We Would Rather Text than Talk: Personality, Identity, and Anonymity in Modern Virtual Environments. In B. Baggio (Ed.), *Analyzing Digital Discourse and Human Behavior in Modern Virtual Environments* (pp. 110–125). Hershey, PA: IGI Global. doi:10.4018/978-1-4666-9899-4.ch006

Başal, B. (2017). Actor Effect: A Study on Historical Figures Who Have Shaped the Advertising Narration. In R. Yılmaz (Ed.), *Narrative Advertising Models and Conceptualization in the Digital Age* (pp. 34–60). Hershey, PA: IGI Global. doi:10.4018/978-1-5225-2373-4.ch003

Behjati, M., & Cosmas, J. (2017). Self-Organizing Network Solutions: A Principal Step Towards Real 4G and Beyond. In D. Singh (Ed.), *Routing Protocols and Architectural Solutions for Optimal Wireless Networks and Security* (pp. 241–253). Hershey, PA: IGI Global. doi:10.4018/978-1-5225-2342-0.ch011

Bekafigo, M., & Pingley, A. C. (2017). Do Campaigns "Go Negative" on Twitter? In Y. Ibrahim (Ed.), *Politics, Protest, and Empowerment in Digital Spaces* (pp. 178–191). Hershey, PA: IGI Global. doi:10.4018/978-1-5225-1862-4.ch011

Bender, S., & Dickenson, P. (2016). Utilizing Social Media to Engage Students in Online Learning: Building Relationships Outside of the Learning Management System. In P. Dickenson & J. Jaurez (Eds.), *Increasing Productivity and Efficiency in Online Teaching* (pp. 84–105). Hershey, PA: IGI Global. doi:10.4018/978-1-5225-0347-7.ch005

Bermingham, N., & Prendergast, M. (2016). Bespoke Mobile Application Development: Facilitating Transition of Foundation Students to Higher Education. In L. Briz-Ponce, J. Juanes-Méndez, & F. García-Peñalvo (Eds.), *Handbook of Research on Mobile Devices and Applications in Higher Education Settings* (pp. 222–249). Hershey, PA: IGI Global. doi:10.4018/978-1-5225-0256-2.ch010

Bishop, J. (2017). Developing and Validating the "This Is Why We Can't Have Nice Things Scale": Optimising Political Online Communities for Internet Trolling. In Y. Ibrahim (Ed.), *Politics, Protest, and Empowerment in Digital Spaces* (pp. 153–177). Hershey, PA: IGI Global. doi:10.4018/978-1-5225-1862-4.ch010

Bolat, N. (2017). The Functions of the Narrator in Digital Advertising. In R. Yılmaz (Ed.), *Narrative Advertising Models and Conceptualization in the Digital Age* (pp. 184–201). Hershey, PA: IGI Global. doi:10.4018/978-1-5225-2373-4.ch011

Bowen, G., & Bowen, D. (2016). Social Media: Strategic Decision Making Tool. In W. Ozuem & G. Bowen (Eds.), *Competitive Social Media Marketing Strategies* (pp. 94–111). Hershey, PA: IGI Global. doi:10.4018/978-1-4666-9776-8.ch005

Brown, M. A. Sr. (2017). SNIP: High Touch Approach to Communication. In *Solutions for High-Touch Communications in a High-Tech World* (pp. 71–88). Hershey, PA: IGI Global. doi:10.4018/978-1-5225-1897-6.ch004

Brown, M. A. Sr. (2017). Comparing FTF and Online Communication Knowledge. In *Solutions for High-Touch Communications in a High-Tech World* (pp. 103–113). Hershey, PA: IGI Global. doi:10.4018/978-1-5225-1897-6.ch006

Brown, M. A. Sr. (2017). Where Do We Go from Here? In *Solutions for High-Touch Communications in a High-Tech World* (pp. 137–159). Hershey, PA: IGI Global. doi:10.4018/978-1-5225-1897-6.ch008

Brown, M. A. Sr. (2017). Bridging the Communication Gap. In *Solutions for High-Touch Communications in a High-Tech World* (pp. 1–22). Hershey, PA: IGI Global. doi:10.4018/978-1-5225-1897-6.ch001

Brown, M. A. Sr. (2017). Key Strategies for Communication. In *Solutions for High-Touch Communications in a High-Tech World* (pp. 179–202). Hershey, PA: IGI Global. doi:10.4018/978-1-5225-1897-6.ch010

Bryant, K. N. (2017). WordUp!: Student Responses to Social Media in the Technical Writing Classroom. In K. Bryant (Ed.), *Engaging 21st Century Writers with Social Media* (pp. 231–245). Hershey, PA: IGI Global. doi:10.4018/978-1-5225-0562-4.ch014

Buck, E. H. (2017). Slacktivism, Supervision, and #Selfies: Illuminating Social Media Composition through Reception Theory. In K. Bryant (Ed.), *Engaging 21st Century Writers with Social Media* (pp. 163–178). Hershey, PA: IGI Global. doi:10.4018/978-1-5225-0562-4.ch010

Bucur, B. (2016). Sociological School of Bucharest's Publications and the Romanian Political Propaganda in the Interwar Period. In A. Fox (Ed.), *Global Perspectives on Media Events in Contemporary Society* (pp. 106–120). Hershey, PA: IGI Global. doi:10.4018/978-1-4666-9967-0.ch008

Bull, R., & Pianosi, M. (2017). Social Media, Participation, and Citizenship: New Strategic Directions. In V. Benson, R. Tuninga, & G. Saridakis (Eds.), *Analyzing the Strategic Role of Social Networking in Firm Growth and Productivity* (pp. 76–94). Hershey, PA: IGI Global. doi:10.4018/978-1-5225-0559-4.ch005

Camillo, A. A., & Camillo, I. C. (2016). The Ethics of Strategic Managerial Communication in the Global Context. In A. Normore, L. Long, & M. Javidi (Eds.), *Handbook of Research on Effective Communication, Leadership, and Conflict Resolution* (pp. 566–590). Hershey, PA: IGI Global. doi:10.4018/978-1-4666-9970-0.ch030

Cassard, A., & Sloboda, B. W. (2016). Faculty Perception of Virtual 3-D Learning Environment to Assess Student Learning. In D. Choi, A. Dailey-Hebert, & J. Simmons Estes (Eds.), *Emerging Tools and Applications of Virtual Reality in Education* (pp. 48–74). Hershey, PA: IGI Global. doi:10.4018/978-1-4666-9837-6.ch003

Castellano, S., & Khelladi, I. (2017). Play It Like Beckham!: The Influence of Social Networks on E-Reputation – The Case of Sportspeople and Their Online Fan Base. In A. Mesquita (Ed.), *Research Paradigms and Contemporary Perspectives on Human-Technology Interaction* (pp. 43–61). Hershey, PA: IGI Global. doi:10.4018/978-1-5225-1868-6.ch003

Castellet, A. (2016). What If Devices Take Command: Content Innovation Perspectives for Smart Wearables in the Mobile Ecosystem. *International Journal of Handheld Computing Research*, *7*(2), 16–33. doi:10.4018/IJHCR.2016040102

Chugh, R., & Joshi, M. (2017). Challenges of Knowledge Management amidst Rapidly Evolving Tools of Social Media. In R. Chugh (Ed.), *Harnessing Social Media as a Knowledge Management Tool* (pp. 299–314). Hershey, PA: IGI Global. doi:10.4018/978-1-5225-0495-5.ch014

Cockburn, T., & Smith, P. A. (2016). Leadership in the Digital Age: Rhythms and the Beat of Change. In A. Normore, L. Long, & M. Javidi (Eds.), *Handbook of Research on Effective Communication, Leadership, and Conflict Resolution* (pp. 1–20). Hershey, PA: IGI Global. doi:10.4018/978-1-4666-9970-0.ch001

Cole, A. W., & Salek, T. A. (2017). Adopting a Parasocial Connection to Overcome Professional Kakoethos in Online Health Information. In M. Folk & S. Apostel (Eds.), *Establishing and Evaluating Digital Ethos and Online Credibility* (pp. 104–120). Hershey, PA: IGI Global. doi:10.4018/978-1-5225-1072-7.ch006

Cossiavelou, V. (2017). ACTA as Media Gatekeeping Factor: The EU Role as Global Negotiator. *International Journal of Interdisciplinary Telecommunications and Networking*, *9*(1), 26–37. doi:10.4018/IJITN.2017010103

Costanza, F. (2017). Social Media Marketing and Value Co-Creation: A System Dynamics Approach. In S. Rozenes & Y. Cohen (Eds.), *Handbook of Research on Strategic Alliances and Value Co-Creation in the Service Industry* (pp. 205–230). Hershey, PA: IGI Global. doi:10.4018/978-1-5225-2084-9.ch011

Cross, D. E. (2016). Globalization and Media's Impact on Cross Cultural Communication: Managing Organizational Change. In A. Normore, L. Long, & M. Javidi (Eds.), *Handbook of Research on Effective Communication, Leadership, and Conflict Resolution* (pp. 21–41). Hershey, PA: IGI Global. doi:10.4018/978-1-4666-9970-0.ch002

Damásio, M. J., Henriques, S., Teixeira-Botelho, I., & Dias, P. (2016). Mobile Media and Social Interaction: Mobile Services and Content as Drivers of Social Interaction. In J. Aguado, C. Feijóo, & I. Martínez (Eds.), *Emerging Perspectives on the Mobile Content Evolution* (pp. 357–379). Hershey, PA: IGI Global. doi:10.4018/978-1-4666-8838-4.ch018

Davis, A., & Foley, L. (2016). Digital Storytelling. In B. Guzzetti & M. Lesley (Eds.), *Handbook of Research on the Societal Impact of Digital Media* (pp. 317–342). Hershey, PA: IGI Global. doi:10.4018/978-1-4666-8310-5.ch013

Davis, S., Palmer, L., & Etienne, J. (2016). The Geography of Digital Literacy: Mapping Communications Technology Training Programs in Austin, Texas. In B. Passarelli, J. Straubhaar, & A. Cuevas-Cerveró (Eds.), *Handbook of Research on Comparative Approaches to the Digital Age Revolution in Europe and the Americas* (pp. 371–384). Hershey, PA: IGI Global. doi:10.4018/978-1-4666-8740-0.ch022

Delello, J. A., & McWhorter, R. R. (2016). New Visual Literacies and Competencies for Education and the Workplace. In B. Guzzetti & M. Lesley (Eds.), *Handbook of Research on the Societal Impact of Digital Media* (pp. 127–162). Hershey, PA: IGI Global. doi:10.4018/978-1-4666-8310-5.ch006

Di Virgilio, F., & Antonelli, G. (2018). Consumer Behavior, Trust, and Electronic Word-of-Mouth Communication: Developing an Online Purchase Intention Model. In F. Di Virgilio (Ed.), *Social Media for Knowledge Management Applications in Modern Organizations* (pp. 58–80). Hershey, PA: IGI Global. doi:10.4018/978-1-5225-2897-5.ch003

Dixit, S. K. (2016). eWOM Marketing in Hospitality Industry. In A. Singh, & P. Duhan (Eds.), Managing Public Relations and Brand Image through Social Media (pp. 266-280). Hershey, PA: IGI Global. doi:10.4018/978-1-5225-0332-3.ch014

Duhan, P., & Singh, A. (2016). Facebook Experience Is Different: An Empirical Study in Indian Context. In S. Rathore & A. Panwar (Eds.), *Capturing, Analyzing, and Managing Word-of-Mouth in the Digital Marketplace* (pp. 188–212). Hershey, PA: IGI Global. doi:10.4018/978-1-4666-9449-1.ch011

Dunne, D. J. (2016). The Scholar's Ludo-Narrative Game and Multimodal Graphic Novel: A Comparison of Fringe Scholarship. In A. Connor & S. Marks (Eds.), *Creative Technologies for Multidisciplinary Applications* (pp. 182–207). Hershey, PA: IGI Global. doi:10.4018/978-1-5225-0016-2.ch008

DuQuette, J. L. (2017). Lessons from Cypris Chat: Revisiting Virtual Communities as Communities. In G. Panconesi & M. Guida (Eds.), *Handbook of Research on Collaborative Teaching Practice in Virtual Learning Environments* (pp. 299–316). Hershey, PA: IGI Global. doi:10.4018/978-1-5225-2426-7.ch016

Ekhlassi, A., Niknejhad Moghadam, M., & Adibi, A. (2018). The Concept of Social Media: The Functional Building Blocks. In *Building Brand Identity in the Age of Social Media: Emerging Research and Opportunities* (pp. 29–60). Hershey, PA: IGI Global. doi:10.4018/978-1-5225-5143-0.ch002

Ekhlassi, A., Niknejhad Moghadam, M., & Adibi, A. (2018). Social Media Branding Strategy: Social Media Marketing Approach. In *Building Brand Identity in the Age of Social Media: Emerging Research and Opportunities* (pp. 94–117). Hershey, PA: IGI Global. doi:10.4018/978-1-5225-5143-0.ch004

Ekhlassi, A., Niknejhad Moghadam, M., & Adibi, A. (2018). The Impact of Social Media on Brand Loyalty: Achieving "E-Trust" Through Engagement. In *Building Brand Identity in the Age of Social Media: Emerging Research and Opportunities* (pp. 155–168). Hershey, PA: IGI Global. doi:10.4018/978-1-5225-5143-0.ch007

Elegbe, O. (2017). An Assessment of Media Contribution to Behaviour Change and HIV Prevention in Nigeria. In O. Nelson, B. Ojebuyi, & A. Salawu (Eds.), *Impacts of the Media on African Socio-Economic Development* (pp. 261–280). Hershey, PA: IGI Global. doi:10.4018/978-1-5225-1859-4.ch017

Endong, F. P. (2018). Hashtag Activism and the Transnationalization of Nigerian-Born Movements Against Terrorism: A Critical Appraisal of the #BringBackOurGirls Campaign. In F. Endong (Ed.), *Exploring the Role of Social Media in Transnational Advocacy* (pp. 36–54). Hershey, PA: IGI Global. doi:10.4018/978-1-5225-2854-8.ch003

Erragcha, N. (2017). Using Social Media Tools in Marketing: Opportunities and Challenges. In M. Brown Sr., (Ed.), *Social Media Performance Evaluation and Success Measurements* (pp. 106–129). Hershey, PA: IGI Global. doi:10.4018/978-1-5225-1963-8.ch006

Ezeh, N. C. (2018). Media Campaign on Exclusive Breastfeeding: Awareness, Perception, and Acceptability Among Mothers in Anambra State, Nigeria. In A. Salawu & T. Owolabi (Eds.), *Exploring Journalism Practice and Perception in Developing Countries* (pp. 172–193). Hershey, PA: IGI Global. doi:10.4018/978-1-5225-3376-4.ch009

Fawole, O. A., & Osho, O. A. (2017). Influence of Social Media on Dating Relationships of Emerging Adults in Nigerian Universities: Social Media and Dating in Nigeria. In M. Wright (Ed.), *Identity, Sexuality, and Relationships among Emerging Adults in the Digital Age* (pp. 168–177). Hershey, PA: IGI Global. doi:10.4018/978-1-5225-1856-3.ch011

Fayoyin, A. (2017). Electoral Polling and Reporting in Africa: Professional and Policy Implications for Media Practice and Political Communication in a Digital Age. In N. Mhiripiri & T. Chari (Eds.), *Media Law, Ethics, and Policy in the Digital Age* (pp. 164–181). Hershey, PA: IGI Global. doi:10.4018/978-1-5225-2095-5.ch009

Fayoyin, A. (2018). Rethinking Media Engagement Strategies for Social Change in Africa: Context, Approaches, and Implications for Development Communication. In A. Salawu & T. Owolabi (Eds.), *Exploring Journalism Practice and Perception in Developing Countries* (pp. 257–280). Hershey, PA: IGI Global. doi:10.4018/978-1-5225-3376-4.ch013

Fechine, Y., & Rêgo, S. C. (2018). Transmedia Television Journalism in Brazil: Jornal da Record News as Reference. In R. Gambarato & G. Alzamora (Eds.), *Exploring Transmedia Journalism in the Digital Age* (pp. 253–265). Hershey, PA: IGI Global. doi:10.4018/978-1-5225-3781-6.ch015

Feng, J., & Lo, K. (2016). Video Broadcasting Protocol for Streaming Applications with Cooperative Clients. In D. Kanellopoulos (Ed.), *Emerging Research on Networked Multimedia Communication Systems* (pp. 205–229). Hershey, PA: IGI Global. doi:10.4018/978-1-4666-8850-6.ch006

Fiore, C. (2017). The Blogging Method: Improving Traditional Student Writing Practices. In K. Bryant (Ed.), *Engaging 21st Century Writers with Social Media* (pp. 179–198). Hershey, PA: IGI Global. doi:10.4018/978-1-5225-0562-4.ch011

Fleming, J., & Kajimoto, M. (2016). The Freedom of Critical Thinking: Examining Efforts to Teach American News Literacy Principles in Hong Kong, Vietnam, and Malaysia. In M. Yildiz & J. Keengwe (Eds.), *Handbook of Research on Media Literacy in the Digital Age* (pp. 208–235). Hershey, PA: IGI Global. doi:10.4018/978-1-4666-9667-9.ch010

Gambarato, R. R., Alzamora, G. C., & Tárcia, L. P. (2018). 2016 Rio Summer Olympics and the Transmedia Journalism of Planned Events. In R. Gambarato & G. Alzamora (Eds.), *Exploring Transmedia Journalism in the Digital Age* (pp. 126–146). Hershey, PA: IGI Global. doi:10.4018/978-1-5225-3781-6.ch008

Ganguin, S., Gemkow, J., & Haubold, R. (2017). Information Overload as a Challenge and Changing Point for Educational Media Literacies. In R. Marques & J. Batista (Eds.), *Information and Communication Overload in the Digital Age* (pp. 302–328). Hershey, PA: IGI Global. doi:10.4018/978-1-5225-2061-0.ch013

Gao, Y. (2016). Reviewing Gratification Effects in Mobile Gaming. In X. Xu (Ed.), *Handbook of Research on Human Social Interaction in the Age of Mobile Devices* (pp. 406–428). Hershey, PA: IGI Global. doi:10.4018/978-1-5225-0469-6.ch017

Gardner, G. C. (2017). The Lived Experience of Smartphone Use in a Unit of the United States Army. In F. Topor (Ed.), *Handbook of Research on Individualism and Identity in the Globalized Digital Age* (pp. 88–117). Hershey, PA: IGI Global. doi:10.4018/978-1-5225-0522-8.ch005

Giessen, H. W. (2016). The Medium, the Content, and the Performance: An Overview on Media-Based Learning. In B. Khan (Ed.), *Revolutionizing Modern Education through Meaningful E-Learning Implementation* (pp. 42–55). Hershey, PA: IGI Global. doi:10.4018/978-1-5225-0466-5.ch003

Giltenane, J. (2016). Investigating the Intention to Use Social Media Tools Within Virtual Project Teams. In G. Silvius (Ed.), *Strategic Integration of Social Media into Project Management Practice* (pp. 83–105). Hershey, PA: IGI Global. doi:10.4018/978-1-4666-9867-3.ch006

Golightly, D., & Houghton, R. J. (2018). Social Media as a Tool to Understand Behaviour on the Railways. In S. Kohli, A. Kumar, J. Easton, & C. Roberts (Eds.), *Innovative Applications of Big Data in the Railway Industry* (pp. 224–239). Hershey, PA: IGI Global. doi:10.4018/978-1-5225-3176-0.ch010

Goovaerts, M., Nieuwenhuysen, P., & Dhamdhere, S. N. (2016). VLIR-UOS Workshop 'E-Info Discovery and Management for Institutes in the South': Presentations and Conclusions, Antwerp, 8-19 December, 2014. In E. de Smet, & S. Dhamdhere (Eds.), E-Discovery Tools and Applications in Modern Libraries (pp. 1-40). Hershey, PA: IGI Global. doi:10.4018/978-1-5225-0474-0.ch001

Grützmann, A., Carvalho de Castro, C., Meireles, A. A., & Rodrigues, R. C. (2016). Organizational Architecture and Online Social Networks: Insights from Innovative Brazilian Companies. In G. Jamil, J. Poças Rascão, F. Ribeiro, & A. Malheiro da Silva (Eds.), *Handbook of Research on Information Architecture and Management in Modern Organizations* (pp. 508–524). Hershey, PA: IGI Global. doi:10.4018/978-1-4666-8637-3.ch023

Gundogan, M. B. (2017). In Search for a "Good Fit" Between Augmented Reality and Mobile Learning Ecosystem. In G. Kurubacak & H. Altinpulluk (Eds.), *Mobile Technologies and Augmented Reality in Open Education* (pp. 135–153). Hershey, PA: IGI Global. doi:10.4018/978-1-5225-2110-5.ch007

Gupta, H. (2018). Impact of Digital Communication on Consumer Behaviour Processes in Luxury Branding Segment: A Study of Apparel Industry. In S. Dasgupta, S. Biswal, & M. Ramesh (Eds.), *Holistic Approaches to Brand Culture and Communication Across Industries* (pp. 132–157). Hershey, PA: IGI Global. doi:10.4018/978-1-5225-3150-0.ch008

Hai-Jew, S. (2017). Creating "(Social) Network Art" with NodeXL. In S. Hai-Jew (Ed.), *Social Media Data Extraction and Content Analysis* (pp. 342–393). Hershey, PA: IGI Global. doi:10.4018/978-1-5225-0648-5.ch011

Hai-Jew, S. (2017). Employing the Sentiment Analysis Tool in NVivo 11 Plus on Social Media Data: Eight Initial Case Types. In N. Rao (Ed.), *Social Media Listening and Monitoring for Business Applications* (pp. 175–244). Hershey, PA: IGI Global. doi:10.4018/978-1-5225-0846-5.ch010

Hai-Jew, S. (2017). Conducting Sentiment Analysis and Post-Sentiment Data Exploration through Automated Means. In S. Hai-Jew (Ed.), *Social Media Data Extraction and Content Analysis* (pp. 202–240). Hershey, PA: IGI Global. doi:10.4018/978-1-5225-0648-5.ch008

Hai-Jew, S. (2017). Applied Analytical "Distant Reading" using NVivo 11 Plus. In S. Hai-Jew (Ed.), *Social Media Data Extraction and Content Analysis* (pp. 159–201). Hershey, PA: IGI Global. doi:10.4018/978-1-5225-0648-5.ch007

Hai-Jew, S. (2017). Flickering Emotions: Feeling-Based Associations from Related Tags Networks on Flickr. In S. Hai-Jew (Ed.), *Social Media Data Extraction and Content Analysis* (pp. 296–341). Hershey, PA: IGI Global. doi:10.4018/978-1-5225-0648-5.ch010

Hai-Jew, S. (2017). Manually Profiling Egos and Entities across Social Media Platforms: Evaluating Shared Messaging and Contents, User Networks, and Metadata. In V. Benson, R. Tuninga, & G. Saridakis (Eds.), *Analyzing the Strategic Role of Social Networking in Firm Growth and Productivity* (pp. 352–405). Hershey, PA: IGI Global. doi:10.4018/978-1-5225-0559-4.ch019

Hai-Jew, S. (2017). Exploring "User," "Video," and (Pseudo) Multi-Mode Networks on YouTube with NodeXL. In S. Hai-Jew (Ed.), *Social Media Data Extraction and Content Analysis* (pp. 242–295). Hershey, PA: IGI Global. doi:10.4018/978-1-5225-0648-5.ch009

Hai-Jew, S. (2018). Exploring "Mass Surveillance" Through Computational Linguistic Analysis of Five Text Corpora: Academic, Mainstream Journalism, Microblogging Hashtag Conversation, Wikipedia Articles, and Leaked Government Data. In *Techniques for Coding Imagery and Multimedia: Emerging Research and Opportunities* (pp. 212–286). Hershey, PA: IGI Global. doi:10.4018/978-1-5225-2679-7.ch004

Hai-Jew, S. (2018). Exploring Identity-Based Humor in a #Selfies #Humor Image Set From Instagram. In *Techniques for Coding Imagery and Multimedia: Emerging Research and Opportunities* (pp. 1–90). Hershey, PA: IGI Global. doi:10.4018/978-1-5225-2679-7.ch001

Hai-Jew, S. (2018). See Ya!: Exploring American Renunciation of Citizenship Through Targeted and Sparse Social Media Data Sets and a Custom Spatial-Based Linguistic Analysis Dictionary. In *Techniques for Coding Imagery and Multimedia: Emerging Research and Opportunities* (pp. 287–393). Hershey, PA: IGI Global. doi:10.4018/978-1-5225-2679-7.ch005

Han, H. S., Zhang, J., Peikazadi, N., Shi, G., Hung, A., Doan, C. P., & Filippelli, S. (2016). An Entertaining Game-Like Learning Environment in a Virtual World for Education. In S. D'Agustino (Ed.), *Creating Teacher Immediacy in Online Learning Environments* (pp. 290–306). Hershey, PA: IGI Global. doi:10.4018/978-1-4666-9995-3.ch015

Harrin, E. (2016). Barriers to Social Media Adoption on Projects. In G. Silvius (Ed.), *Strategic Integration of Social Media into Project Management Practice* (pp. 106–124). Hershey, PA: IGI Global. doi:10.4018/978-1-4666-9867-3.ch007

Harvey, K. E. (2016). Local News and Mobile: Major Tipping Points. In X. Xu (Ed.), *Handbook of Research on Human Social Interaction in the Age of Mobile Devices* (pp. 171–199). Hershey, PA: IGI Global. doi:10.4018/978-1-5225-0469-6.ch009

Harvey, K. E., & An, Y. (2016). Marketing and Mobile: Increasing Integration. In X. Xu (Ed.), *Handbook of Research on Human Social Interaction in the Age of Mobile Devices* (pp. 220–247). Hershey, PA: IGI Global. doi:10.4018/978-1-5225-0469-6.ch011

Harvey, K. E., Auter, P. J., & Stevens, S. (2016). Educators and Mobile: Challenges and Trends. In X. Xu (Ed.), *Handbook of Research on Human Social Interaction in the Age of Mobile Devices* (pp. 61–95). Hershey, PA: IGI Global. doi:10.4018/978-1-5225-0469-6.ch004

Hasan, H., & Linger, H. (2017). Connected Living for Positive Ageing. In S. Gordon (Ed.), *Online Communities as Agents of Change and Social Movements* (pp. 203–223). Hershey, PA: IGI Global. doi:10.4018/978-1-5225-2495-3.ch008

Hashim, K., Al-Sharqi, L., & Kutbi, I. (2016). Perceptions of Social Media Impact on Social Behavior of Students: A Comparison between Students and Faculty. *International Journal of Virtual Communities and Social Networking*, *8*(2), 1–11. doi:10.4018/IJVCSN.2016040101

Henriques, S., & Damasio, M. J. (2016). The Value of Mobile Communication for Social Belonging: Mobile Apps and the Impact on Social Interaction. *International Journal of Handheld Computing Research, 7*(2), 44–58. doi:10.4018/IJHCR.2016040104

Hersey, L. N. (2017). CHOICES: Measuring Return on Investment in a Nonprofit Organization. In M. Brown Sr., (Ed.), *Social Media Performance Evaluation and Success Measurements* (pp. 157–179). Hershey, PA: IGI Global. doi:10.4018/978-1-5225-1963-8.ch008

Heuva, W. E. (2017). Deferring Citizens' "Right to Know" in an Information Age: The Information Deficit in Namibia. In N. Mhiripiri & T. Chari (Eds.), *Media Law, Ethics, and Policy in the Digital Age* (pp. 245–267). Hershey, PA: IGI Global. doi:10.4018/978-1-5225-2095-5.ch014

Hopwood, M., & McLean, H. (2017). Social Media in Crisis Communication: The Lance Armstrong Saga. In V. Benson, R. Tuninga, & G. Saridakis (Eds.), *Analyzing the Strategic Role of Social Networking in Firm Growth and Productivity* (pp. 45–58). Hershey, PA: IGI Global. doi:10.4018/978-1-5225-0559-4.ch003

Hotur, S. K. (2018). Indian Approaches to E-Diplomacy: An Overview. In S. Bute (Ed.), *Media Diplomacy and Its Evolving Role in the Current Geopolitical Climate* (pp. 27–35). Hershey, PA: IGI Global. doi:10.4018/978-1-5225-3859-2.ch002

Ibadildin, N., & Harvey, K. E. (2016). Business and Mobile: Rapid Restructure Required. In X. Xu (Ed.), *Handbook of Research on Human Social Interaction in the Age of Mobile Devices* (pp. 312–350). Hershey, PA: IGI Global. doi:10.4018/978-1-5225-0469-6.ch014

Iwasaki, Y. (2017). Youth Engagement in the Era of New Media. In M. Adria & Y. Mao (Eds.), *Handbook of Research on Citizen Engagement and Public Participation in the Era of New Media* (pp. 90–105). Hershey, PA: IGI Global. doi:10.4018/978-1-5225-1081-9.ch006

Jamieson, H. V. (2017). We have a Situation!: Cyberformance and Civic Engagement in Post-Democracy. In R. Shin (Ed.), *Convergence of Contemporary Art, Visual Culture, and Global Civic Engagement* (pp. 297–317). Hershey, PA: IGI Global. doi:10.4018/978-1-5225-1665-1.ch017

Jimoh, J., & Kayode, J. (2018). Imperative of Peace and Conflict-Sensitive Journalism in Development. In A. Salawu & T. Owolabi (Eds.), *Exploring Journalism Practice and Perception in Developing Countries* (pp. 150–171). Hershey, PA: IGI Global. doi:10.4018/978-1-5225-3376-4.ch008

Johns, R. (2016). Increasing Value of a Tangible Product through Intangible Attributes: Value Co-Creation and Brand Building within Online Communities – Virtual Communities and Value. In R. English & R. Johns (Eds.), *Gender Considerations in Online Consumption Behavior and Internet Use* (pp. 112–124). Hershey, PA: IGI Global. doi:10.4018/978-1-5225-0010-0.ch008

Kanellopoulos, D. N. (2018). Group Synchronization for Multimedia Systems. In M. Khosrow-Pour, D.B.A. (Ed.), Encyclopedia of Information Science and Technology, Fourth Edition (pp. 6435-6446). Hershey, PA: IGI Global. doi:10.4018/978-1-5225-2255-3.ch559

Kapepo, M. I., & Mayisela, T. (2017). Integrating Digital Literacies Into an Undergraduate Course: Inclusiveness Through Use of ICTs. In C. Ayo & V. Mbarika (Eds.), *Sustainable ICT Adoption and Integration for Socio-Economic Development* (pp. 152–173). Hershey, PA: IGI Global. doi:10.4018/978-1-5225-2565-3.ch007

Karahoca, A., & Yengin, İ. (2018). Understanding the Potentials of Social Media in Collaborative Learning. In M. Khosrow-Pour, D.B.A. (Ed.), Encyclopedia of Information Science and Technology, Fourth Edition (pp. 7168-7180). Hershey, PA: IGI Global. doi:10.4018/978-1-5225-2255-3.ch623

Karataş, S., Ceran, O., Ülker, Ü., Gün, E. T., Köse, N. Ö., Kılıç, M., ... Tok, Z. A. (2016). A Trend Analysis of Mobile Learning. In D. Parsons (Ed.), *Mobile and Blended Learning Innovations for Improved Learning Outcomes* (pp. 248–276). Hershey, PA: IGI Global. doi:10.4018/978-1-5225-0359-0.ch013

Kasemsap, K. (2016). Role of Social Media in Brand Promotion: An International Marketing Perspective. In A. Singh & P. Duhan (Eds.), *Managing Public Relations and Brand Image through Social Media* (pp. 62–88). Hershey, PA: IGI Global. doi:10.4018/978-1-5225-0332-3.ch005

Kasemsap, K. (2016). The Roles of Social Media Marketing and Brand Management in Global Marketing. In W. Ozuem & G. Bowen (Eds.), *Competitive Social Media Marketing Strategies* (pp. 173–200). Hershey, PA: IGI Global. doi:10.4018/978-1-4666-9776-8.ch009

Kasemsap, K. (2017). Professional and Business Applications of Social Media Platforms. In V. Benson, R. Tuninga, & G. Saridakis (Eds.), *Analyzing the Strategic Role of Social Networking in Firm Growth and Productivity* (pp. 427–450). Hershey, PA: IGI Global. doi:10.4018/978-1-5225-0559-4.ch021

Kasemsap, K. (2017). Mastering Social Media in the Modern Business World. In N. Rao (Ed.), *Social Media Listening and Monitoring for Business Applications* (pp. 18–44). Hershey, PA: IGI Global. doi:10.4018/978-1-5225-0846-5.ch002

Kato, Y., & Kato, S. (2016). Mobile Phone Use during Class at a Japanese Women's College. In M. Yildiz & J. Keengwe (Eds.), *Handbook of Research on Media Literacy in the Digital Age* (pp. 436–455). Hershey, PA: IGI Global. doi:10.4018/978-1-4666-9667-9.ch021

Kaufmann, H. R., & Manarioti, A. (2017). Consumer Engagement in Social Media Platforms. In *Encouraging Participative Consumerism Through Evolutionary Digital Marketing: Emerging Research and Opportunities* (pp. 95–123). Hershey, PA: IGI Global. doi:10.4018/978-1-68318-012-8.ch004

Kavoura, A., & Kefallonitis, E. (2018). The Effect of Social Media Networking in the Travel Industry. In M. Khosrow-Pour, D.B.A. (Ed.), Encyclopedia of Information Science and Technology, Fourth Edition (pp. 4052-4063). Hershey, PA: IGI Global. doi:10.4018/978-1-5225-2255-3.ch351

Kawamura, Y. (2018). Practice and Modeling of Advertising Communication Strategy: Sender-Driven and Receiver-Driven. In T. Ogata & S. Asakawa (Eds.), *Content Generation Through Narrative Communication and Simulation* (pp. 358–379). Hershey, PA: IGI Global. doi:10.4018/978-1-5225-4775-4.ch013

Kell, C., & Czerniewicz, L. (2017). Visibility of Scholarly Research and Changing Research Communication Practices: A Case Study from Namibia. In A. Esposito (Ed.), *Research 2.0 and the Impact of Digital Technologies on Scholarly Inquiry* (pp. 97–116). Hershey, PA: IGI Global. doi:10.4018/978-1-5225-0830-4.ch006

Khalil, G. E. (2016). Change through Experience: How Experiential Play and Emotional Engagement Drive Health Game Success. In D. Novák, B. Tulu, & H. Brendryen (Eds.), *Handbook of Research on Holistic Perspectives in Gamification for Clinical Practice* (pp. 10–34). Hershey, PA: IGI Global. doi:10.4018/978-1-4666-9522-1.ch002

Kılınç, U. (2017). Create It! Extend It!: Evolution of Comics Through Narrative Advertising. In R. Yılmaz (Ed.), *Narrative Advertising Models and Conceptualization in the Digital Age* (pp. 117–132). Hershey, PA: IGI Global. doi:10.4018/978-1-5225-2373-4.ch007

Kim, J. H. (2016). Pedagogical Approaches to Media Literacy Education in the United States. In M. Yildiz & J. Keengwe (Eds.), *Handbook of Research on Media Literacy in the Digital Age* (pp. 53–74). Hershey, PA: IGI Global. doi:10.4018/978-1-4666-9667-9.ch003

Kirigha, J. M., Mukhongo, L. L., & Masinde, R. (2016). Beyond Web 2.0. Social Media and Urban Educated Youths Participation in Kenyan Politics. In L. Mukhongo & J. Macharia (Eds.), *Political Influence of the Media in Developing Countries* (pp. 156–174). Hershey, PA: IGI Global. doi:10.4018/978-1-4666-9613-6.ch010

Krochmal, M. M. (2016). Training for Mobile Journalism. In D. Mentor (Ed.), *Handbook of Research on Mobile Learning in Contemporary Classrooms* (pp. 336–362). Hershey, PA: IGI Global. doi:10.4018/978-1-5225-0251-7.ch017

Kumar, P., & Sinha, A. (2018). Business-Oriented Analytics With Social Network of Things. In H. Bansal, G. Shrivastava, G. Nguyen, & L. Stanciu (Eds.), *Social Network Analytics for Contemporary Business Organizations* (pp. 166–187). Hershey, PA: IGI Global. doi:10.4018/978-1-5225-5097-6.ch009

Kunock, A. I. (2017). Boko Haram Insurgency in Cameroon: Role of Mass Media in Conflict Management. In N. Mhiripiri & T. Chari (Eds.), *Media Law, Ethics, and Policy in the Digital Age* (pp. 226–244). Hershey, PA: IGI Global. doi:10.4018/978-1-5225-2095-5.ch013

Labadie, J. A. (2018). Digitally Mediated Art Inspired by Technology Integration: A Personal Journey. In A. Ursyn (Ed.), *Visual Approaches to Cognitive Education With Technology Integration* (pp. 121–162). Hershey, PA: IGI Global. doi:10.4018/978-1-5225-5332-8.ch008

Lefkowith, S. (2017). Credibility and Crisis in Pseudonymous Communities. In M. Folk & S. Apostel (Eds.), *Establishing and Evaluating Digital Ethos and Online Credibility* (pp. 190–236). Hershey, PA: IGI Global. doi:10.4018/978-1-5225-1072-7.ch010

Lemoine, P. A., Hackett, P. T., & Richardson, M. D. (2016). The Impact of Social Media on Instruction in Higher Education. In L. Briz-Ponce, J. Juanes-Méndez, & F. García-Peñalvo (Eds.), *Handbook of Research on Mobile Devices and Applications in Higher Education Settings* (pp. 373–401). Hershey, PA: IGI Global. doi:10.4018/978-1-5225-0256-2.ch016

Liampotis, N., Papadopoulou, E., Kalatzis, N., Roussaki, I. G., Kosmides, P., Sykas, E. D., ... Taylor, N. K. (2016). Tailoring Privacy-Aware Trustworthy Cooperating Smart Spaces for University Environments. In A. Panagopoulos (Ed.), *Handbook of Research on Next Generation Mobile Communication Systems* (pp. 410–439). Hershey, PA: IGI Global. doi:10.4018/978-1-4666-8732-5.ch016

Luppicini, R. (2017). Technoethics and Digital Democracy for Future Citizens. In R. Luppicini & R. Baarda (Eds.), *Digital Media Integration for Participatory Democracy* (pp. 1–21). Hershey, PA: IGI Global. doi:10.4018/978-1-5225-2463-2.ch001

Mahajan, I. M., Rather, M., Shafiq, H., & Qadri, U. (2016). Media Literacy Organizations. In M. Yildiz & J. Keengwe (Eds.), *Handbook of Research on Media Literacy in the Digital Age* (pp. 236–248). Hershey, PA: IGI Global. doi:10.4018/978-1-4666-9667-9.ch011

Maher, D. (2018). Supporting Pre-Service Teachers' Understanding and Use of Mobile Devices. In J. Keengwe (Ed.), *Handbook of Research on Mobile Technology, Constructivism, and Meaningful Learning* (pp. 160–177). Hershey, PA: IGI Global. doi:10.4018/978-1-5225-3949-0.ch009

Makhwanya, A. (2018). Barriers to Social Media Advocacy: Lessons Learnt From the Project "Tell Them We Are From Here". In F. Endong (Ed.), *Exploring the Role of Social Media in Transnational Advocacy* (pp. 55–72). Hershey, PA: IGI Global. doi:10.4018/978-1-5225-2854-8.ch004

Manli, G., & Rezaei, S. (2017). Value and Risk: Dual Pillars of Apps Usefulness. In S. Rezaei (Ed.), *Apps Management and E-Commerce Transactions in Real-Time* (pp. 274–292). Hershey, PA: IGI Global. doi:10.4018/978-1-5225-2449-6.ch013

Manrique, C. G., & Manrique, G. G. (2017). Social Media's Role in Alleviating Political Corruption and Scandals: The Philippines during and after the Marcos Regime. In K. Demirhan & D. Çakır-Demirhan (Eds.), *Political Scandal, Corruption, and Legitimacy in the Age of Social Media* (pp. 205–222). Hershey, PA: IGI Global. doi:10.4018/978-1-5225-2019-1.ch009

Manzoor, A. (2016). Cultural Barriers to Organizational Social Media Adoption. In A. Goel & P. Singhal (Eds.), *Product Innovation through Knowledge Management and Social Media Strategies* (pp. 31–45). Hershey, PA: IGI Global. doi:10.4018/978-1-4666-9607-5.ch002

Manzoor, A. (2016). Social Media for Project Management. In G. Silvius (Ed.), *Strategic Integration of Social Media into Project Management Practice* (pp. 51–65). Hershey, PA: IGI Global. doi:10.4018/978-1-4666-9867-3.ch004

Marovitz, M. (2017). Social Networking Engagement and Crisis Communication Considerations. In M. Brown Sr., (Ed.), *Social Media Performance Evaluation and Success Measurements* (pp. 130–155). Hershey, PA: IGI Global. doi:10.4018/978-1-5225-1963-8.ch007

Mathur, D., & Mathur, D. (2016). Word of Mouth on Social Media: A Potent Tool for Brand Building. In S. Rathore & A. Panwar (Eds.), *Capturing, Analyzing, and Managing Word-of-Mouth in the Digital Marketplace* (pp. 45–60). Hershey, PA: IGI Global. doi:10.4018/978-1-4666-9449-1.ch003

Maulana, I. (2018). Spontaneous Taking and Posting Selfie: Reclaiming the Lost Trust. In S. Hai-Jew (Ed.), *Selfies as a Mode of Social Media and Work Space Research* (pp. 28–50). Hershey, PA: IGI Global. doi:10.4018/978-1-5225-3373-3.ch002

Mayo, S. (2018). A Collective Consciousness Model in a Post-Media Society. In M. Khosrow-Pour (Ed.), *Enhancing Art, Culture, and Design With Technological Integration* (pp. 25–49). Hershey, PA: IGI Global. doi:10.4018/978-1-5225-5023-5.ch002

Mazur, E., Signorella, M. L., & Hough, M. (2018). The Internet Behavior of Older Adults. In M. Khosrow-Pour, D.B.A. (Ed.), Encyclopedia of Information Science and Technology, Fourth Edition (pp. 7026-7035). Hershey, PA: IGI Global. doi:10.4018/978-1-5225-2255-3.ch609

McGuire, M. (2017). Reblogging as Writing: The Role of Tumblr in the Writing Classroom. In K. Bryant (Ed.), *Engaging 21st Century Writers with Social Media* (pp. 116–131). Hershey, PA: IGI Global. doi:10.4018/978-1-5225-0562-4.ch007

McKee, J. (2018). Architecture as a Tool to Solve Business Planning Problems. In M. Khosrow-Pour, D.B.A. (Ed.), Encyclopedia of Information Science and Technology, Fourth Edition (pp. 573-586). Hershey, PA: IGI Global. doi:10.4018/978-1-5225-2255-3.ch050

McMahon, D. (2017). With a Little Help from My Friends: The Irish Radio Industry's Strategic Appropriation of Facebook for Commercial Growth. In V. Benson, R. Tuninga, & G. Saridakis (Eds.), *Analyzing the Strategic Role of Social Networking in Firm Growth and Productivity* (pp. 157–171). Hershey, PA: IGI Global. doi:10.4018/978-1-5225-0559-4.ch009

McPherson, M. J., & Lemon, N. (2017). The Hook, Woo, and Spin: Academics Creating Relations on Social Media. In A. Esposito (Ed.), *Research 2.0 and the Impact of Digital Technologies on Scholarly Inquiry* (pp. 167–187). Hershey, PA: IGI Global. doi:10.4018/978-1-5225-0830-4.ch009

Melro, A., & Oliveira, L. (2018). Screen Culture. In M. Khosrow-Pour, D.B.A. (Ed.), Encyclopedia of Information Science and Technology, Fourth Edition (pp. 4255-4266). Hershey, PA: IGI Global. doi:10.4018/978-1-5225-2255-3.ch369

Merwin, G. A. Jr, McDonald, J. S., Bennett, J. R. Jr, & Merwin, K. A. (2016). Social Media Applications Promote Constituent Involvement in Government Management. In G. Silvius (Ed.), *Strategic Integration of Social Media into Project Management Practice* (pp. 272–291). Hershey, PA: IGI Global. doi:10.4018/978-1-4666-9867-3.ch016

Mhiripiri, N. A., & Chikakano, J. (2017). Criminal Defamation, the Criminalisation of Expression, Media and Information Dissemination in the Digital Age: A Legal and Ethical Perspective. In N. Mhiripiri & T. Chari (Eds.), *Media Law, Ethics, and Policy in the Digital Age* (pp. 1–24). Hershey, PA: IGI Global. doi:10.4018/978-1-5225-2095-5.ch001

Miliopoulou, G., & Cossiavelou, V. (2016). Brands and Media Gatekeeping in the Social Media: Current Trends and Practices – An Exploratory Research. *International Journal of Interdisciplinary Telecommunications and Networking*, *8*(4), 51–64. doi:10.4018/IJITN.2016100105

Miron, E., Palmor, A., Ravid, G., Sharon, A., Tikotsky, A., & Zirkel, Y. (2017). Principles and Good Practices for Using Wikis within Organizations. In R. Chugh (Ed.), *Harnessing Social Media as a Knowledge Management Tool* (pp. 143–176). Hershey, PA: IGI Global. doi:10.4018/978-1-5225-0495-5.ch008

Mishra, K. E., Mishra, A. K., & Walker, K. (2016). Leadership Communication, Internal Marketing, and Employee Engagement: A Recipe to Create Brand Ambassadors. In A. Normore, L. Long, & M. Javidi (Eds.), *Handbook of Research on Effective Communication, Leadership, and Conflict Resolution* (pp. 311–329). Hershey, PA: IGI Global. doi:10.4018/978-1-4666-9970-0.ch017

Moeller, C. L. (2018). Sharing Your Personal Medical Experience Online: Is It an Irresponsible Act or Patient Empowerment? In S. Sekalala & B. Niezgoda (Eds.), *Global Perspectives on Health Communication in the Age of Social Media* (pp. 185–209). Hershey, PA: IGI Global. doi:10.4018/978-1-5225-3716-8.ch007

Mosanako, S. (2017). Broadcasting Policy in Botswana: The Case of Botswana Television. In O. Nelson, B. Ojebuyi, & A. Salawu (Eds.), *Impacts of the Media on African Socio-Economic Development* (pp. 217–230). Hershey, PA: IGI Global. doi:10.4018/978-1-5225-1859-4.ch014

Nazari, A. (2016). Developing a Social Media Communication Plan. In G. Silvius (Ed.), *Strategic Integration of Social Media into Project Management Practice* (pp. 194–217). Hershey, PA: IGI Global. doi:10.4018/978-1-4666-9867-3.ch012

Neto, B. M. (2016). From Information Society to Community Service: The Birth of E-Citizenship. In B. Passarelli, J. Straubhaar, & A. Cuevas-Cerveró (Eds.), *Handbook of Research on Comparative Approaches to the Digital Age Revolution in Europe and the Americas* (pp. 101–123). Hershey, PA: IGI Global. doi:10.4018/978-1-4666-8740-0.ch007

Noguti, V., Singh, S., & Waller, D. S. (2016). Gender Differences in Motivations to Use Social Networking Sites. In R. English & R. Johns (Eds.), *Gender Considerations in Online Consumption Behavior and Internet Use* (pp. 32–49). Hershey, PA: IGI Global. doi:10.4018/978-1-5225-0010-0.ch003

Noor, R. (2017). Citizen Journalism: News Gathering by Amateurs. In M. Adria & Y. Mao (Eds.), *Handbook of Research on Citizen Engagement and Public Participation in the Era of New Media* (pp. 194–229). Hershey, PA: IGI Global. doi:10.4018/978-1-5225-1081-9.ch012

Nwagbara, U., Oruh, E. S., & Brown, C. (2016). State Fragility and Stakeholder Engagement: New Media and Stakeholders' Voice Amplification in the Nigerian Petroleum Industry. In W. Ozuem & G. Bowen (Eds.), *Competitive Social Media Marketing Strategies* (pp. 136–154). Hershey, PA: IGI Global. doi:10.4018/978-1-4666-9776-8.ch007

Obermayer, N., Csepregi, A., & Kővári, E. (2017). Knowledge Sharing Relation to Competence, Emotional Intelligence, and Social Media Regarding Generations. In A. Bencsik (Ed.), *Knowledge Management Initiatives and Strategies in Small and Medium Enterprises* (pp. 269–290). Hershey, PA: IGI Global. doi:10.4018/978-1-5225-1642-2.ch013

Obermayer, N., Gaál, Z., Szabó, L., & Csepregi, A. (2017). Leveraging Knowledge Sharing over Social Media Tools. In R. Chugh (Ed.), *Harnessing Social Media as a Knowledge Management Tool* (pp. 1–24). Hershey, PA: IGI Global. doi:10.4018/978-1-5225-0495-5.ch001

Ogwezzy-Ndisika, A. O., & Faustino, B. A. (2016). Gender Responsive Election Coverage in Nigeria: A Score Card of 2011 General Elections. In L. Mukhongo & J. Macharia (Eds.), *Political Influence of the Media in Developing Countries* (pp. 234–249). Hershey, PA: IGI Global. doi:10.4018/978-1-4666-9613-6.ch015

Okoroafor, O. E. (2018). New Media Technology and Development Journalism in Nigeria. In A. Salawu & T. Owolabi (Eds.), *Exploring Journalism Practice and Perception in Developing Countries* (pp. 105–125). Hershey, PA: IGI Global. doi:10.4018/978-1-5225-3376-4.ch006

Olaleye, S. A., Sanusi, I. T., & Ukpabi, D. C. (2018). Assessment of Mobile Money Enablers in Nigeria. In F. Mtenzi, G. Oreku, D. Lupiana, & J. Yonazi (Eds.), *Mobile Technologies and Socio-Economic Development in Emerging Nations* (pp. 129–155). Hershey, PA: IGI Global. doi:10.4018/978-1-5225-4029-8.ch007

Ozuem, W., Pinho, C. A., & Azemi, Y. (2016). User-Generated Content and Perceived Customer Value. In W. Ozuem & G. Bowen (Eds.), *Competitive Social Media Marketing Strategies* (pp. 50–63). Hershey, PA: IGI Global. doi:10.4018/978-1-4666-9776-8.ch003

Pacchiega, C. (2017). An Informal Methodology for Teaching Through Virtual Worlds: Using Internet Tools and Virtual Worlds in a Coordinated Pattern to Teach Various Subjects. In G. Panconesi & M. Guida (Eds.), *Handbook of Research on Collaborative Teaching Practice in Virtual Learning Environments* (pp. 163–180). Hershey, PA: IGI Global. doi:10.4018/978-1-5225-2426-7.ch009

Pase, A. F., Goss, B. M., & Tietzmann, R. (2018). A Matter of Time: Transmedia Journalism Challenges. In R. Gambarato & G. Alzamora (Eds.), *Exploring Transmedia Journalism in the Digital Age* (pp. 49–66). Hershey, PA: IGI Global. doi:10.4018/978-1-5225-3781-6.ch004

Passarelli, B., & Paletta, F. C. (2016). Living inside the NET: The Primacy of Interactions and Processes. In B. Passarelli, J. Straubhaar, & A. Cuevas-Cerveró (Eds.), *Handbook of Research on Comparative Approaches to the Digital Age Revolution in Europe and the Americas* (pp. 1–15). Hershey, PA: IGI Global. doi:10.4018/978-1-4666-8740-0.ch001

Patkin, T. T. (2017). Social Media and Knowledge Management in a Crisis Context: Barriers and Opportunities. In R. Chugh (Ed.), *Harnessing Social Media as a Knowledge Management Tool* (pp. 125–142). Hershey, PA: IGI Global. doi:10.4018/978-1-5225-0495-5.ch007

Pavlíček, A. (2017). Social Media and Creativity: How to Engage Users and Tourists. In A. Kiráľová (Ed.), *Driving Tourism through Creative Destinations and Activities* (pp. 181–202). Hershey, PA: IGI Global. doi:10.4018/978-1-5225-2016-0.ch009

Pillay, K., & Maharaj, M. (2017). The Business of Advocacy: A Case Study of Greenpeace. In V. Benson, R. Tuninga, & G. Saridakis (Eds.), *Analyzing the Strategic Role of Social Networking in Firm Growth and Productivity* (pp. 59–75). Hershey, PA: IGI Global. doi:10.4018/978-1-5225-0559-4.ch004

Piven, I. P., & Breazeale, M. (2017). Desperately Seeking Customer Engagement: The Five-Sources Model of Brand Value on Social Media. In V. Benson, R. Tuninga, & G. Saridakis (Eds.), *Analyzing the Strategic Role of Social Networking in Firm Growth and Productivity* (pp. 283–313). Hershey, PA: IGI Global. doi:10.4018/978-1-5225-0559-4.ch016

Pokharel, R. (2017). New Media and Technology: How Do They Change the Notions of the Rhetorical Situations? In B. Gurung & M. Limbu (Eds.), *Integration of Cloud Technologies in Digitally Networked Classrooms and Learning Communities* (pp. 120–148). Hershey, PA: IGI Global. doi:10.4018/978-1-5225-1650-7.ch008

Popoola, I. S. (2016). The Press and the Emergent Political Class in Nigeria: Media, Elections, and Democracy. In L. Mukhongo & J. Macharia (Eds.), *Political Influence of the Media in Developing Countries* (pp. 45–58). Hershey, PA: IGI Global. doi:10.4018/978-1-4666-9613-6.ch004

Porlezza, C., Benecchi, E., & Colapinto, C. (2018). The Transmedia Revitalization of Investigative Journalism: Opportunities and Challenges of the Serial Podcast. In R. Gambarato & G. Alzamora (Eds.), *Exploring Transmedia Journalism in the Digital Age* (pp. 183–201). Hershey, PA: IGI Global. doi:10.4018/978-1-5225-3781-6.ch011

Ramluckan, T., Ally, S. E., & van Niekerk, B. (2017). Twitter Use in Student Protests: The Case of South Africa's #FeesMustFall Campaign. In M. Korstanje (Ed.), *Threat Mitigation and Detection of Cyber Warfare and Terrorism Activities* (pp. 220–253). Hershey, PA: IGI Global. doi:10.4018/978-1-5225-1938-6.ch010

Rao, N. R. (2017). Social Media: An Enabler for Governance. In N. Rao (Ed.), *Social Media Listening and Monitoring for Business Applications* (pp. 151–164). Hershey, PA: IGI Global. doi:10.4018/978-1-5225-0846-5.ch008

Rathore, A. K., Tuli, N., & Ilavarasan, P. V. (2016). Pro-Business or Common Citizen?: An Analysis of an Indian Woman CEO's Tweets. *International Journal of Virtual Communities and Social Networking*, *8*(1), 19–29. doi:10.4018/IJVCSN.2016010102

Redi, F. (2017). Enhancing Coopetition Among Small Tourism Destinations by Creativity. In A. Kiráľová (Ed.), *Driving Tourism through Creative Destinations and Activities* (pp. 223–244). Hershey, PA: IGI Global. doi:10.4018/978-1-5225-2016-0.ch011

Reeves, M. (2016). Social Media: It Can Play a Positive Role in Education. In R. English & R. Johns (Eds.), *Gender Considerations in Online Consumption Behavior and Internet Use* (pp. 82–95). Hershey, PA: IGI Global. doi:10.4018/978-1-5225-0010-0.ch006

Reis, Z. A. (2016). Bring the Media Literacy of Turkish Pre-Service Teachers to the Table. In M. Yildiz & J. Keengwe (Eds.), *Handbook of Research on Media Literacy in the Digital Age* (pp. 405–422). Hershey, PA: IGI Global. doi:10.4018/978-1-4666-9667-9.ch019

Resuloğlu, F., & Yılmaz, R. (2017). A Model for Interactive Advertising Narration. In R. Yılmaz (Ed.), *Narrative Advertising Models and Conceptualization in the Digital Age* (pp. 1–20). Hershey, PA: IGI Global. doi:10.4018/978-1-5225-2373-4.ch001

Ritzhaupt, A. D., Poling, N., Frey, C., Kang, Y., & Johnson, M. (2016). A Phenomenological Study of Games, Simulations, and Virtual Environments Courses: What Are We Teaching and How? *International Journal of Gaming and Computer-Mediated Simulations*, *8*(3), 59–73. doi:10.4018/IJGCMS.2016070104

Ross, D. B., Eleno-Orama, M., & Salah, E. V. (2018). The Aging and Technological Society: Learning Our Way Through the Decades. In V. Bryan, A. Musgrove, & J. Powers (Eds.), *Handbook of Research on Human Development in the Digital Age* (pp. 205–234). Hershey, PA: IGI Global. doi:10.4018/978-1-5225-2838-8.ch010

Rusko, R., & Merenheimo, P. (2017). Co-Creating the Christmas Story: Digitalizing as a Shared Resource for a Shared Brand. In I. Oncioiu (Ed.), *Driving Innovation and Business Success in the Digital Economy* (pp. 137–157). Hershey, PA: IGI Global. doi:10.4018/978-1-5225-1779-5.ch010

Sabao, C., & Chikara, T. O. (2018). Social Media as Alternative Public Sphere for Citizen Participation and Protest in National Politics in Zimbabwe: The Case of #thisflag. In F. Endong (Ed.), *Exploring the Role of Social Media in Transnational Advocacy* (pp. 17–35). Hershey, PA: IGI Global. doi:10.4018/978-1-5225-2854-8.ch002

Samarthya-Howard, A., & Rogers, D. (2018). Scaling Mobile Technologies to Maximize Reach and Impact: Partnering With Mobile Network Operators and Governments. In S. Takavarasha Jr & C. Adams (Eds.), *Affordability Issues Surrounding the Use of ICT for Development and Poverty Reduction* (pp. 193–211). Hershey, PA: IGI Global. doi:10.4018/978-1-5225-3179-1.ch009

Sandoval-Almazan, R. (2017). Political Messaging in Digital Spaces: The Case of Twitter in Mexico's Presidential Campaign. In Y. Ibrahim (Ed.), *Politics, Protest, and Empowerment in Digital Spaces* (pp. 72–90). Hershey, PA: IGI Global. doi:10.4018/978-1-5225-1862-4.ch005

Schultz, C. D., & Dellnitz, A. (2018). Attribution Modeling in Online Advertising. In K. Yang (Ed.), *Multi-Platform Advertising Strategies in the Global Marketplace* (pp. 226–249). Hershey, PA: IGI Global. doi:10.4018/978-1-5225-3114-2.ch009

Schultz, C. D., & Holsing, C. (2018). Differences Across Device Usage in Search Engine Advertising. In K. Yang (Ed.), *Multi-Platform Advertising Strategies in the Global Marketplace* (pp. 250–279). Hershey, PA: IGI Global. doi:10.4018/978-1-5225-3114-2.ch010

Senadheera, V., Warren, M., Leitch, S., & Pye, G. (2017). Facebook Content Analysis: A Study into Australian Banks' Social Media Community Engagement. In S. Hai-Jew (Ed.), *Social Media Data Extraction and Content Analysis* (pp. 412–432). Hershey, PA: IGI Global. doi:10.4018/978-1-5225-0648-5.ch013

Sharma, A. R. (2018). Promoting Global Competencies in India: Media and Information Literacy as Stepping Stone. In M. Yildiz, S. Funk, & B. De Abreu (Eds.), *Promoting Global Competencies Through Media Literacy* (pp. 160–174). Hershey, PA: IGI Global. doi:10.4018/978-1-5225-3082-4.ch010

Sillah, A. (2017). Nonprofit Organizations and Social Media Use: An Analysis of Nonprofit Organizations' Effective Use of Social Media Tools. In M. Brown Sr., (Ed.), *Social Media Performance Evaluation and Success Measurements* (pp. 180–195). Hershey, PA: IGI Global. doi:10.4018/978-1-5225-1963-8.ch009

Škorić, M. (2017). Adaptation of Winlink 2000 Emergency Amateur Radio Email Network to a VHF Packet Radio Infrastructure. In A. El Oualkadi & J. Zbitou (Eds.), *Handbook of Research on Advanced Trends in Microwave and Communication Engineering* (pp. 498–528). Hershey, PA: IGI Global. doi:10.4018/978-1-5225-0773-4.ch016

Skubida, D. (2016). Can Some Computer Games Be a Sport?: Issues with Legitimization of eSport as a Sporting Activity. *International Journal of Gaming and Computer-Mediated Simulations*, *8*(4), 38–52. doi:10.4018/IJGCMS.2016100103

Sonnenberg, C. (2016). Mobile Content Adaptation: An Analysis of Techniques and Frameworks. In J. Aguado, C. Feijóo, & I. Martínez (Eds.), *Emerging Perspectives on the Mobile Content Evolution* (pp. 177–199). Hershey, PA: IGI Global. doi:10.4018/978-1-4666-8838-4.ch010

Sonnevend, J. (2016). More Hope!: Ceremonial Media Events Are Still Powerful in the Twenty-First Century. In A. Fox (Ed.), *Global Perspectives on Media Events in Contemporary Society* (pp. 132–140). Hershey, PA: IGI Global. doi:10.4018/978-1-4666-9967-0.ch010

Sood, T. (2017). Services Marketing: A Sector of the Current Millennium. In T. Sood (Ed.), *Strategic Marketing Management and Tactics in the Service Industry* (pp. 15–42). Hershey, PA: IGI Global. doi:10.4018/978-1-5225-2475-5.ch002

Stairs, G. A. (2016). The Amplification of the Sunni-Shia Divide through Contemporary Communications Technology: Fear and Loathing in the Modern Middle East. In S. Gibson & A. Lando (Eds.), *Impact of Communication and the Media on Ethnic Conflict* (pp. 214–231). Hershey, PA: IGI Global. doi:10.4018/978-1-4666-9728-7.ch013

Stokinger, E., & Ozuem, W. (2016). The Intersection of Social Media and Customer Retention in the Luxury Beauty Industry. In W. Ozuem & G. Bowen (Eds.), *Competitive Social Media Marketing Strategies* (pp. 235–258). Hershey, PA: IGI Global. doi:10.4018/978-1-4666-9776-8.ch012

Sudarsanam, S. K. (2017). Social Media Metrics. In N. Rao (Ed.), *Social Media Listening and Monitoring for Business Applications* (pp. 131–149). Hershey, PA: IGI Global. doi:10.4018/978-1-5225-0846-5.ch007

Swiatek, L. (2017). Accessing the Finest Minds: Insights into Creativity from Esteemed Media Professionals. In N. Silton (Ed.), *Exploring the Benefits of Creativity in Education, Media, and the Arts* (pp. 240–263). Hershey, PA: IGI Global. doi:10.4018/978-1-5225-0504-4.ch012

Switzer, J. S., & Switzer, R. V. (2016). Virtual Teams: Profiles of Successful Leaders. In B. Baggio (Ed.), *Analyzing Digital Discourse and Human Behavior in Modern Virtual Environments* (pp. 1–24). Hershey, PA: IGI Global. doi:10.4018/978-1-4666-9899-4.ch001

Tabbane, R. S., & Debabi, M. (2016). Electronic Word of Mouth: Definitions and Concepts. In S. Rathore & A. Panwar (Eds.), *Capturing, Analyzing, and Managing Word-of-Mouth in the Digital Marketplace* (pp. 1–27). Hershey, PA: IGI Global. doi:10.4018/978-1-4666-9449-1.ch001

Tellería, A. S. (2016). The Role of the Profile and the Digital Identity on the Mobile Content. In J. Aguado, C. Feijóo, & I. Martínez (Eds.), *Emerging Perspectives on the Mobile Content Evolution* (pp. 263–282). Hershey, PA: IGI Global. doi:10.4018/978-1-4666-8838-4.ch014

Teurlings, J. (2017). What Critical Media Studies Should Not Take from Actor-Network Theory. In M. Spöhrer & B. Ochsner (Eds.), *Applying the Actor-Network Theory in Media Studies* (pp. 66–78). Hershey, PA: IGI Global. doi:10.4018/978-1-5225-0616-4.ch005

Tomé, V. (2018). Assessing Media Literacy in Teacher Education. In M. Yildiz, S. Funk, & B. De Abreu (Eds.), *Promoting Global Competencies Through Media Literacy* (pp. 1–19). Hershey, PA: IGI Global. doi:10.4018/978-1-5225-3082-4.ch001

Toscano, J. P. (2017). Social Media and Public Participation: Opportunities, Barriers, and a New Framework. In M. Adria & Y. Mao (Eds.), *Handbook of Research on Citizen Engagement and Public Participation in the Era of New Media* (pp. 73–89). Hershey, PA: IGI Global. doi:10.4018/978-1-5225-1081-9.ch005

Trauth, E. (2017). Creating Meaning for Millennials: Bakhtin, Rosenblatt, and the Use of Social Media in the Composition Classroom. In K. Bryant (Ed.), *Engaging 21st Century Writers with Social Media* (pp. 151–162). Hershey, PA: IGI Global. doi:10.4018/978-1-5225-0562-4.ch009

Ugangu, W. (2016). Kenya's Difficult Political Transitions Ethnicity and the Role of Media. In L. Mukhongo & J. Macharia (Eds.), *Political Influence of the Media in Developing Countries* (pp. 12–24). Hershey, PA: IGI Global. doi:10.4018/978-1-4666-9613-6.ch002

Uprety, S. (2018). Print Media's Role in Securitization: National Security and Diplomacy Discourses in Nepal. In S. Bute (Ed.), *Media Diplomacy and Its Evolving Role in the Current Geopolitical Climate* (pp. 56–82). Hershey, PA: IGI Global. doi:10.4018/978-1-5225-3859-2.ch004

Van der Merwe, L. (2016). Social Media Use within Project Teams: Practical Application of Social Media on Projects. In G. Silvius (Ed.), *Strategic Integration of Social Media into Project Management Practice* (pp. 139–159). Hershey, PA: IGI Global. doi:10.4018/978-1-4666-9867-3.ch009

van der Vyver, A. G. (2018). A Model for Economic Development With Telecentres and the Social Media: Overcoming Affordability Constraints. In S. Takavarasha Jr & C. Adams (Eds.), *Affordability Issues Surrounding the Use of ICT for Development and Poverty Reduction* (pp. 112–140). Hershey, PA: IGI Global. doi:10.4018/978-1-5225-3179-1.ch006

van Dokkum, E., & Ravesteijn, P. (2016). Managing Project Communication: Using Social Media for Communication in Projects. In G. Silvius (Ed.), *Strategic Integration of Social Media into Project Management Practice* (pp. 35–50). Hershey, PA: IGI Global. doi:10.4018/978-1-4666-9867-3.ch003

van Niekerk, B. (2018). Social Media Activism From an Information Warfare and Security Perspective. In F. Endong (Ed.), *Exploring the Role of Social Media in Transnational Advocacy* (pp. 1–16). Hershey, PA: IGI Global. doi:10.4018/978-1-5225-2854-8.ch001

Varnali, K., & Gorgulu, V. (2017). Determinants of Brand Recall in Social Networking Sites. In W. Al-Rabayah, R. Khasawneh, R. Abu-shamaa, & I. Alsmadi (Eds.), *Strategic Uses of Social Media for Improved Customer Retention* (pp. 124–153). Hershey, PA: IGI Global. doi:10.4018/978-1-5225-1686-6.ch007

Varty, C. T., O'Neill, T. A., & Hambley, L. A. (2017). Leading Anywhere Workers: A Scientific and Practical Framework. In Y. Blount & M. Gloet (Eds.), *Anywhere Working and the New Era of Telecommuting* (pp. 47–88). Hershey, PA: IGI Global. doi:10.4018/978-1-5225-2328-4.ch003

Vatikiotis, P. (2016). Social Media Activism: A Contested Field. In T. Deželan & I. Vobič (Eds.), *R)evolutionizing Political Communication through Social Media* (pp. 40–54). Hershey, PA: IGI Global. doi:10.4018/978-1-4666-9879-6.ch003

Velikovsky, J. T. (2018). The Holon/Parton Structure of the Meme, or The Unit of Culture. In M. Khosrow-Pour, D.B.A. (Ed.), Encyclopedia of Information Science and Technology, Fourth Edition (pp. 4666-4678). Hershey, PA: IGI Global. doi:10.4018/978-1-5225-2255-3.ch405

Venkatesh, R., & Jayasingh, S. (2017). Transformation of Business through Social Media. In N. Rao (Ed.), *Social Media Listening and Monitoring for Business Applications* (pp. 1–17). Hershey, PA: IGI Global. doi:10.4018/978-1-5225-0846-5.ch001

Vesnic-Alujevic, L. (2016). European Elections and Facebook: Political Advertising and Deliberation? In T. Deželan & I. Vobič (Eds.), *R)evolutionizing Political Communication through Social Media* (pp. 191–209). Hershey, PA: IGI Global. doi:10.4018/978-1-4666-9879-6.ch010

Virkar, S. (2017). Trolls Just Want to Have Fun: Electronic Aggression within the Context of E-Participation and Other Online Political Behaviour in the United Kingdom. In M. Korstanje (Ed.), *Threat Mitigation and Detection of Cyber Warfare and Terrorism Activities* (pp. 111–162). Hershey, PA: IGI Global. doi:10.4018/978-1-5225-1938-6.ch006

Wakabi, W. (2017). When Citizens in Authoritarian States Use Facebook for Social Ties but Not Political Participation. In Y. Ibrahim (Ed.), *Politics, Protest, and Empowerment in Digital Spaces* (pp. 192–214). Hershey, PA: IGI Global. doi:10.4018/978-1-5225-1862-4.ch012

Weisberg, D. J. (2016). Methods and Strategies in Using Digital Literacy in Media and the Arts. In M. Yildiz & J. Keengwe (Eds.), *Handbook of Research on Media Literacy in the Digital Age* (pp. 456–471). Hershey, PA: IGI Global. doi:10.4018/978-1-4666-9667-9.ch022

Weisgerber, C., & Butler, S. H. (2016). Debranding Digital Identity: Personal Branding and Identity Work in a Networked Age. *International Journal of Interactive Communication Systems and Technologies, 6*(1), 17–34. doi:10.4018/IJICST.2016010102

Wijngaard, P., Wensveen, I., Basten, A., & de Vries, T. (2016). Projects without Email, Is that Possible? In G. Silvius (Ed.), *Strategic Integration of Social Media into Project Management Practice* (pp. 218–235). Hershey, PA: IGI Global. doi:10.4018/978-1-4666-9867-3.ch013

Wright, K. (2018). “Show Me What You Are Saying”: Visual Literacy in the Composition Classroom. In A. August (Ed.), *Visual Imagery, Metadata, and Multimodal Literacies Across the Curriculum* (pp. 24–49). Hershey, PA: IGI Global. doi:10.4018/978-1-5225-2808-1.ch002

Yang, K. C. (2018). Understanding How Mexican and U.S. Consumers Decide to Use Mobile Social Media: A Cross-National Qualitative Study. In K. Yang (Ed.), *Multi-Platform Advertising Strategies in the Global Marketplace* (pp. 168–198). Hershey, PA: IGI Global. doi:10.4018/978-1-5225-3114-2.ch007

Yang, K. C., & Kang, Y. (2016). Exploring Female Hispanic Consumers’ Adoption of Mobile Social Media in the U.S. In R. English & R. Johns (Eds.), *Gender Considerations in Online Consumption Behavior and Internet Use* (pp. 185–207). Hershey, PA: IGI Global. doi:10.4018/978-1-5225-0010-0.ch012

Yao, Q., & Wu, M. (2016). Examining the Role of WeChat in Advertising. In X. Xu (Ed.), *Handbook of Research on Human Social Interaction in the Age of Mobile Devices* (pp. 386–405). Hershey, PA: IGI Global. doi:10.4018/978-1-5225-0469-6.ch016

Yarchi, M., Wolfsfeld, G., Samuel-Azran, T., & Segev, E. (2017). Invest, Engage, and Win: Online Campaigns and Their Outcomes in an Israeli Election. In M. Brown Sr., (Ed.), *Social Media Performance Evaluation and Success Measurements* (pp. 225–248). Hershey, PA: IGI Global. doi:10.4018/978-1-5225-1963-8.ch011

Yeboah-Banin, A. A., & Amoakohene, M. I. (2018). The Dark Side of Multi-Platform Advertising in an Emerging Economy Context. In K. Yang (Ed.), *Multi-Platform Advertising Strategies in the Global Marketplace* (pp. 30–53). Hershey, PA: IGI Global. doi:10.4018/978-1-5225-3114-2.ch002

Yılmaz, R., Çakır, A., & Resuloğlu, F. (2017). Historical Transformation of the Advertising Narration in Turkey: From Stereotype to Digital Media. In R. Yılmaz (Ed.), *Narrative Advertising Models and Conceptualization in the Digital Age* (pp. 133–152). Hershey, PA: IGI Global. doi:10.4018/978-1-5225-2373-4.ch008

Yusuf, S., Hassan, M. S., & Ibrahim, A. M. (2018). Cyberbullying Among Malaysian Children Based on Research Evidence. In M. Khosrow-Pour, D.B.A. (Ed.), Encyclopedia of Information Science and Technology, Fourth Edition (pp. 1704-1722). Hershey, PA: IGI Global. doi:10.4018/978-1-5225-2255-3.ch149

Zervas, P., & Alexandraki, C. (2016). Facilitating Open Source Software and Standards to Assembly a Platform for Networked Music Performance. In D. Kanellopoulos (Ed.), *Emerging Research on Networked Multimedia Communication Systems* (pp. 334–365). Hershey, PA: IGI Global. doi:10.4018/978-1-4666-8850-6.ch011

About the Contributors

Serpil Kır was born in Alaşehir, Manisa, Turkey. She completed her primary, secondary, and high school education in Alaşehir. She started her college education at Selçuk University Faculty of Communication Department of Public Relations and Publicity and graduated in 2011. In 2014, she completed her master's degree in Public Relations and Publicity in Selcuk University Institute of Social Sciences with her thesis titled, "Test Drives in the Context of Experiential Marketing". In 2018, she completed her doctorate degree in Public Relations and Publicity in Selcuk University Institute of Social Sciences with her thesis titled "Investigation of Factors Affecting Online Shopping Tendency in the Context of Sensory Activation Technology Acceptance Model". She works as assistant professor in Hatay Mustafa Kemal University.

* * *

Bahar Akbulak is a Teaching Assistant Faculty Member in the Faculty of Communication at Bolu Abant Izzet Baysal University. She graduated from the Graphic Arts department at Abant İzzet Baysal University, Faculty of Education, in 2005. In 2012, she completed a master's degree in the Graphic Design programme at Abant İzzet Baysal University Institute educational Sciences. Her research interests are art, graphic design, communication design, and visual communication design studies.

Zuhal Akmeşe was born in Mardin, Turkey, in 1982. After she graduated from İzmir Kız Lisesi (High School), she taught Social Sciences for the Kocatepe University Faculty of Education. Later in 2010, she received her bachelor degree, ranking first in class from Ege University Communication Faculty Radio-TV Cinema, and then she gained Master of Arts degree with the thesis, "After 2000 Representation of Youth in Turkish Cinema". İn 2017, she gained her PhD degree at İstanbul University Communication Faculty Radio-TV Cinema department with the thesis, "Framing on Television and Create The Social Point of View". She is working at Dicle University as a Research Assistant. She wrote many articles about communication, documentary, and photography. She produced eight short films and five documentaries.

Paulo M. Barroso is an Assistant professor at the Polytechnic Institute of Viseu, Portugal (College of Education, Department of Communication and Art), teaching Advertising Semiotics, Sociology of Communication, and Communication Ethics and Deontology. He is an integrated researcher (ORCID 0000-0001-7638-5064) at the Investigation Centre in Communication, Information, and Digital Culture

(CIC-Digital) of the Faculty of Social Sciences and Humanities, New University of Lisbon. Research interests are in Semiotics, Argumentation and Rhetoric, Ethics, Media Languages, and Theories and Models of Communication. He earned a BA and MA in Communication Sciences; BA, MA and PhD in Philosophy (in the scientific area of Philosophy of Language); post-doctorate researcher (6 years) in Communication Sciences, having participated in international conferences and published several articles and books in these fields (e.g. Grammar, Expressiveness, and Inter-subjective Meanings: Wittgenstein's Philosophy of Psychology. Newcastle-Upon-Tyne, UK: Cambridge Scholars Publishing, 2015).

Mikail Batu is Associate Professor at the Department of Public Relations of Ege University, where he completed his PhD program. His areas of interest include Social Media, Corporate Communication, Gender, Communication Sciences, and New Media.

Selçuk Bazarcı has been working as a research assistant in the Communication Faculty of Ege University since 2015. He earned a master's degree in advertising in 2017, and then started his Ph.D. in the same field. He mainly studies digital advertising.

Arzu Kalafat Cat is a research assistant at Abant İzzet Baysal University Faculty of Communication. She graduated from the department of Publıc Administration at Anadolu Unıversity, Faculty of Economics and Administrative Sciences, in 2010. In 2013, she completed a master's degree in Public Relations and Publicity at Atatürk Unıversity Institute Of Social Sciences. In 2019, She completed her Ph.D. in the department of Basic Communication Sciences at Atatürk University Institute Of Social Sciences. Her research interests are consumption culture, marketing, brand management, consumer behavior, marketing, public relations, and new media studies.

Christopher Chepken is a seasoned trainer and researcher in the field of Computer Science, specializing in ICT4D. He has done ICT4D/ICTD research and consultancy in urban commuting challenges-telecommuting; Software Engineering; Health; and Agriculture and Business process outsourcing, Christopher brings considerable experience in carrying out ICT and development-related work. He has experience both as a practitioner and as a researcher from Eastern and Southern African countries. Besides teaching research and consultancy, Christopher is involved in administrative tasks of the University, the current one being the Coordinator for the four Masters programmes at the School of Computing and Informatics, University of Nairobi. He also gets involved in conference organizations, including being the general conference co-chair for AfriCHI (africhi.net) held in Nairobi, Kenya (2016), and ICTD 2013, held in South Arica. Christopher's other areas of specialization include Computer and cyber security, including mobile technology, where he has experience in cyber security practice and training at academic and industry levels.

Naziat Choudhury is an Associate Professor at the Department of Mass Communication and Journalism, University of Rajshahi, Bangladesh. She has completed her Masters from University of Calgary, Canada, and PhD from Monash University, Australia. She is currently working on social media in the context of China.

Şadiye Deniz has been an Associate Professor at Ege University since 2014. She finished her Phd in 2009. Her professions are digital journalism, political communication, and social politics.

Türker Elitaş was born in Erzurum, Turkey. He completed his primary, secondary, and high school education in Erzurum. He started his college education at Atatürk University Faculty of Communication Department of Radio, Television, and Cinema and graduated in 2005. In 2014, he completed his master's degree in Radio, Television, and Cinema at Atatürk University Institute of Social Sciences. In 2017, he completed his doctorate's degree in Radio, Television, and Cinema in Marmara University Institute of Social Sciences with his thesis titled, "New communication technologies in distance education license period: Ataturk University Distance Education Center". He is still working in Communication faculty at Manas University in the Kyrgyz Republic as an assistant professor. The author is studying new media, new communication technologies, distance education, and communication technology.

Unanza Gulzar is an assistant professor in the School of Law, The NorthCap University, Gurugram, India. Unanza Gulzar has done her LLB and LLM from University of Kashmir specializing in Business laws, and holds a Bachelor's degree in science. She is pursuing a PhD in e-commerce. She has qualified the U.G.C. NET twice and State eligibility test for lectureship in 2013. She has to her credit a Book on "Money Laundering in India: Legal Perspective". She has published more than 21 research papers in various national and international reputed law journals. She has received a number of letters of appreciation from law journals for her contributions. Her articles have appeared in various newspapers/web portals on a wide range of subjects from politics to typical legal matters. She has presented 23 research papers in various Seminars and conferences. She has over four years of teaching experience and has taught in Central University of Kashmir. She has received Professional Development Training organized by National Law School University, Bangalore and Menon institute of legal advocacy training in association with SLS, CUK.

Sefer Kalaman is Assistant Professor at the Department of Radio, Television, and Film of Yozgat Bozok University, Turkey. He completed his PhD program at the Radio and Television Department of Ege University. His areas of interest include New Media, Digital Culture, Gender, Digimodernism, Privacy, Immigration, Middle East, Asia, and Research Methods.

Savaş Keskin was born in Diyarbakır, Turkey, in 1990. He has a Bachelor's degree in Department of Public Relations and Advertising in the Faculty of Communication of Fırat State University/Elazığ-Turkey. He has a Master's Degree in the Department of Communication Science in the Institute of Social Science of Fırat State University/Elazığ-Turkey. He continues Doctoral Education in the Department of Radio, Television, and Cinema in the Institute of Social Science of Istanbul University/Istanbul-Turkey. His working area includes communication sociology, social media, identity, otherness, subaltern identity, virtual community, minority, representation, and self-representation. He has had published many manuscripts, proceedings, and book chapters about his working area. He is lecturer in the Health Services Vocational School in Bayburt University.

Rengim Sine Nazlı is an Assistant Professor Dr. Faculty Member for Bolu Abant Izzet Baysal University, Faculty of Communication, Department of Journalism. She graduated from the department of Journalism at Selcuk Unıversity, Faculty of Communication, in 2007. In 2010, she completed master's degree in Journalism programme. at Selcuk University Institute of Social Sciences. In 2017, she completed PhD degree in the department of Journalism programme at Selcuk University Institute of Social Sciences. Her research interests are consumption culture, new media, critical media theories, and journalistic studies.

Özen Okat was born in İzmir, Turkey. She got her undergraduate degree at Ege University, Faculty of Communication, Department of Journalism, and Ph.D. from Ege University Social Sciences Institute, Department of Advertising. She worked as a graphic design supervisor at Ege University, Rectorate, and as a lecturer at Ege University, Ege Vocational School. Okat is a faculty member at Ege University Faculty of Communication. She has numerous articles in various books, international journals and numerous international symposiums. Her research interests are marketing management, digital marketing, neuromarketing, advertising design, and visual design of advertising.

Emel Özdemir was born in 1982. After completing high school in Antalya Karatay Super High School, she graduated from Ege University, English Language and Literature Department, in 2004. In 2007, she received a master's degree from the Department of Translation and Interpreting at Muğla University. She began doctoral studies at Ege University, Department of General Journalism, and received her Ph.D. degree in 2012 with her thesis, "The Evaluation of The Turkish Image That is Constructed in The Globalized World in The Foreign Press". She works in Akdeniz University, The Faculty of Communication, Department of Journalism, in Antalya.

Bahadır Burak Solak was born in Çorum, Turkey. He received his BA degree in the Department of Public Relations and Publicity at the Communication Faculty of Akdeniz University. He also holds a master's degree in the field of public relations and publicity at the same university's social sciences institute. The Ph.D. degree is continuing at Ege University Social Sciences Institute, Department of Advertising. He has been working as a research assistant at the Public Relations and Advertising Department of the Faculty of Communication at Trabzon University since 2014. Solak has numerous articles published in international journals and presented in international symposiums. His research interests are marketing management, digital marketing, neuromarketing, advertising design, and visual design of advertising.

Gurur Sönmez was born in İstanbul in 1990. He started a one-year psychology foundation program in Coventry University (UK) in 2009, then began his university education in Marmara University, Department of Radio, Television, and Cinema, in 2010. He worked as director assistant and editor in several productions. He directed two short films that were screened, finalized, and awarded at various festivals. He graduated from a master's degree program in 2017 at Marmara University, Department of Radio, Television. During that period, he worked as an Editor-in-Chief in a respectable monthly technology magazine called Newtech. He has led a team comprised of junior and senior editors, writers, and other professionals in charge of creating original content to be published. He has started his Ph.D. in Radio, Television, and Cinema at Istanbul University. He has been working as a research assistant in Radio, Television, and Cinema in Istanbul Medipol University since 2018.

Mehmet Ferhat Sönmez was born in Elazığ, Turkey. He completed his primary, secondary, and high school education in Elazığ. He started his college education at Gazi University Faculty of Communication Department of Journalism and graduated in 2005. In 2013, he completed his master's degree in communication at Fırat University Institute of Social Sciences. Sönmez is a phD student in Radio, Television, and Cinema in İstanbul University Institute of Social Sciences. He is working in Communication faculty, Fırat University, Turkey, as a research assistant. His research interests are identity, social media and identities, study in new media, and new communication technologies.

Index

W

Ensure Quality Research is Introduced to the Academic Community

Become an IGI Global Reviewer for Authored Book Projects

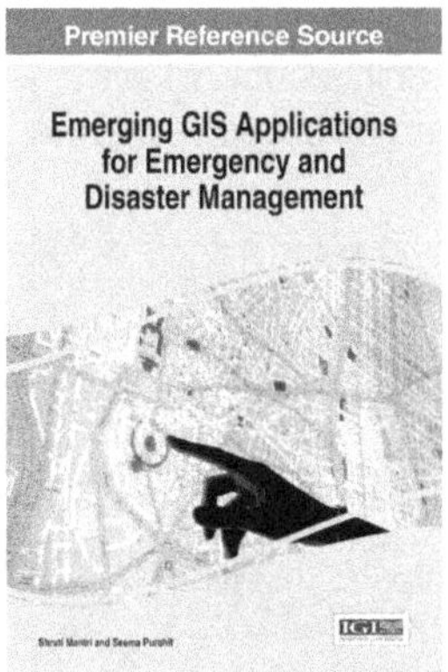

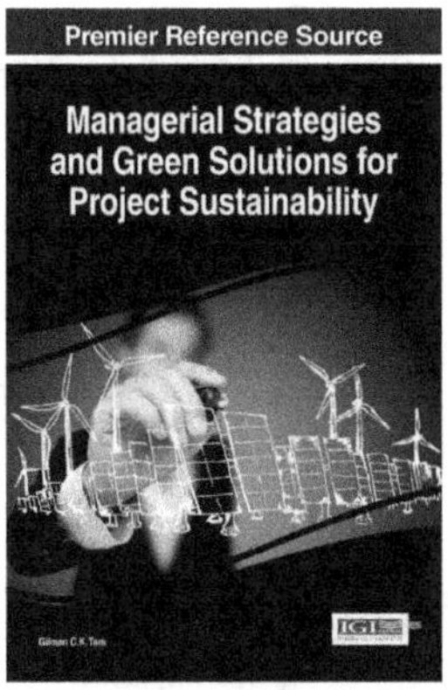

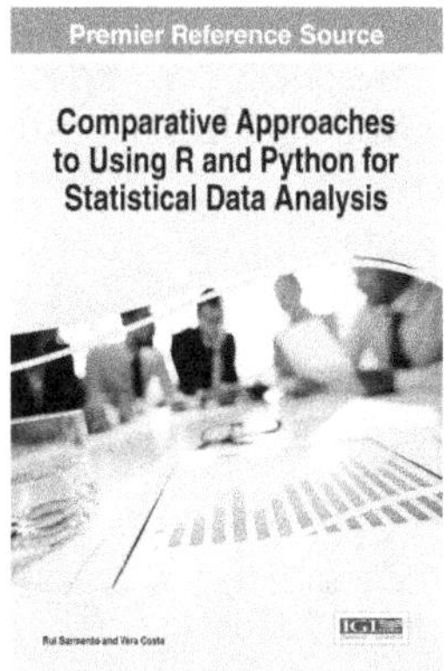

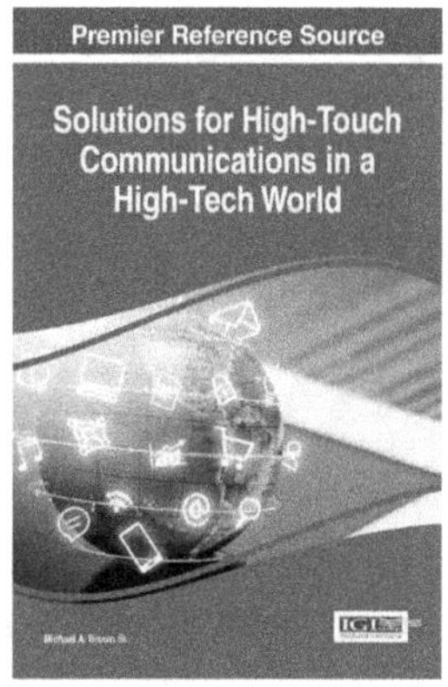

The overall success of an authored book project is dependent on quality and timely reviews.

In this competitive age of scholarly publishing, constructive and timely feedback significantly expedites the turnaround time of manuscripts from submission to acceptance, allowing the publication and discovery of forward-thinking research at a much more expeditious rate. Several IGI Global authored book projects are currently seeking highly-qualified experts in the field to fill vacancies on their respective editorial review boards:

Applications and Inquiries may be sent to:
development@igi-global.com

Applicants must have a doctorate (or an equivalent degree) as well as publishing and reviewing experience. Reviewers are asked to complete the open-ended evaluation questions with as much detail as possible in a timely, collegial, and constructive manner. All reviewers' tenures run for one-year terms on the editorial review boards and are expected to complete at least three reviews per term. Upon successful completion of this term, reviewers can be considered for an additional term.

If you have a colleague that may be interested in this opportunity, we encourage you to share this information with them.

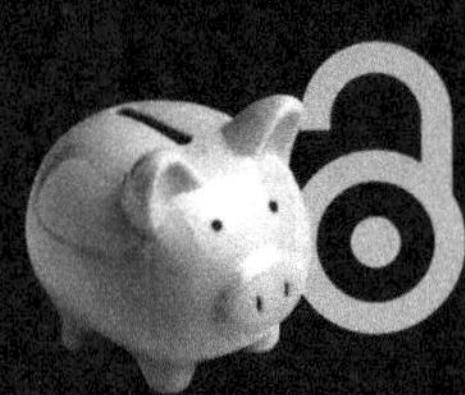

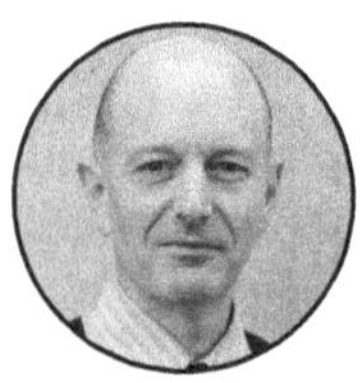

www.ingramcontent.com/pod-product-compliance
Lightning Source LLC
LaVergne TN
LVHW080017210926
848481LV00011B/3046